MATHEMATICAL FINANCE

BICENTENNIAL
1807
⊕WILEY
2007
BICENTENNIAL

THE WILEY BICENTENNIAL–KNOWLEDGE FOR GENERATIONS

*E*ach generation has its unique needs and aspirations. When Charles Wiley first opened his small printing shop in lower Manhattan in 1807, it was a generation of boundless potential searching for an identity. And we were there, helping to define a new American literary tradition. Over half a century later, in the midst of the Second Industrial Revolution, it was a generation focused on building the future. Once again, we were there, supplying the critical scientific, technical, and engineering knowledge that helped frame the world. Throughout the 20th Century, and into the new millennium, nations began to reach out beyond their own borders and a new international community was born. Wiley was there, expanding its operations around the world to enable a global exchange of ideas, opinions, and know-how.

For 200 years, Wiley has been an integral part of each generation's journey, enabling the flow of information and understanding necessary to meet their needs and fulfill their aspirations. Today, bold new technologies are changing the way we live and learn. Wiley will be there, providing you the must-have knowledge you need to imagine new worlds, new possibilities, and new opportunities.

Generations come and go, but you can always count on Wiley to provide you the knowledge you need, when and where you need it!

WILLIAM J. PESCE
PRESIDENT AND CHIEF EXECUTIVE OFFICER

PETER BOOTH WILEY
CHAIRMAN OF THE BOARD

MATHEMATICAL FINANCE
Theory, Modeling, Implementation

Christian Fries
University of Frankfurt
Department of Mathematics
Frankfurt, Germany

BICENTENNIAL
BICENTENNIAL
1807
WILEY
2007
BICENTENNIAL
BICENTENNIAL

WILEY-INTERSCIENCE
A John Wiley & Sons, Inc., Publication

Published by John Wiley & Sons, Inc., Hoboken, New Jersey.
Published simultaneously in Canada.

For general information on our other products and services or for technical support, please contact our Customer Care Department within the United States at (800) 762-2974, outside the United States at (317) 572-3993 or fax (317) 572-4002.

Wiley also publishes its books in a variety of electronic formats. Some content that appears in print may not be available in electronic format. For information about Wiley products, visit our web site at www.wiley.com.

Wiley Bicentennial Logo: Richard J. Pacifico

Library of Congress Cataloging-in-Publication Data:

Mathematical finance : theory, modeling, implementation / Christian Fries.
    p. cm.
  "Published simultaneously in Canada."
  Includes bibliographical references and index.
  ISBN 978-0-470-04722-4 (cloth : alk. paper)
  1. Derivative securities—Prices—Mathematical models. 2. Securities—Mathematical models. 3. Investments—Mathematical models. I. Title.
  HG6024.A3F75 2007
  332.601'5195—dc22   2007011325

Printed in the United States of America.

10 9 8 7 6 5 4 3 2 1

Nowadays people know the price of everything
and the value of nothing.

Oscar Wilde
The Picture of Dorian Gray [38]

Typeset by the author using TeXShop for Mac OS X™. Drawings by the author using OmniGraffle for Mac OS X™. Charts created using Java™ code by the author.

Version 1.5.12. Build 20070702.

Picture Credits

All figures are © copyright Christian Fries except the special section icons (see Section 1.3.3) licensed through iStockphoto.com.

Note

This book is also available in German. See
`http://www.christian-fries.de/finmath/book`

Acknowledgment

I am grateful to Andreas Bahmer, Hans-Josef Beauvisage, Michael Belledin, Dr. Urs Braun, Oliver Dauben, Peter Dellbrügger, Dr. Jörg Dickersbach, Dr. Holger Dietz, Sinan Dikmen, Dr. Dirk Ebmeyer, Fabian Eckstädt, Dr. Lydia Fechner, Christian Ferber, Frank Genheimer, Dr. Gido Herbers, Dr. Ansgar Jüngel, Dr. Jörg Kampen, Dr. Christoph Kiehn, Dr. Christoph Kühn, Dr. Jürgen Linde, Markus Meister, Dr. Sean Matthews, Dilys and Bill McCann, Michael Paulsen, Matthias Peter, Dr. Erwin Pier-Ribbert, Rosemarie Philippi, Dr. Radu Tunaru, Frank Ritz, Marius Rott, Oliver Schlüter, Thomas Schwiertz, Arndt Unterweger, Benedikt Wilbertz, Andre Woker, Polina Zeydis, and Jörg Zinnegger.

Their support and their feedback as well as the stimulating discussions we had contributed significantly to this work.

I am most grateful to my wife and my family. I thank them for their continuous support and generous tolerance.

Contents

III Interest Rate Structures, Interest Rate Products, and Analytic Pricing Formulas   119

8 Interest Rate Structures   123

9 Simple Interest Rate Products   133

10 The Black Model for a Caplet   147

V   Pricing Models for Interest Rate Derivatives   293

19 LIBOR Market Model   297

20 Swap Rate Market Models   329

VIII Appendices   463

List of Listings   500

Bibliography   503

Index   511

CHAPTER 1

Introduction

1.1 Theory, Modeling, and Implementation

This book tries to give a balanced representation of the theoretical foundations of mathematical finance, especially derivative pricing, state-of-the-art models, which are actually used in practice, and their implementation.

In practice, none of the three aspects—*theory*, *modeling*, and *implementation*—can be considered alone. Knowledge of the theory is worthless if it isn't applied. Theory provides the tools for consistent modeling. A model without implementation is essentially worthless. Good implementation requires a deep understanding of the model and the underlying theory.

With this in mind, the book tries to build a bridge from academia to practice and from theory to object-oriented implementation.

1.2 Interest Rate Models and Interest Rate Derivatives

The text concentrates on the modeling of interest rates as stochastic (undetermined) quantities and the evaluation of interest rate derivatives under such models. *However, this is not a specialization!* Although the mathematical modeling of stock prices was the historical starting point and interest rates were assumed to be constant, some important theoretical aspects are significant only for stochastic interest rates (e.g. the change of numéraire technique). So for didactic reasons it is meaningful to start with interest rate models. Another reason to start with interest rate models is that interest rate models are the foundation of hybrid models. Since the numéraire, the reference asset, is most likely an interest-rate-related product, a need for stochastic interest rates implies the need to build upon an interest rate model; see Figures 1.1 and 1.2. We will do so in Chapter 29. Nevertheless, the first model studied will be, of course, the Black-Scholes model for a single stock, after which we will move to stochastic interest rates.

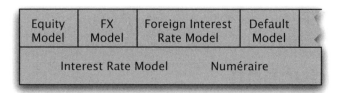

Equity Model	FX Model	Foreign Interest Rate Model	Default Model
Interest Rate Model		Numéraire	

Figure 1.1. *Hybrid Models: The numéraire, the reference asset in the modeling of price processes, is most likely an interest rate product. This choice is not mathematically necessary but common for almost all models. Interest rate processes are the natural starting point for the modeling of price processes.*

Black–Scholes Model

Equity Model	$dS = \mu S\, dt + \sigma S\, dW(t)$
Interest Rate Model	$dB = r B\, dt$

Equity Hybrid LIBOR Market Model

Equity Model	$dS = \mu S\, dt + \sigma S\, dW(t)$
Interest Rate Model	$dL_i = \mu_i L_i\, dt + \sigma_i L_i\, dW_i(t)$

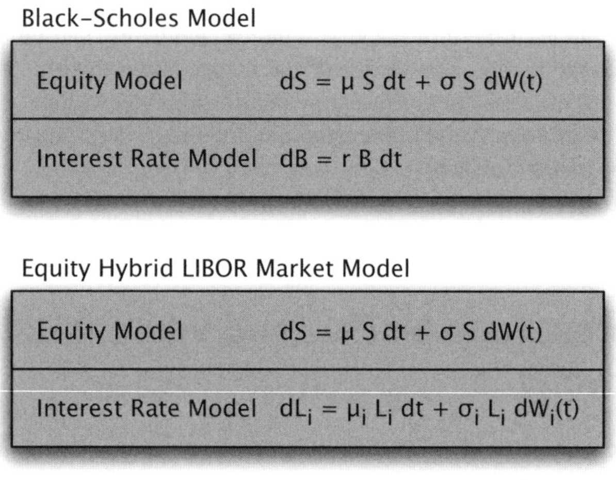

Figure 1.2. *The Black-Scholes model may be interpreted as a hybrid model with deterministic interest rates. The solution of* $dB(t) = rB(t)dt$ *is* $B(0)\exp(r\,t)$, *i.e. it is deterministic and given in closed form. Thus the interest rate component is trivial. Within a LIBOR market model the interest rate is a stochastic quantity. This also changes properties of the stock process.*

1.3 About This Book

1.3.1 How to Read This Book

The text may be read in a *nonlinear* way, i.e., the chapters have been kept as free-standing as possible. Chapter 2 provides the foundations in the order of their dependence. The reader familiar with the concepts of stochastic processes and martingales may skip the chapter and use it as reference only. To get a feeling for the mathematical concepts, one should read the special sections *Interpretation* and *Motivation*. Readers familiar with programming and implementation may prefer Chapter 13 as an illustration of the basic concepts.

The appendix gives a selection of the results and techniques from diverse areas (linear algebra, calculus, optimization), which are used in the text and in the implementation, but which are less important for understanding the essential concepts.

1.3.2 Abridged Versions

For a *crash course* focusing on particular aspects some chapters may be skipped. What follows are a few suggestions in this direction.

1.3.2.1 Abridged version "Monte Carlo Pricing"

Foundations (Chapter 2) → Replication (Chapter 3)
→ Black-Scholes Model (Chapter 4)
→ Discretization / Monte-Carlo Simulation (Chapter 13)

1.3.2.2 Abridged version "LIBOR Market Model"

Foundations (Chapter 2) → Replication (Chapter 3)
→ Interest Rate Structures (Chapter 8) → Black Model (Chapter 10)
→ LIBOR Market Model (Chapter 19)
→ Instantaneous and Terminal Correlation (Chapter 21)
→ Shape of the Interest Rate Curve (Chapter 25)

1.3.2.3 Abridged version "Markov Functional Model"

Foundations (Chapter 2) → Replication (Chapter 3)
→ Interest Rate Structures (Chapter 8) → Black Model (Chapter 10)
→ The Density of the Underlying of a European Option (Chapter 5)
→ Markov Functional Models (Chapter 27)

1.3.3 Special Sections

The text contains special sections giving notes on interpretation, motivation, and practical aspects. These are marked by the following symbols:

 Interpretation: *Provides an interpretation of the preceding topic. Casts light on purposes and practical aspects.* ◁|

 Motivation: *Provides motivation for the following topic. Sometimes notes deficiencies in the previous results.* ◁|

 Further Reading: *Suggested literature and associated topics.* ◁|

 Experiment: *Guide for a software experiment where aspects of the preceding topic can be explored.* ◁|

 Tip: *Hints for practical use and software implementation of the preceding topics.* ◁|

1.3.4 Notation

We will model the time evolution of stocks or interest rates with random variables parametrized through a time parameter t. Such stochastic processes may depend on other parameters like maturity or interest rate period. We will separate these two different kinds of parameters by a semicolon—see Figure 1.3.

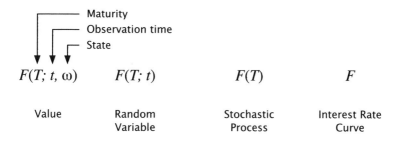

Figure 1.3. *On the notation.*

1.3.5 Feedback

Please help to improve this work!   Please send error reports and suggestions to

Christian Fries <email@christian-fries.de>.

Thank you.

1.3.6 Resources

In connection with this book the following resources are available:

- Interactive experiments and exercises:
  `http://www.christian-fries.de/finmath/applets`

- Java™ source code:
  `http://www.finmath.net/`

- Figures (in Color): The figures in this book are reproduced in black and white.
  The original color figures may be obtained from
  `http://www.christian-fries.de/finmath/book`

- Updates: For updates and error corrections see
  `http://www.christian-fries.de/finmath/book/errata`

Part I

Foundations: Probability Theory, Stochastic Processes, and Risk-Neutral Evaluation

CHAPTER 2

Foundations

2.1 Probability Theory

Definition 1 (Probability Space, σ Algebra):
Let Ω denote a set and $\mathcal{F}$ a family of subsets of Ω. $\mathcal{F}$ is a $\sigma\text{-}algebra$ if

1. $\emptyset \in \mathcal{F}$.

2. $F \in \mathcal{F} \implies \Omega \setminus F \in \mathcal{F}$.

3. $F_1, F_2, F_3, \ldots \in \mathcal{F} \implies \bigcup_{i=1}^{\infty} F_i \in \mathcal{F}$.

The pair $(\Omega, \mathcal{F})$ is a *measurable space*. A function $P : \mathcal{F} \to [0, \infty)$ is a *probability measure* if

1. $P(\emptyset) = 0$, $P(\Omega) = 1$.

2. For $F_1, F_2, F_3, \ldots \in \mathcal{F}$ mutually disjoint (i.e. $i \neq j \implies F_i \cap F_j = \emptyset$), we have

$$P\left(\bigcup_{i=1}^{\infty} F_i\right) = \sum_{i=1}^{\infty} P(F_i).$$

The triple $(\Omega, \mathcal{F}, P)$ is called *probability space* (if instead of 1 we require only $P(\emptyset) = 0$, then P is called *measure* and $(\Omega, \mathcal{F}, P)$ is called measure space).

Interpretation: The set Ω may be interpreted as the set of elementary events. Only one such event may occur. The subset $F \subset \Omega$ may then be interpreted as an event configuration, e.g. as if one asked only for a specific property of an event, a property that might be shared by more than one event. Then the complement of a set of events corresponds to the negation of the property in question, and the union of two subsets $F_1, F_2 \subset \Omega$ corresponds to combining the questions for the two corresponding properties with an "or". Likewise the intersection corresponds to an "and": only those events that share both properties are part of the intersection. A σ-algebra may then be interpreted as a set of properties, e.g., the set of properties by which we may distinguish the events or the set of properties on which we may base decisions and answer questions. Thus the σ-algebra may be interpreted as information (on properties of events).

Thus a probability space $(\Omega, \mathcal{F}, P)$ may be interpreted as a set of elementary events, a family of properties of the events, and a map that assigns a probability to each property of the events, the probability that an event with the respective properties will occur. ◁|

Since conditional expectation will be one of the central concepts, we remind the reader of the notions of *conditional probability* and *independence*.

Definition 2 (Independence, Conditional Probability):
Let $(\Omega, \mathcal{F}, P)$ denote a probability space and $A, B \in \mathcal{F}$.

1. We say that A and B are *independent*, if

$$P(A \cap B) = P(A)\, P(B).$$

2. For $P(B) > 0$ we define the *conditional probability of A under the hypothesis B* as

$$P(A|B) := \frac{P(A \cap B)}{P(B)}.$$

The Borel σ-algebra $\mathcal{B}(\mathbb{R})$ or $\mathcal{B}(\mathbb{R}^n)$ plays a special role in integration theory. We define it next.

Definition 3 (Borel σ-Algebra, Lebesgue Measure):
Let $n \in \mathbb{N}$ and $a_i < b_i$ ($i = 1, \ldots, n$). By $\mathcal{B}(\mathbb{R}^n)$ we denote the smallest σ-algebra for which

$$(a_1, b_1) \times \cdots \times (a_n, b_n) \in \mathcal{B}(\mathbb{R}^n).$$

$\mathcal{B}(\mathbb{R}^n)$ is called the *Borel σ-algebra*. The measure λ defined on $\mathcal{B}(\mathbb{R}^n)$ with

$$\lambda((a_1, b_1) \times \cdots \times (a_n, b_n)) := \prod_{i=1}^{n} (b_i - a_i)$$

is called a *Lebesgue measure* on $\mathcal{B}(\mathbb{R}^n)$.

Remark 4 (Lebesgue Measure): Obviously the Lebesgue measure is *not* a probability measure on $(\mathbb{R}^n, \mathcal{B}(\mathbb{R}^n))$ since $\lambda(\mathbb{R}^n) = \infty$. It will be needed in the discussion of Lebesgue integration and we give the definition merely for completeness.[1]

Definition 5 (Measurable, Random Variable):
Let $(\Omega, \mathcal{F})$ and $(S, \mathcal{S})$ denote two measurable spaces.

1. A map $T : \Omega \mapsto S$ is called $(\mathcal{F}, \mathcal{S})$-measurable if[2]

$$T^{-1}(A) \in \mathcal{F} \text{ for all } A \in \mathcal{S}.$$

   If $T : \Omega \mapsto S$ is a $(\mathcal{F}, \mathcal{S})$-measurable map we write more concisely

$$T : (\Omega, \mathcal{F}) \mapsto (S, \mathcal{S}).$$

2. A measurable map $X : (\Omega, \mathcal{F}) \mapsto (S, \mathcal{S})$ is also called a *random variable*. A random variable $X : (\Omega, \mathcal{F}) \mapsto (S, \mathcal{S})$ is called a *n-dimensional real-valued random variable* if $S = \mathbb{R}^n$ and $\mathcal{S} = \mathcal{B}(\mathbb{R}^n)$.

We are interested in the probability for which a given random variable attains a certain value or range of values. This is given by the following definition.

Definition 6 (Image Measure):
Let $X : (\Omega, \mathcal{F}) \mapsto (S, \mathcal{S})$ denote a random variable and P a measure on the measureable space $(\Omega, \mathcal{F})$. Then

$$P_X(A) := P(X^{-1}(A)) \qquad \forall A \in \mathcal{S}$$

defines a probability measure on $(S, \mathcal{S})$, which we call the *image measure* of P with respect to X.

 Interpretation: A real-valued random variable assigns a real value (or vector of values) to each elementary event ω. This value may be interpreted as the result of an experiment, depending on the events. In our context the random variables mostly stand for payments or values of financial products depending on the *state of the world*. How random a random variable is depends on the random variable itself. The random variable that assigns the same value to all events ω exhibits no randomness at all. If we could observe

---

[1] The Lebesgue measure measures intervals ($n = 1$) according to their length, rectangles ($n = 2$) according to their area, and cubes ($n = 3$) according to their volume.

[2] We define $T^{-1}(A) := \{\omega \in \Omega \mid T(\omega) \in A\}$.

11

only the result of such an experiment (random variable), we would not be able to say anything about the state of the world ω that led to the result.

The image measure is the probability measure induced by the probability measure P (a probability measure on $(\Omega, \mathcal{F})$) and the map X on the image space $(S, \mathcal{S})$.

The property of being measurable may be interpreted as the property that the distinguishable events in the image space $(S, \mathcal{S})$ are not finer (better distinguishable) than the events in the preimage space $(\Omega, \mathcal{F})$. Only then it is possible to use the probability measure P on $(\Omega, \mathcal{F})$ to define a probability measure on $(S, \mathcal{S})$, see Figure 2.1.   ◁|

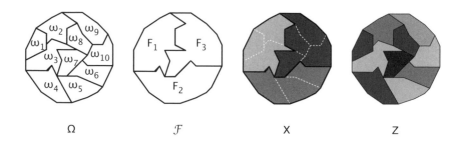

Ω $\mathcal{F}$ X   Z

Figure 2.1. *Illustration of measurability: The random variables X and Z assign a gray value to each elementary event $\omega_1, \ldots, \omega_{10}$ as shown. The σ-algebra $\mathcal{F}$ is generated by the sets $F_1 = \{\omega_1, \omega_2, \omega_3\}$, $F_2 = \{\omega_4, \omega_5, \omega_6\}$, $F_3 = \{\omega_7, \ldots, \omega_{10}\}$. The random variable X is measurable with respect to $\mathcal{F}$, the random variable Z is not measurable with respect to $\mathcal{F}$.*

Exercise: Let X be as in Definition 6. Show that

$$\{X^{-1}(A) \mid A \in \mathcal{S}\}$$

is a σ-algebra. What would be an interpretation of $X^{-1}(A)$?

Motivation: We will now define the *Lebesgue integral* and give an interpretation and a comparison to the (possibly more familiar) Riemann integral.

The definition of the Lebesgue integral is not only given to prepare the definition of the conditional expectation (Definition 15). The definition will also show the construction of the Lebesgue integral and we will later use similar steps to construct the Itô integral.   ◁|

12

Definition 7 (Integral, Lebesgue Integral):

Let $(\Omega, \mathcal{F}, \mu)$ denote a measure space.

1. Let f denote a $(\mathcal{F}, \mathcal{B}(\mathbb{R}))$-measurable real-valued, nonnegative map. f is called an *elementary function* if f takes on only a finite number of values $a_1, \ldots, a_n$. For an elementary function we define

$$\int_\Omega f(\omega) \, d\mu(\omega) := \sum_{i=1}^n a_i \mu(A_i)$$

where $A_i := f^{-1}(\{a_i\})$ ($\Rightarrow A_i \in \mathcal{F}$)[3] as the *(Lebesgue) integral* of f.

2. Let f denote a nonnegative map defined on Ω, such that a monotonically increasing sequence $(u_k)_{k \in \mathbb{N}}$ of elementary maps with $f := \sup_{k \in \mathbb{N}} u_k$ exists. Then

$$\int_\Omega f(\omega) \, d\mu(\omega) := \sup_{k \in \mathbb{N}} \int_\Omega u_k(\omega) d\mu(\omega)$$

is unique and is called the (Lebesgue) integral of f.

3. Let f denote a map on Ω such that we have for $f^+ := \max(f, 0)$ and $f^- := \max(-f, 0)$, respectively, a monotone increasing sequence of elementary maps as in the previous definition. Furthermore we require that $\int_\Omega f^\pm \, d\mu < \infty$. Then f is called *integrable* with respect to μ and we define

$$\int_\Omega f \, d\mu := \int_\Omega f^+ \, d\mu - \int_\Omega f^- \, d\mu$$

as the (Lebesgue) integral of f.

Remark 8 ($\int f(x) \, dx$, $\int f(t) \, dt$): If the measure μ is the Lebesgue measure $\mu = \lambda$ we use the shortened notation

$$\int_A f(x) \, d\lambda(x) =: \int_A f(x) \, dx.$$

In this case $\Omega = \mathbb{R}^n$ and we denote the elements of Ω by latin letters, e.g. x (instead of ω). If the elements of $\Omega = \mathbb{R}$ have the interpretation of a time we usually denote them by t.

---

[3] We have $(a_i - \frac{1}{n}, a_i + \frac{1}{n}) \in \mathcal{B}(\mathbb{R})$ by definition. Then $\{a_i\} = \bigcap_{n=1}^{\infty} (a_i - \frac{1}{n}, a_i + \frac{1}{n}) \in \mathcal{B}(\mathbb{R})$.

13

Theorem 9 (*measurable* ⇔ *integrable* **for nonnegative maps**): For a non-negative map f on Ω a monotone increasing sequence $(u_k)_{k \in \mathbb{N}}$ of elementary maps with $f = \sup_{k \in \mathbb{N}} u_k$ exists if and only if f is $\mathcal{F}$-measurable.

Proof: See [1], §12. □|

Interpretation: To develop an understanding of the (Lebesgue) integral we consider its definition for elementary maps:

$$\int_\Omega f(\omega)\, d\mu(\omega) = \sum_{i=1}^{n} a_i \mu(A_i), \qquad A_i := f^{-1}(\{a_i\}).$$

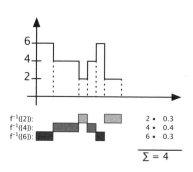

$f^{-1}(\{2\})$:	2 •	0.3
$f^{-1}(\{4\})$:	4 •	0.4
$f^{-1}(\{6\})$:	6 •	0.3
	$\Sigma = 4$	

The Lebesgue integral of an (elementary) map f is the weighted sum of the function values a_i of f, each weighted by the measure $\mu(A_i)$ of the set on which this value is attained (i.e., $A_i = f^{-1}(\{a_i\})$).

If in addition we have $\mu(\Omega) = 1$, where Ω is the domain of f, then the integral is a weighted average of the function values a_i of f.

For a real-valued (elementary) function of a real-valued argument, e.g., $f : [a, b] \mapsto \mathbb{R}$, and the Lebesgue measure, the integral corresponds to the naive concept of an integral as being the sum all rectangles given by

base area (Lebesgue measure of the interval) × height (function value)

which is also the concept behind the Riemann integral.

Part 2 and 3 of Definition 7 extend this concept via a limit approximation to more general functions. ◁|

Excursus: On the Difference between Lebesgue and Riemann Integrals[4]

The construction of the Lebesgue integral differs from the construction of the Riemann integral (which is perhaps more familiar) in the way the sets A_i are chosen. The Riemann integral starts from a given partition of the domain and multiplies the

---

[4] An understanding of the difference between the Lebesgue and Riemann integrals does not play a major role in the following text. The excursus can safely be skipped. It should serve to satisfy curiosity, e.g., if the concept of a Riemann integral is more familiar.

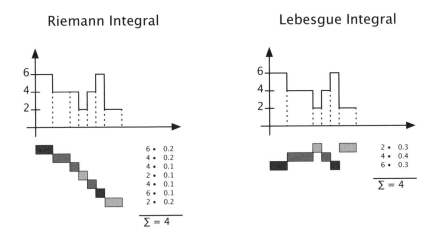

Figure 2.2. *Lebesgue integral versus Riemann integral.*

size of each subinterval by a corresponding functional value of (any) chosen point belonging to that interval (e.g., the center point). The Lebesgue integral chooses the partition as preimage $f^{-1}(\{a_i\})$ of given function values a_i. In short: the Riemann integral partitions the domain of f, the Lebesgue integral partitions the range of f. For elementary functions both approaches give the same integral value; see Figure 2.2. For general functions the corresponding integrals are defined as the limit of a sequence of approximating elementary functions (if it exists). Here, the two concepts are different: In the limit, all Riemann integrable functions are Lebesgue integrable, and the two limits give the same value for the integral. However, there exist Lebesgue integrable functions for which the Riemann integral is not defined (its limit construction does not converge).

Definition 10 (Distribution): ⌐

Let P denote a probability measure on $(\mathbb{R}, \mathcal{B}(\mathbb{R}))$ (e.g., the image measure of a random variable). The function

$$F_P(x) := P(\,(-\infty, x)\,)$$

is called the *distribution function*[5] of P. If P denotes a probability measure on $(\mathbb{R}^n, \mathcal{B}(\mathbb{R}^n))$ the *n-dimensional distribution function* is defined as

$$F_P(x_1, \ldots, x_n) := P(\,(-\infty, x_1) \times \cdots \times (-\infty, x_n)\,).$$

---

[5] We have $(-\infty, x) \in \mathcal{B}(\mathbb{R})$ since $(-\infty, x) = \cup_{i=1}^{\infty}(x - i, x)$, see Definition 1.

The distribution function of a random variable X is defined as the distribution function of its image measure P_X (see Definition 6).

Definition 11 (Density):
Let F_P denote the distribution function of a probability measure P. If F_P is differentiable, we define

$$\phi(x) := \frac{\partial}{\partial x} F_P(x)$$

as the *density* of P. If P is the image measure of some random variable X, we also say that ϕ is the *density* of X.

Remark 12 (Integration Using a Known Density/Distribution): To calculate the integral of a function of a random variable (e.g., to calculate expectation or variance), it is sufficient to know the density or distribution function of the random variable. Let g denote a sufficiently smooth function and X a random variable on $(\Omega, \mathcal{F}, P)$; then we have

$$\int_\Omega g(X(\omega)) \, dP(\omega) = \int_{-\infty}^{\infty} g(x) \, dF_{P_X}(x) = \int_{-\infty}^{\infty} g(x)\phi(x) \, dx,$$

where F_{P_X} denotes the distribution function of X and ϕ the density of X (i.e., of P_X). In this case it is neither necessary to know the underlying space Ω, the measure P, nor how X is modeled (i.e., defined) on this space.

Definition 13 (Independence of Random Variables):
Let $X : (\Omega, \mathcal{F}) \mapsto (S, \mathcal{S})$ and $Y : (\Omega, \mathcal{F}) \mapsto (S, \mathcal{S})$ denote two random variables. X and Y are called *independent*, if for all $A, B \in \mathcal{S}$ the events $X^{-1}(A)$ and $Y^{-1}(B)$ are independent in the sense of Definition 2.

Remark 14 (Independence): For $i = 1, \ldots, n$ let $X_i : \Omega \mapsto \mathbb{R}$ denote random variables with distribution functions F_{X_i} and let $F_{(X_1, \ldots, X_n)}$ denote the distribution function of $(X_1, \ldots, X_n) : \Omega \mapsto \mathbb{R}^n$. Then the X_i are pairwise independent if and only if

$$F_{(X_1, \ldots, X_n)}(x_1, \ldots, x_n) = F_{X_1}(x_1) \cdot \ldots \cdot F_{X_n}(x_n).$$

Definition 15 (Expectation, Conditional Expectation):
Let X denote a real-valued random variable on the probability space $(\Omega, \mathcal{F}, P)$.

1. If X is P-integrable, we define

$$E^P(X) := \int_\Omega X \, dP$$

as the *expectation* of X.

2. Furthermore, let $F_i \in \mathcal{F}$ with $P(F_i) > 0$. Then

$$E^P(X|F_i) := \frac{1}{P(F_i)} \int_{F_i} X \, dP$$

is called the *conditional expectation of X under (the hypothesis)* F_i.

Theorem 16 (Conditional Expectation[6]): Let X denote a real-valued random variable on $(\Omega, \mathcal{F}, P)$, either nonnegative or integrable. Then we have for each σ-algebra $\mathcal{C} \subset \mathcal{F}$ a nonnegative or integrable real-valued random variable $X_{|C}$ on Ω, unique in the sense of almost sure equality,[7] such that $X_{|C}$ is $\mathcal{C}$-measurable and

$$\forall \, C \in \mathcal{C} \, : \quad \int_C X_{|C} \, dP = \int_C X \, dP, \quad \text{i.e.,} \quad E^P(X|C) = E^P(X_C|C).$$

We will discuss the interpretation of this theorem after giving a name to X_C:

Definition 17 (Conditional Expectation (continued)):
Under the assumptions and with the notation of Theorem 16 we define:

1. The random variable $X_{|C}$ is called the *conditional expectation of X under (the hypothesis)* C and is denoted by

$$E^P(X|C) := X_{|C}. \tag{2.1}$$

2. Let Y denote another random variable on the same measure space. We define:

$$E^P(X|Y) := E(X|\sigma(Y)), \tag{2.2}$$

where $\sigma(Y)$ is the σ-algebra generated by Y, i.e., the smallest σ-algebra, with respect to which Y is measurable, i.e., $\sigma(Y) := \sigma(Y^{-1}(S))$.

 Interpretation: First note that the two concepts of expectations from Definition 15 are just special cases of the conditional expectation defined in Definition 17, namely:

- Let $C = \{\emptyset, \Omega\}$. Then $E(X \mid C) = X_{|C}$ where $X_{|C}(\omega) = E(X) \quad \forall \, \omega \in \Omega$.

---

[6] See [2], Chapter 15
[7] A property holds *P-almost surely* if the set of $\omega \in \Omega$ for which the property does not hold has measure zero.

- For $C = \{\emptyset, F, \Omega \setminus F, \Omega\}$ we have $X_{|C}(\omega) = \begin{cases} E(X \mid F) & \text{if } \omega \in F \\ E(X \mid \Omega \setminus F) & \text{if } \omega \in \Omega \setminus F \end{cases}$,

with $E(X) = P(F)\, E(X|F) + (1 - P(F))\, E(X|\Omega \setminus F)$

The conditional expectation is a random variable that is derived from X such that only events (sets) in C can be distinguished. In the first case we have a very coarse C and the image of $X_{|C}$ contains only the expectation $E(X)$. This is the smallest piece of information on X. As C becomes finer, more and more information about X becomes visible in $X_{|C}$. Furthermore, if X itself is C-measurable, then X and $X_{|C}$ are (P-almost surely) indistinguishable.

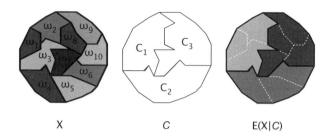

$$X \qquad\qquad C \qquad\qquad E(X|C)$$

Figure 2.3. *Conditional expectation: Let the σ-algebra C be generated by the sets $C_1 = \{\omega_1, \omega_2, \omega_3\}$, $C_2 = \{\omega_4, \omega_5, \omega_6\}$, $C_3 = \{\omega_7, \ldots, \omega_{10}\}$.*

In this sense C may be interpreted as an information set and $X_{|C}$ as a *filtered* version of X. If it is only possible to make statements about events in C, then we can only make statements about X which could also be made about $X_{|C}$, see Figure 2.3.   ◁|

2.2 Stochastic Processes

Definition 18 (Stochastic Process):   ⌐

A family $X = \{X_t \mid 0 \leq t < \infty\}$ of random variables

$$X_t : (\Omega, \mathcal{F}) \to (S, \mathcal{S})$$

is called *(time continuous) stochastic process*. If $(S, \mathcal{S}) = (\mathbb{R}^d, \mathcal{B}(\mathbb{R}^d))$, we say that X is a *d-dimensional stochastic process*. The family X may also be interpreted as a $X : [0, \infty) \times \Omega \to S$:

$$X(t, \omega) := X_t(\omega) \qquad \forall\, (t, \omega) \in [0, \infty) \times \Omega.$$

If the range $(S, \mathcal{S})$ is not given explicitly, we assume $(S, \mathcal{S}) = (\mathbb{R}^d, \mathcal{B}(\mathbb{R}^d))$.   ⌐

Interpretation: The parameter t obviously refers to time. For fixed $t \in [0, \infty)$ we view $X(t)$ as the outcome of an experiment at time t. Note that all random variables $X(t)$ are modeled over the *same* measurable space $(\Omega, \mathcal{F})$. Thus we do not assume a family $(\Omega_t, \mathcal{F}_t)$ of measurable spaces, one for each X_t. The stochastic process X assigns a *path* to each $\omega \in \Omega$: For a fixed $\omega \in \Omega$ the *path* $X(\cdot, \omega) := \{(t, X(t, \omega)) \mid t \in [0, \infty)\}$ is a sequence of outcomes of the random experiments X_t (a *trajectory*) associated with a state ω. Knowledge about $\omega \in \Omega$ implies knowledge of the whole history (past, present, and future) $X(\omega)$.

To model the different levels of knowledge and thus distinguish between past and future, we will define in Section 2.3 the concept of a filtration and an adapted process.

◁|

Definition 19 (Path):

Let X denote a stochastic process. For a fixed $\omega \in \Omega$ the mapping $t \mapsto X(t, \omega)$ is called the path of X (in state ω).

Definition 20 (Equality of Stochastic Processes):
We define three notions of equality of stochastic processes:

1. Two stochastic processes X and Y are called *indistinguishable* if

$$P(X_t = Y_t : \forall\, 0 \le t < \infty) = 1.$$

2. A stochastic process Y is a *modification* of X if

$$P(X_t = Y_t) = 1 : \forall\, 0 \le t < \infty.$$

3. Two stochastic processes X and Y *have the same finite-dimensional distributions*, if

$$\forall\, n : \forall\, 0 \le t_1 < t_2 < \cdots < t_n < \infty : \forall\, A \in \mathcal{B}(S^n) :$$
$$P((X_{t_1}, \ldots, X_{t_n}) \in A) = P((Y_{t_1}, \ldots, Y_{t_n}) \in A).$$

Remark 21 (On the Equality of Stochastic Processes): While in Definition 20.3 only the distributions generated by the processes are considered, Definitions 20.1 and 20.2 consider the pointwise differences between the processes. The difference between 20.1 and 20.2 will become apparent in the following example:

Let $Z : (\mathbb{R}, \mathcal{B}(\mathbb{R})) \to ([-1, 1], \mathcal{B}([-1, 1]))$ be a random variable on $(\Omega, \mathcal{F}, P) = (\mathbb{R}, \mathcal{B}, \lambda)$ and $t \mapsto X(t) := t \cdot Z$ be a stochastic process.[8] Let $P(\{Z \in A\}) = P(\{-Z \in A\})$

---

[8] An interpretation of this process would be the position of a moving particle, having at time 0 the position 0 and the random speed Z.

$\forall\, A \in \mathcal{B}([-1, 1])$ and $P(\{Z = x\}) = 0\ \forall\, x \in [-1, 1]$, e.g., an equally distributed Z. Furthermore let $\forall\, (t, \omega) \in [0, \infty) \times \mathbb{R}$

$$Y_1(t, \omega) := \begin{cases} X(t, \omega) & \text{for } t \neq \omega \\ -X(t, \omega) & \text{for } t = \omega \end{cases}, \qquad Y_2(t, \omega) := -X(t, \omega).$$

The Y_1 is a modification of X, since $Y_1(t)$ differs from $X(t)$ (for fixed t) only on a set with probability 0. However, X and Y_1 are not indistinguishable, since $P(X(t) = Y_1(t)\ :\ \forall\, 0 \leq t < \infty) = \frac{1}{2}$ (the two processes are different on 50% of all paths). Y_2 is neither indistinguishable nor a modification of X, but due to $P(\{Z \in A\}) = P(\{Z \in -A\})\ \forall A \in \mathcal{B}([-1, 1])$ it fulfills condition 3 in Definition 20.

To summarize, condition 1 in Definition 20 considers the equality of the processes X, Y, condition 2 in Definition 20 considers equality of the random variables $X(t)$, $Y(t)$ for fixed t, and condition 3 in Definition 20 considers the equality of distributions. In our applications we are interested only in the distributions of processes.

2.3 Filtration

Definition 22 (Filtration): ⌐

Let $(\Omega, \mathcal{F})$ denote a measurable space. A family of σ-algebras $\{\mathcal{F}_t \mid t \geq 0\}$, where

$$\mathcal{F}_s \subseteq \mathcal{F}_t \subseteq \mathcal{F} \qquad \text{for } 0 \leq s \leq t,$$

is called a *filtration* on $(\Omega, \mathcal{F})$. ⌟

Definition 23 (Generated Filtration): ⌐

Let X denote a stochastic process on $(\Omega, \mathcal{F})$. We define

$\mathcal{F}_t^X := \sigma(X_s; 0 \leq s \leq t)$

:= the smallest σ-algebra with respect to which X_s is measurable $\forall\, s \in [0, t]$.

⌟

Definition 24 (Adapted Process): ⌐

Let X denote a stochastic process on $(\Omega, \mathcal{F})$ and $\{\mathcal{F}_t\}$ a filtration on $(\Omega, \mathcal{F})$. The process X is called $\{\mathcal{F}_t\}$-adapted, if X_t is $\mathcal{F}_t$-measurable for all $t \geq 0$. ⌟

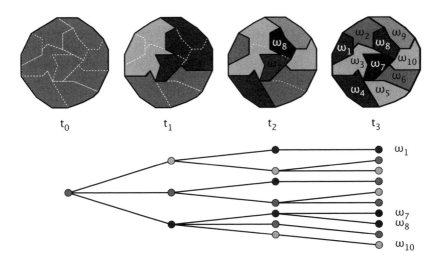

Figure 2.4. *Illustration of a filtration and an adapted process.*

 Interpretation: In Figure 2.4 we depict a filtration of four σ-algebras with increasing refinement (left to right). The black borders surround the generators of the corresponding σ-algebra. If a stochastic process maps a gray value for each elementary event (or path) ω_i of Ω (left), then the process is adapted if it takes a constant gray value on the generators of the respective σ-algebra. If at time t_2 the process assigns to ω_7 the same dark gray as to ω_8, then the process is adapted, otherwise it is not.

By means of the conditional expectation (see Theorem 16 and the interpretation of Figure 2.3) we may create an adapted process from a given filtration $\{\mathcal{F}_t \mid t \geq 0\}$ and an $\mathcal{F}$-measurable random variable Z:

Lemma 25 (Process of the Conditional Expectation): Let $\{\mathcal{F}_t \mid t \geq 0\}$ denote a filtration $\mathcal{F}_s \subseteq \mathcal{F}_t \subseteq \mathcal{F}$ and Z an $\mathcal{F}$ measurable random variable. Then

$$X(t) := \mathrm{E}(Z \mid \mathcal{F}_t)$$

is a $\{\mathcal{F}_t\}$-adapted process.

This lemma shows how the filtration (and the corresponding adapted process) may be viewed as a model for *information*: The random variable $X(t)$ in Lemma 25 allows with increasing t more and more specific statements about the nature of Z. Compare this to the illustrations in Figure 2.1 and 2.3. ◁|

The concepts of an adapted process only links random variables $X(t)$ to σ-algebras $\mathcal{F}_t$ for any t. It does not necessarily imply that the stochastic process X (interpreted as a random variable on $[0, \infty) \times \Omega$) is measurable. A stronger requirement is given by the following Definition.

Definition 26 (Progressively Measurable):
An (n-dimensional) stochastic process X is called *progessively measurable* with respect to the filtration $\{\mathcal{F}_t\}$ if for each $T > 0$ the mapping

$$X : ([0,T] \times \Omega \, , \, \mathcal{B}([0,T] \otimes \mathcal{F}_t)) \;\to\; (\mathbb{R}, \mathcal{B}(\mathbb{R}^n))$$

is measurable.

Remark 27: Any progressively measurable process is measurable and adapted. Conversly a measurable and adapted process has a progressively measurable modification; see [20].

Another regularity requirement for stochastic processes is that of being *previsible*:

Definition 28 (Previsible Process):
Let X denote a (real-valued) stochastic process on $(\Omega, \mathcal{F})$ and $\{\mathcal{F}_t\}$ a filtration on $(\Omega, \mathcal{F})$. The process X is called $\{\mathcal{F}_t\}$-*previsible*, if X is $\{\mathcal{F}_t\}$-adapted and bounded with left continuous paths.

2.4 Brownian Motion

Definition 29 (Brownian Motion):
Let $W : [0, \infty) \times \Omega \to \mathbb{R}^n$ denote a stochastic process with the following properties:

1. $W(0) = 0$ (*P*-almost surely).

2. The map $t \mapsto W(t)$ is continuous (*P*-almost surely).

3. For given $t_0 < t_1 < \cdots < t_k$ the increments $W(t_1) - W(t_0), \ldots, W(t_k) - W(t_{k-1})$ are mutually independent.

4. For all $0 \le s \le t$ we have $W(t) - W(s) \sim \mathcal{N}(0, (t-s)I_n)$, i.e., the increment is normally distributed with mean 0 and covariance matrix $(t-s)I_n$, where I_n denotes the $n \times n$ identity matrix.

Then W is called *(n-dimensional) P-Brownian motion* or a *(n-dimensional) P-Wiener process*.

We have not yet discussed the question of whether a process with such properties exists (it does). The question for its existence is nontrivial. For example, if we want to replace normally distributed by lognormally distributed in property 4 in Definition 29 there would be no such process.[9] If we set $s = 0$ in property 4, we see that we have prescribed the distribution of $W(t)$ as well as the distribution of the increments $W(t) - W(s)$.

Remark 30 (Brownian Motion): Property 4 is less axiomatic than one might assume: The central requirement is the independence of the increments together with the requirement that increments of the same time step size $t - s$ have the same nonnegative variance (here $t - s$) and mean 0. That the increments are normally distributed is more a consequence than an requirement, see Theorem 31. This theorem also gives a construction of the Brownian motion.

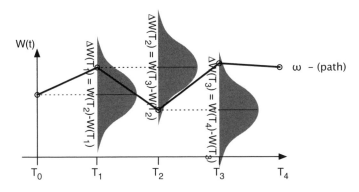

Figure 2.5. *Time discretization of a Brownian motion: The transition $\Delta W(T_i) := W(T_{i+1}) - W(T_i)$ from time T_i to T_{i+1} is normally distributed. The mean of the transition is 0, i.e., under the condition that at time T_i the state $W(T_i) = x^*$ was attained, the (conditional) expectation of $W(T_{i+1})$ is x^*: $E(W(T_{i+1}) \mid W(T_i) = x^*) = x^*$.*

 Tip (Time-Discrete Realizations): In the following we will often consider the realizations of a stochastic process at discrete times $0 = T_0 < T_1 < \cdots < T_N$ only (e.g., this will be the case when we consider the implementation). If we need only the realizations $W(T_i)$, we may generate them by the time-discrete increments $\Delta W(T_i) := W(T_{i+1}) - W(T_i)$ since from Definition 29 we have $W(T_i) = \sum_{k=0}^{i-1} \Delta W(T_k)$, $W(T_0) := 0$. See Figure 2.4. ◁|

---

[9] Note that the sum of two (independent) normally distributed random variables is normally distributed, but the sum of two lognormally distributed random variables is not lognormal.

2.5 Wiener Measure, Canonical Setup

The following theorem gives a construction (or approximation) of a Brownian motion. It defines the *Wiener measure* and shows that the properties of a Brownian motion are less axiomatic than one might assume from Definition 29; rather they are consequences of independence.

Theorem 31 (Invariance Principle of Donsker (1951); see [20] §2): Let $(\Omega, \mathcal{F}, P)$ denote a probability space and $(Y_j)_{j=1}^{\infty}$ a sequence of independent identically distributed random variables (not necessarily normally distributed!) with mean 0 and variance $\sigma^2 > 0$. Define $S_0 := 0$ and $S_k := \sum_{j=1}^{k} Y_j$. Let X^n denote a stochastic process defined as the (scaled) linear interpolation of the S_k's at time steps of size $\frac{1}{n}$:

$$X^n(t) := \frac{1}{\sigma \sqrt{n}} ((t \cdot n - [t \cdot n]) S_{[t \cdot n]+1} + ([t \cdot n] + 1 - t \cdot n)) S_{[t \cdot n]}),$$

where $[x]$ denotes the largest integer number less or equal to x.

A path $X^n(\omega)$, $\omega \in \Omega$, defines a continuous map $[0, \infty) \mapsto \mathbb{R}$ and X^n is $(\Omega, \mathcal{F}) \mapsto (C([0, \infty)), \mathcal{B}(C([0, \infty))))$-measurable[10]. Let P^n denote the image measure of X^n defined on $(C([0, \infty)), \mathcal{B}(C([0, \infty))))$. Then we have:

- $\{P_n\}_{n=1}^{\infty}$ converges on $(C([0, \infty)), \mathcal{B}(C([0, \infty))))$ to a measure P^* in the weak sense[11].

- The process W defined on $(C([0, \infty)), \mathcal{B}(C([0, \infty))))$ by

$$W(t, \omega) := \omega(t)$$

is a P^* Brownian motion.

Proof: See [20] §2. □|

Definition 32 (Wiener Measure): ⌐

The measure P^* from Theorem 31 is called *Wiener measure*. ⌐

Definition 33 (Canonical Setup): ⌐

The space

$$(C([0, \infty)), \mathcal{B}(C([0, \infty))), P^*)$$

---

[10] With $C([0, \infty))$ denoting the space of continuous maps $[0, \infty) \mapsto \mathbb{R}$ endowed with the metric of equicontinuous convergence $d(f, g) = \sum_{i=1}^{\infty} \frac{1}{2^n} \frac{d_n(f,g)}{1+d_n(f,g)}$ with $d_n(f, g) = \sup_{0 \le t \le n} |f(t) - g(t)|$ (then $(C([0, \infty)), d)$ is a complete metric space) and $\mathcal{B}(C([0, \infty)))$ denoting the Borel σ-algebra induced by that metric, i.e., the smallest σ-algebra containing the d-open sets.

[11] A sequence of probability measures $\{P_n\}_{n=1}^{\infty}$ *converges in the weak sense* to a measure P^*, if $\int f \, dP_n \to \int f \, dP^*$ for all continuous bounded maps $f : \Omega \mapsto \mathbb{R}$.

(as defined in Theorem 31) is called the *canonical setup* for a Brownian motion W defined by $W(t, \omega) := \omega(t)$, $\omega \in C([0, \infty))$. ⌐

Remark 34: A more detailed discussion of Theorem 31 may be found in [20]. A less formal discussion of properties of the Brownian motion may be found in [13].

2.6 Itô Calculus

 Motivation: The Brownian motion W is our first encounter with an important continuous stochastic process. The Brownian motion may be viewed as the limit of a scaled *random walk*.[12] If we interpret the Brownian motion W in this sense as a model for the movement of a particle, then $W(T)$ denotes the position of the particle at time T and $W(T+\Delta T)-W(T)$ the position change that occurs from T to $T + \Delta T$; to be precise, $W(T)$ models the probability distribution of the particle position.

The model of a Brownian motion is that position changes are normally distributed with mean 0 and standard deviation $\sqrt{\Delta T}$. Requiring mean 0 corresponds to requiring that the position change has no directional preference. The standard deviation $\sqrt{\Delta T}$ is, apart from a constant which we assume to be 1, a consequence of the requirement that position changes are independent of the position and time at which they occur.

To motivate the class of Itô processes we consider the Brownian motion at discrete times $0 = T_0 < T_1 < \cdots < T_N$. The random variable $W(T_i)$ (position of the particle) may be expressed through the increments $\Delta W(T_i) := W(T_{i+1}) - W(T_i)$:

$$W(T_i) = \sum_{j=0}^{i-1} \Delta W(T_j).$$

Using the increments $\Delta W(T_j)$ we may define a whole family of discrete stochastic processes (Figure 2.6). We give a step by step introduction and use the illustrative interpretation of a particle movement: First we assume that the particle may lose energy over time (for example). Then the increments may still be normally distributed but their standard deviation no longer will be $\sqrt{T_{j+1} - T_j}$. Instead it might be a time-dependent scaling thereof, e.g., $e^{-T_j} \sqrt{T_{j+1} - T_j}$ where the standard deviation decays exponentially. Multiplying the increments $\Delta W(T_j)$ by a factor gives normally

---

[12] In a (one-dimensional) random walk a particle changes position at discrete time steps by a (constant) distance (say 1) in either direction with equal probability. In other words, we have binomial distributed Y_i in Theorem 31.

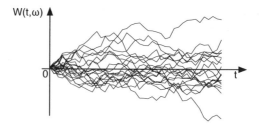

Figure 2.6. *Brownian motion: Paths of (a discretization of) a Brownian motion*

Figure 2.7. *Paths of (a discretization of) a Brownian motion with time-dependent instantaneous volatility.*

distributed increments with arbitrary standard deviations. Thus we consider a process of the form

$$X(T_i) = \sum_{j=0}^{i-1} \sigma(T_j)\Delta W(T_j),$$

where in our example we would use $\sigma(T_j) := e^{-T_j}$ (Figure 2.7).

Next we consider the case where the particle has a preference for a certain direction, i.e., a *drift* (Figure 2.8). This is modeled by increments having a mean different from zero. The addition of a constant μ to a normally distributed random variable with mean zero will give a normally distributed random variable with mean μ. We want μ to be the drift per time unit and allow that μ may change over time. Thus we add $\mu(T_j)(T_{j+1} - T_j)$ to the corresponding increment over period T_j to T_{j+1}. If we also consider the starting point to be random, modeled by a random variable $X(0)$, we then

26

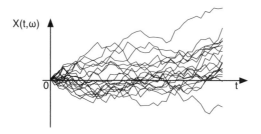

Figure 2.8. *Paths of (a discretization of) a Brownian motion with drift.*

consider processes of the form

$$X(T_i) = X(0) + \sum_{j=0}^{i-1} \mu(T_j) \underbrace{(T_{j+1} - T_j)}_{=:\Delta T_j} + \sigma(T_j) \Delta W(T_j) \quad ,$$

$$\underbrace{\text{Normal distributed with mean } \mu(T_j) \Delta T_j \text{ and}}$$
standard deviation $\sigma(T_j) \sqrt{\Delta T_j}$

i.e.,

$$X(T_{j+1}) - X(T_j) = \mu(T_j) \underbrace{(T_{j+1} - T_j)}_{=: \Delta T_j} + \sigma(T_j) \Delta W(T_j).$$

Our next generalization of the process is that the parameters $\mu(T_j)$ and $\sigma(T_j)$ could depend on the paths, i.e., are assumed to be random variables. This might appear odd since then one could create *any* time-discrete stochastic processes.[13] However, it would make sense to allow the parameters $\mu(T_j)$ and/or $\sigma(T_j)$ (used in the increment from $X(T_j)$ to $X(T_{j+1})$) to depend on the current state of $X(T_j)$ as in

$$\underbrace{X(T_{j+1}) - X(T_j)}_{=: \Delta X(T_j)} = X(T_j) \underbrace{(T_{j+1} - T_j)}_{=: \Delta T_j} + \sigma(T_j) \Delta W(T_j).$$

Here we would have $\mu(T_j) = \mu(T_j, X(T_j)) = X(T_j)$, i.e., a drift that is a random variable. It is an important fact that the drift for the increment from T_j to T_{j+1} is known in T_j. More generally, we allow μ and σ to be stochastic processes if they are $\{\mathcal{F}_t\}$-adapted.[14]

---

[13] If $T_i \mapsto S(T_i)$ is an arbitrary time-discrete stochastic process, we set $\sigma(T_j) := 0$ and $\mu(T_j) := (S(T_{j+1}) - S(T_j))/(T_{j+1} - T_j)$ and have $X(T_i) = S(T_i)$.

[14] The increment $\Delta W(T_j)$ is not $\mathcal{F}_{T_j}$-measurable. It is only $\mathcal{F}_{T_{j+1}}$-measurable. The requirements that $\mu(T_j)$ is $\mathcal{F}_{T_j}$-measurable excludes the example in footnote 13.

27

The continuous analog to the time-discrete processes considered above are Itô processes, i.e., processes of the form

$$X(T) = X(0) + \int_0^T \mu(t) \, dt + \int_0^T \sigma(t) \, dW(t), \qquad (2.3)$$

i.e.,

$$X(T) - X(0) = \int_0^T \mu(t) \, dt + \int_0^T \sigma(t) \, dW(t)$$

and as a short-hand we will write

$$dX(t) = \mu(t)dt + \sigma(t)dW(t).$$

While the dt part of (2.3) may be (and will be) understood pathwise as a Lebesgue (or even Riemann) integral, we need to define the $dW(t)$ part (as we will see), the *Itô integral*.[15] ◁|

2.6.1 Itô Integral

In this section we define the Itô integral $\int_S^T f(t, \omega) \, dW(t, \omega)$. We do not present the mathematical theory in full detail. For a more detailed discussion of the Itô integral see, e.g, [13, 20, 21, 27].

Definition 35 (The Filtration $\{\mathcal{F}_t\}$ generated by W): ⌐
Let $(\Omega, \mathcal{F}, P)$ denote a probability space and $W(t)$ a Brownian motion defined on $(\Omega, \mathcal{F}, P)$ (e.g., by the canonical setup). We define $\mathcal{F}_t$ as the σ-algebra generated by $W(s)$, $s \leq t$, i.e., the smallest σ-algebra, which contains sets of the form

$$\{\omega; W(t_1, \omega) \in F_1, \ldots, W(t_k, \omega) \in F_k, \} = \bigcap_{i=1}^k W(t_i)^{-1}(F_i)$$

for arbitrary $t_i < t$ and $F_i \subset \mathbb{R}$, $F_i \in \mathcal{B}(\mathbb{R})$ ($j \leq k$) and arbitrary $k \in \mathbb{N}$. Furthermore we assume that all sets of measure zero belong to $\mathcal{F}_t$. Then $\{\mathcal{F}_t\}$ is a filtration which we call the *filtration generated by W*. ⌐

Remark 36: W is a $\{\mathcal{F}_t\}$-adapted process.

---

[15] The $dW(t)$ part may not be interpreted as a Lebesgue-Stieltjes integral through $\sum f(\tau_j)(W(t_{j+1}) - W(t_j))$, $\tau_j \in [t_j, t_{j+1}]$, since $t \mapsto W(t, \omega)$ is not of bounded variation. Thus the limit will depend on the specific choice of $\tau_j \in [t_j, t_{j+1}]$; see Exercise 7.

Definition 37 (Itô Integral for Elementary Processes):
A stochastic process ϕ is called elementary, if

$$\phi(t, \omega) = \sum_{j \in \mathbb{N} \cup \{0\}} e_j(\omega) \, 1_{(t_j, t_{j+1}]}(t),$$

where $\{t_j \mid j \in \mathbb{N} \cup \{0\}\}$ is a strictly monotone sequence in $[0, \infty)$ with $t_0 := 0$ and $\{e_j \mid j \in \mathbb{N} \cup \{0\}\}$ a sequence of $\mathcal{F}_{t_j}$-measurable random variables and $1_{(t_j, t_{j+1}]}$ denotes the indicator function[16].[17]

For an elementary process we define the Itô integral as

$$\int_0^\infty \phi(t, \omega) \, dW(t) := \sum_{j \in \mathbb{N} \cup \{0\}} e_j(\omega) \, (W(t_{j+1}, \omega) - W(t_j, \omega)),$$

$$\int_S^T \phi(t, \omega) \, dW(t) := \int_0^\infty \phi(t, \omega) \, 1_{(S,T]}(t) \, dW(t).$$

Remark 38 (On the One-Sided Continuity of the Integrand): In some textbooks (e.g., [27]) the elementary integral is defined using the indicator function $1_{[t_j, t_{j+1})}$ in place of $1_{(t_j, t_{j+1}]}$. For continuous integrators, as we consider here ($W(t)$), it makes no difference which variant we use. However, if jump processes are considered (see, e.g., [29]), and also with respect to the interpretation of the integral as a trading strategy (see page 62), our definition is the better suited.

Lemma 39 (Itô Isometry): Let ϕ denote an elementary process such that $\phi(\cdot, \omega)$ is bounded. Then we have

$$E\left[\left(\int_S^T \phi(t, \cdot) \, dW(t)\right)^2\right] = E\left[\int_S^T \phi(t, \cdot)^2 \, dt\right].$$

Definition 40 (Itô Integral):
The *class of integrands of the Itô integral* is defined as the set of maps

$$f : [0, \infty) \times \Omega \mapsto \mathbb{R},$$

for which

1. f is a $\mathcal{B} \times \mathcal{F}$-measurable map,

---

[16] We define $1_{(t_j, t_{j+1}]}(t) = 1$, if $t \in (t_j, t_{j+1}]$ and $= 0$ else.
[17] Note that by this definition every path is elementary in the sense of Definition 7.

2. f is an $\mathcal{F}_t$-adapted process,

3. f is P-almost surely of finite quadratic variation, i.e., we have $E[\int_S^T f(t, \omega)^2 \, dt] < \infty$.

If f belongs to this class, then there exists an approximating sequence $\{\phi_n\}$ of elementary processes with

$$E^P\left[\int_S^T (f(t, \omega) - \phi_n(t, \omega))^2 \, dt\right] \to 0 \quad (\text{for } n \to \infty),$$

and the *Itô integral* is defined as the (unique) L_2 limit

$$\int_S^T f(t, \omega) \, dW(t) := \lim_{n \to \infty} \int_S^T \phi_n(t, \omega) dW(t).$$

Remark 41: For a proof of the statements made in this definition (e.g., the existence and uniqueness of the limit), see [27].

2.6.2 Itô Process

Definition 42 (Itô Process):
Let σ denote a stochastic process belonging to the class of integrands of the Itô integrals (see Definition 40) with

$$P\left(\int_0^t \sigma(\tau, \omega)^2 \, d\tau < \infty \; \forall t \geq 0\right) = 1$$

and μ an $\{\mathcal{F}_t\}$-adapted process with

$$P\left(\int_0^t |\mu(\tau, \omega)| \, d\tau < \infty \; \forall t \geq 0\right) = 1.$$

Then the process X defined through

$$X(t, \omega) = X(0, \omega) + \int_0^t \mu(s, \omega) \, ds + \int_0^t \sigma(s, \omega) \, dW(s, \omega),$$

where $X(0, \cdot)$ is $(\mathcal{F}_0, \mathcal{B}(\mathbb{R}))$-measurable is called *Itô process* (Remark: X is $\mathcal{F}_t$-adapted).

This definition is generalized by the m-dimensional Brownian motion as Definition 43.

Definition 43 (Itô Process (m-factorial, n-dimensional)[18]):

Let $W = (W_1, \ldots, W_m)^\top$ denote an m-dimensional Brownian motion defined on $(\Omega, \mathcal{F}, P)$. Let $\sigma_{i,j}$ $(i = 1, \ldots, n, \ j = 1, \ldots, m)$ denote stochastic processes belonging to the class of integrands of the Itô integral (compare Definition 40) with

$$P\left(\int_0^t \sigma_{i,j}(\tau, \omega)^2 \, d\tau < \infty \ \forall t \geq 0 \right) = 1, \quad i = 1, \ldots, n, \ j = 1, \ldots, m,$$

and μ_i $(i = 1, \ldots, n)$ an $\{\mathcal{F}_t\}$-adapted process with

$$P\left(\int_0^t |\mu_i(\tau, \omega)| \, d\tau < \infty \ \forall t \geq 0 \right) = 1, \quad i = 1, \ldots, n.$$

Then the (n-dimensional) process $X = (X_1, \ldots, X_n)^\top$ with $X(0, \cdot)$ being $(\mathcal{F}_0, \mathcal{B}(\mathbb{R}^n))$-measurable and

$$X_i(t, \omega) = X_i(0, \omega) + \int_0^t \mu_i(s, \omega) \, ds + \sum_{j=1}^m \int_0^t \sigma_{i,j}(s, \omega) \, dW_j(s, \omega), \quad (i = 1, \ldots, n),$$

is called *(n-dimensional) (m-factorial) Itô process.*[19] We will write X in the shorter matrix notation as

$$X(t, \omega) = X(0, \omega) + \int_0^t \mu(s, \omega) \, ds + \int_0^t \sigma(s, \omega) \cdot dW(s, \omega),$$

with

$$X = \begin{pmatrix} X_1 \\ \vdots \\ X_n \end{pmatrix} \qquad \mu = \begin{pmatrix} \mu_1 \\ \vdots \\ \mu_n \end{pmatrix} \qquad \sigma = \begin{pmatrix} \sigma_{1,1} & \cdots & \sigma_{1,m} \\ \vdots & & \vdots \\ \sigma_{n,1} & \cdots & \sigma_{n,m} \end{pmatrix}.$$

Remark 44 (Itô Process, Differential Notation): For an Itô process

$$X(T, \omega) = X(0, \omega) + \int_0^T \mu(t, \omega) \, dt + \int_0^T \sigma(t, \omega) \cdot dW(t, \omega),$$

as defined by Definitions 42 and 43 we will use the shortened notation

$$dX(t, \omega) = \mu(t, \omega) \, dt + \sigma(t, \omega) \cdot dW(t, \omega).$$

---

[18] Compare [27], Section 4.2

[19] The dimension n denotes the dimension of the image space. The factor dimension m denotes the number of (independent) Brownian motions needed to construct the process.

The stochastic integral

$$\int_{t_1}^{t_2} X(t)\, dY(t)$$

was defined pathwise for integrators Y that are Brownian motions ($dY = dW$) and for integrands X belonging to the class of integrands of the Itô integral. We extend the definition of the stochastic integral to integrators that are Itô processes.

Definition 45 (Integral with Itô Process as Integrator):
Let Y denote an Itô process of the form

$$dY = \mu\, dt + \sigma\, dW$$

and X a stochastic process such that $X(t)\mu(t)$ is integrable with respect to t and $X(t)\sigma(t)$ is integrable with respect to $W(t)$. Then we define

$$\int_{t_1}^{t_2} X(t)\, dY(t) := \int_{t_1}^{t_2} X(t)\mu(t)\, dt + \int_{t_1}^{t_2} X(t)\sigma(t)\, dW(t).$$

As in Remark 44 we will write

$$X(t)\, dY(t) := X(t)\mu(t)\, dt + X(t)\sigma(t)\, dW(t).$$

2.6.3 Itô Lemma and Product Rule

Theorem 46 (Itô Lemma (One Dimensional)[20]): Let X denote an Itô process with

$$dX(t) = \mu\, dt + \sigma\, dW.$$

Let $g(t, x) \in C^2([0, \infty] \times \mathbb{R})$. Then

$$Y(t) := g(t, X(t))$$

is an Itô process with

$$dY = \frac{\partial g}{\partial t}(t, X(t))\, dt + \frac{\partial g}{\partial x}(t, X(t))\, dX + \frac{1}{2}\frac{\partial^2 g}{\partial x^2}(t, X(t))\, (dX)^2, \qquad (2.4)$$

where $(dX)^2 = (dX)\,(dX)$ is given by formal expansion using

$$dt\, dt = 0, \qquad\qquad dt\, dW = 0,$$
$$dW\, dt = 0, \qquad\qquad dW\, dW = dt,$$

---

[20] Compare [27], Section 4.1.

i.e.,

$$(\mathrm{d}X)^2 = (\mathrm{d}X)\,(\mathrm{d}X) = (\mu\mathrm{d}t + \sigma\mathrm{d}W)\,(\mu\mathrm{d}t + \sigma\mathrm{d}W)$$
$$= \mu^2\,\mathrm{d}t\,\mathrm{d}t + \mu\,\sigma\,\mathrm{d}t\,\mathrm{d}W + \mu\,\sigma\,\mathrm{d}W\,\mathrm{d}t + \sigma^2\,\mathrm{d}W\,\mathrm{d}W = \sigma^2\mathrm{d}t$$

Theorem 47 (Itô Lemma[21]): Let X denote an n-dimensional, m-factorial Itô process
with
$$\mathrm{d}X(t) = \mu\,\mathrm{d}t + \sigma\,\mathrm{d}W.$$
Let $g(t, x) \in C^2([0, \infty] \times \mathbb{R}^n; \mathbb{R}^d)$, $g = (g_1, \ldots, g_d)^\mathsf{T}$. Then

$$Y(t) := g(t, X(t))$$

is an d-dimensional, n-factorial Itô process with

$$\mathrm{d}Y_k(t) = \frac{\partial g_k}{\partial t}(t, X(t))\,\mathrm{d}t + \sum_{i=1}^{n}\frac{\partial g_k}{\partial x_i}(t, X(t))\,\mathrm{d}X_i(t)$$
$$+ \frac{1}{2}\sum_{i,j=1}^{n}\frac{\partial^2 g_k}{\partial x_i x_j}(t, X(t))\,(\mathrm{d}X_i(t)\,\mathrm{d}X_j(t)),$$

where $\mathrm{d}X_i(t)\,\mathrm{d}X_j(t)$ is given by formal expansion using

$$\mathrm{d}t\,\mathrm{d}t = 0, \qquad\qquad \mathrm{d}t\,\mathrm{d}W_j = 0,$$
$$\mathrm{d}W_i\,\mathrm{d}t = 0, \qquad\qquad \mathrm{d}W_i\,\mathrm{d}W_j = \begin{cases} \mathrm{d}t & i = j \\ 0 & i \neq j \end{cases}.$$

Theorem 48 (Product Rule): Let X, Y, and $X_1, \ldots, X_n$ denote Itô processes. Then
we have

1. $\mathrm{d}(X\,Y) = Y\,\mathrm{d}X + X\,\mathrm{d}Y + \mathrm{d}X\,\mathrm{d}Y$

2. $\displaystyle \mathrm{d}\left(\prod_{i=1}^{N} X_i\right) = \sum_{i=1}^{N}\prod_{\substack{k=1 \\ k\neq i}}^{N} X_k\,\mathrm{d}X_i + \sum_{\substack{i,j=1 \\ j>i}}^{N}\prod_{\substack{k=1 \\ k\neq i,j}}^{N} X_k\,\mathrm{d}X_i\mathrm{d}X_j$

Proof: We prove only 1 since 2 follows from 1 by induction. We apply the Itô
lemma to the map

$$g : \mathbb{R} \times \mathbb{R} \times \mathbb{R} \to \mathbb{R}, \qquad g(t, x, y) := x \cdot y.$$

---

[21] Compare [27], Section 4.2.

We have

$$\frac{\partial g}{\partial t} = 0 \qquad\qquad \frac{\partial g}{\partial x} = y \qquad\qquad \frac{\partial g}{\partial y} = x$$

$$\frac{\partial g^2}{\partial x^2} = 0 \qquad\qquad \frac{\partial g^2}{\partial y^2} = 0$$

$$\frac{\partial g^2}{\partial xy} = 1 \qquad\qquad \frac{\partial g^2}{\partial yx} = 1$$

and thus

$$d(X\,Y) = d(g(X, Y)) = Y\,dX + X\,dY + dX\,dY.$$

$\square$

Theorem 49 (Quotient Rule): Let X and Y and $Y_1, \ldots, Y_n$ denote Itô processes where, $Y > c$ for a given $c \in \mathbb{R}$. Then we have

1. $$d\left(\frac{X}{Y}\right) = \frac{X}{Y}\left(\frac{dX}{X} - \frac{dY}{Y} - \frac{dX}{X} \cdot \frac{dY}{Y} + \left(\frac{dY}{Y}\right)^2\right)$$

2. $$d\left(\frac{1}{Y}\right) = -\frac{1}{Y^2}dY + \frac{1}{Y^3}dY\,dY$$

3. $$d\left(\prod_{i=1}^{N}\frac{1}{Y_i}\right) = \prod_{i=1}^{N}\frac{1}{Y_i} \cdot \sum_{i=1}^{N}\left(\frac{-dY_i}{Y_i} + \sum_{j\geq i}^{N}\frac{dY_i}{Y_i}\frac{dY_j}{Y_j}\right)$$

Lemma 50 (Drift Adjustment of Lognormal Process): Let $S(t) > 0$ denote an Itô process of the form

$$dS(t) = \mu(t)S(t)\,dt + \sigma(t)S(t)\,dW(t),$$

and $Y(t) := \log(S(t))$. Then we have

$$dY(t) = (\mu(t) - \frac{1}{2}\sigma^2(t))\,dt + \sigma(t)\,dW(t).$$

Proof: See Exercise 8. $\square$

Interpretation: The Itô lemma and its implications such as Lemma 50 may appear unfamiliar. They state that a nonlinear function of a stochastic process will induce a drift of the mean. This may be seen in an elementary example: Consider the time-discrete stochastic process $X(t_i)$ constructed from binomial distributed increments $\Delta B(t_i)$ (instead of Brownian increments $dW(t)$), where $\Delta B(t_i)$ are independent and attain with probability $p = \frac{1}{2}$ the value $+1$ or -1, respectively. Assuming $X(t_0) = 10$, we draw the process

$$X(t_{i+1}) = X(t_i) + \Delta B(t_i)$$

in Figure 2.9 (left), i.e., $\Delta X(t_i) = \Delta B(t_i)$. This process does not exhibit a drift. We have

$$X(t_i) = E(X(t_{i+1}) \mid \mathcal{F}_{t_i}).$$

In other words: In each node in Figure 2.9 the process X attains the mean of the values from the two child nodes.

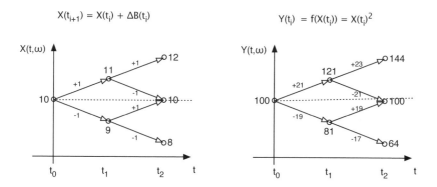

Figure 2.9. *Non-linear functions of stochastic processes induce a drift to the mean.*

As in Figure 2.9 (right) we then consider the process $Y(t_i) = f(X(t_i)) = X(t_i)^2$. This process exhibits in each time step a drift of the mean of $+1$. One can easily check that the increments of the process Y are given by

$$Y(t_{i+1}) = Y(t_i) + 1 + 2X(t_i)\,\Delta B(t_i)$$

35

(check this in Figure 2.9). This corresponds to the result stated by Itô's lemma (see Theorem 46 with $g(t, x) = f(x) = x^2$). Indeed we have

$$
\begin{aligned}
Y(t_{i+1}) &= (X(t_{i+1}))^2 = (X(t_i) + \Delta X(t_i))^2 \\
&= X(t_i)^2 + \Delta X(t_i)^2 + 2X(t_i)\,\Delta X(t_i) \\
&= Y(t_i) + \Delta B(t_i)^2 + 2X(t_i)\,\Delta B(t_i) \\
&= Y(t_i) + 1 + 2X(t_i)\,\Delta B(t_i) \\
&= Y(t_i) + \frac{1}{2}\frac{\partial^2 f}{\partial x^2}(X(t_i))\,(\Delta X(t_i))^2 + \frac{\partial f}{\partial x}(X(t_i))\,\Delta X(t_i).
\end{aligned}
$$

Obviously we may interpret Itô formula (2.4) in Itô's lemma as a (formal) Taylor expansion of $g(X + dX)$ up to the order $(dX)^2$. For the continuous case the higher order increments are (almost surely) 0. For the discrete case this is not the case. For example, consider $(X(t_i) + \Delta X(t_i))^3$ in the example above. ◁|

2.7 Brownian Motion with Instantaneous Correlation

In Definition 29 Brownian motion was defined through normally distributed increments $W(t) - W(s)$, $t > s$ having covariance matrix $(t - s)I_n$. In other words, for $W = (W_1, \ldots, W_n)$ the components are one-dimensional Brownian motions with pairwise independent increments, i.e., for $i \neq j$ we have that $W_i(t) - W_i(s)$ and $W_j(t) - W_j(s)$ are independent (thus uncorrelated).

We define the *Brownian motion with instantaneously correlated increments* as a special Itô process:

Definition 51 (Brownian Motion with Instantaneous Correlated Increments): ⌐
Let U denote an m-dimensional Brownian motion as defined in Definition 29. Let $f_{i,j}$ ($i = 1, \ldots, n$, $j = 1, \ldots, m$) denote stochastic processes belonging to the class of integrands of the Itô integral (see Definitions 40 and 43) with

$$
P\left(\int_0^t f_{i,j}(\tau, \omega)^2\, d\tau < \infty \;\forall\, t \geq 0\right) = 1, \quad i = 1, \ldots, n,\; j = 1, \ldots, m,
$$

furthermore let

$$
F(t) := (f_1, \ldots, f_m) := \begin{pmatrix} f_{1,1}(t) & \cdots & f_{1,m}(t) \\ \vdots & & \vdots \\ f_{n,1}(t) & \cdots & f_{n,m}(t) \end{pmatrix}
$$

denote an $n \times m$ matrix with

$$\sum_{j=1}^{m} f_{i,j}^2(t) = 1, \quad \forall\, i = 1, \ldots, n.$$

Then the Itô process

$$dW(t) = F(t) \cdot dU(t), \qquad W(0) = 0$$

(see Definition 43) is called *m-factorial, n-dimensional Brownian motion with factors* $f_j, j = 1, \ldots, m$. With $R := FF^\top$ we call R the *instantaneous correlation* and W a *Brownian motion with instantaneous correlation R. F is called the *factor matrix.* ⌋

 Interpretation: For simplicity let us consider a constant matrix $F = (f_1, \ldots, f_m)$. Then we have for time-discrete increments

$$\Delta W(T_k) = F \cdot \Delta U(T_k) = f_1\,\Delta U_1(T_k) + \cdots + f_m\,\Delta U_m(T_k). \qquad (2.5)$$

Note that $\Delta U_i(T_k)$ and $\Delta U_j(T_k)$ are independent. They may be interpreted as independent scenarios. If ω is a path with $\Delta U_i(T_k; \omega) \neq 0$ and $\Delta U_j(T_k; \omega) = 0$ for $j \neq i$, then we have from (2.5) that $\Delta W(T_k; \omega) = f_i\,\Delta U_i(T_k; \omega)$, i.e., on the path ω the vector W will receive increments corresponding to the scenario f_i (multiplied by the amplitude $\Delta U_i(T_k; \omega)$). If, for example, $f_1 = (1, \ldots, 1)^\top$, then the scenario corresponds to a parallel shift of W (by the shift size $\Delta U_1(T_k)$).

Our definition of a factor matrix does not allow arbitrary scenarios since we require that $\sum_{j=1}^{m} f_{i,j}^2(t) = 1$, i.e., that $R := FF^\top$ is a correlation matrix. By this assumption we ensure that the components of W_i of W are one-dimensional Brownian motions in the sense of Definition 29.

By means of the factor matrix F we may interpret the implied correlation structure R in a geometrical way. The calculation of F from a given R is a Cholesky decomposition.

We will make use of this construction in the modeling of interest rate curves (Chapter 19: LIBOR Market Model). Here the interpretation of the factors is given by movements of the interest rate curve. The possible shapes of an interest rate curve will then be investigated (Chapter 25)). The question of how to obtain a set of factors or reduce a given set of factors to the relevant ones is discussed in Appendices B.2 and B.3. ◁|

2.8 Martingales

Definition 52 (Martingale):
The stochastic process $\{X(t), \mathcal{F}_t \; ; \; 0 \le t < \infty\}$ is called a *martingale* with respect to the filtration $\{\mathcal{F}_t\}$ and the measure P if

$$X_s \; = \; E(X(t) \,|\, \mathcal{F}_s) \quad P\text{-almost surely}, \quad \forall 0 \le s < t < \infty. \tag{2.6}$$

If (2.6) holds for $\le$ in place of =, then X is a called submartingale. If (2.6) holds for $\ge$ in place of =, then X is called supermartingale.

Lemma 53 (Martingale Itô Processes Are Drift-Free): Let X denote an Itô process of the form

$$dX \; = \; \mu \, dt + \sigma \, dW \quad \text{under } P$$

with $E^P\!\left(\left(\int_0^T \sigma^2(t)\,dt\right)^{1/2}\right) < \infty.$ Then we have

$$X \text{ is a } P\text{-martingale} \quad \Leftrightarrow \quad \mu = 0 \quad (\text{i.e., } X \text{ is drift-free}).$$

2.8.1 Martingale Representation Theorem

Theorem 54 (Martingale Representation Theorem[22]): Let $W(t) = (W_1(t), \ldots, W_m(t))^{\mathsf{T}}$ denote an m-dimensional Brownian motion, $\mathcal{F}_t$ the corresponding filtration. Let $M(t)$ denote a martingale with respect to $\mathcal{F}_t$ with $\int_\Omega |M(t)|^2 \, dP < \infty$ ($\forall \, t \ge 0$).

Then there exists a stochastic process g (with $g(t)$ belonging to the class of integrands of the Itô integral) with

$$M(t) \; = \; M(0) \; + \; \int_0^t g(s) \, dW(s) \quad P\text{-almost surely}, \, \forall \, t \ge 0.$$

---

[22] See [27], Section 4.3 and [20], Theorem 4.15.

2.9 Change of Measure

Definition 55 (Measure with Density): ⌐
Let $(\Omega, \mathcal{F}, P)$ denote a measure space and ϕ a P-integrable nonnegative real-valued random variable. Then

$$Q(A) := \int_A \phi \, dP$$

defines a measure on $(\Omega, \mathcal{F})$, which we call *measure with density ϕ with respect to P*.
⌐

Definition 56 (Equivalent Measure): ⌐
Let P and Q denote two measures on the same measurable space $(\Omega, \mathcal{F})$.

1. Q is called *continuous with respect to P* $\Leftrightarrow$ $(P(A) = 0 \Rightarrow Q(A) = 0 \,\forall A \in \mathcal{F})$.

2. P and Q are called *equivalent* $\Leftrightarrow$ $(P(A) = 0 \Leftrightarrow Q(A) = 0 \,\forall A \in \mathcal{F})$.
⌐

Theorem 57 (Radon-Nikodým Density): Let P and Q denote two measures on a measurable space $(\Omega, \mathcal{F})$. Then we have

Q is continuous with respect to P $\Leftrightarrow$ Q has a density with respect to P.

Proof: See [1]. □

Definition 58 (Radon-Nikodým Density): ⌐
If Q is continuous with respect to P, then we call the density of Q with respect to P the *Radon-Nikodým density* and denote it by $\frac{dQ}{dP}$.
⌐

Theorem 59 (Change of Measure (Girsanov, Cameron, Martin)): Let W denote a (d-dimensional) P-Brownian motion and $\{\mathcal{F}_t\}$ the filtration generated by W fulfilling the *usual conditions*[23]. Let Q denote a measure equivalent to P (w.r.t. $\{\mathcal{F}_t\}$).

1. Then there exists a $\{\mathcal{F}_t\}$-previsible process C with

$$\left.\frac{dQ}{dP}\right|_{\mathcal{F}_t} = \exp\left(\int_0^t C(s)\,dW(s) - \frac{1}{2}\int_0^t |C(s)|^2\,ds\right). \qquad (2.7)$$

---

[23] Given a complete, filtered probability space $(\Omega, \mathcal{F}, \{\mathcal{F}_t \,|\, t \in [0, T]\}, P)$, the filtration $\{\mathcal{F}_t \,|\, t \in [0, T]\}$ satisfies the "usual conditions", if it is right-continuous (i.e., $\mathcal{F}_t = \cap_{\epsilon>0}\mathcal{F}_{t+\epsilon}$) and $\mathcal{F}_0$ (and thus $\mathcal{F}_t$ for every $t \in [0, t]$) contains all P-null sets of $\mathcal{F}$.

2. Let $T > 0$ fixed. Reversed, if ρ denotes a strictly positive P-martingale with respect to $\{\mathcal{F}_t \mid t \in [0, T]\}$ with $E^P(\rho) = 1$, then $\rho(t)$ has the representation (2.7) and defines (as a Radon-Nikodým process) a measure $Q = Q^T$ which is equivalent to P with respect to $\mathcal{F}_T$, given by

$$Q(A) := \int_A \rho(T)\, dP \qquad \forall A \in \mathcal{F}_T. \tag{2.8}$$

In any given case

$$\tilde{W}(t) := W(t) - \int_0^t C(s)\, ds \tag{2.9}$$

is a Q-Brownian motion (with respect to $\{\mathcal{F}_t\}$) (and for $t \leq T$ in the second case).

Remark 60 (Change of Measure–Change of Drift): Equation (2.9) may be written as

$$d\tilde{W}(t) = -C(t)\, dt + dW(t) \tag{2.10}$$

and we see that the *change of measure* (2.7) corresponds to a *change of drift* (2.10). The second case has a restriction on a finite time horizon T, which will be irrelevant in the following applications.

Remark 61 (Radon-Nikodým Process): Note that

$$\rho(t) := \exp\left(\int_0^t C(s)\, dW(s) - \frac{1}{2} \int_0^t |C(s)|^2\, ds \right)$$

is a P-martingale. From this we have that for $A \in \mathcal{F}_t$

$$Q(A) := \int_A \rho(T)\, dP = E^P\left(\mathbf{1}_A\, \rho(T)\right) = E^P\left(E^P\left(\mathbf{1}_A\, \rho(T) \mid \mathcal{F}_t\right)\right)$$

$$\stackrel{(\text{C.1})}{=} E^P\left(\mathbf{1}_A\, E^P\left(\rho(T) \mid \mathcal{F}_t\right)\right) = E^P\left(\mathbf{1}_A\, \rho(t)\right) = \int_A \rho(t)\, dP.$$

Thus, ρ defines a process of consistent Radon-Nikodým densities $\left.\frac{dQ}{dP}\right|_{\mathcal{F}_t}$ on $(\Omega, \mathcal{F}_t)$.

Exercise: (*Change of Measure* in a Binomial Tree): Calculate the probabilities (the measure) such that the process Y depicted in Figure 2.9 is a martingale.

Interpretation: The form of the change of measure (2.7) may be motivated via a simple calculation and derived for a time-discrete Itô process

$$\Delta X(T_i) = \mu(T_i)\,\Delta T_i + \sigma(T_i)\,\Delta W(T_i)$$

by elementary calculations.

At first: Let Z denote a normally distributed random variable with mean 0 and standard deviation σ on a probability space $(\Omega, \mathcal{F}, P)$. Under which measure will Z be normally distributed with mean c and standard deviation σ? The density of a normally distributed random variable with mean c and standard deviation σ is

$$z \mapsto \frac{1}{\sqrt{2\pi}\sigma} \exp\left(-\frac{(z-c)^2}{2\sigma^2}\right).$$

Thus we seek a change of measure $\frac{dQ}{dP}$ such that

$$\underbrace{\frac{1}{2\sqrt{\pi}\sigma}\exp\left(-\frac{(z-c)^2}{2\sigma^2}\right)dz}_{\frac{dQ}{dz}=\text{desired density under }Q} = dQ = \frac{dQ}{dP}dP = \underbrace{\frac{1}{2\sqrt{\pi}\sigma}\exp\left(-\frac{z^2}{2\sigma^2}\right)}_{\frac{dP}{dz}=\text{known density under }P}\underbrace{\frac{dQ}{dP}}_{=\,?}dz.$$

With

$$\exp\left(-\frac{(z-c)^2}{2\sigma^2}\right) = \exp\left(-\frac{z^2-2cz+c^2}{2\sigma^2}\right) = \exp\left(-\frac{z^2}{2\sigma^2}\right)\exp\left(\frac{cz-\frac{1}{2}c^2}{\sigma^2}\right)$$

it follows that the desired change of measure is

$$\frac{dQ}{dP} = \exp\left(\frac{cz-\frac{1}{2}c^2}{\sigma^2}\right). \qquad (2.11)$$

This corresponds to the term in (2.7).

To illustrate this we consider the time-discrete process

$$\Delta X(T_i) = \mu^P(T_i)\,\Delta T_i + \sigma(T_i)\,\Delta W(T_i) \qquad \text{under } P.$$

Under which measure is X a (time-discrete) martingale? We have

$$X \text{ is a } Q\text{-martingale} \;\Leftrightarrow\; \mu^Q(T_i) = 0 \;\Leftrightarrow\; E^Q(\Delta X(T_i)\,|\,\mathcal{F}_{T_i}) = 0$$

$$\Leftrightarrow\; E^Q\left(\frac{1}{\sigma(T_i)}\Delta X(T_i)\,|\,\mathcal{F}_{T_i}\right) = 0.$$

First consider a single increment: The random variable

$$\frac{1}{\sigma(T_i)} \Delta X(T_i) = \frac{\mu(T_i)}{\sigma(T_i)} \Delta T_i + \Delta W(T_i)$$

is (under P) normally distributed with mean $\frac{\mu(T_i)}{\sigma(T_i)} \Delta T_i$ and standard deviation $\sqrt{\Delta T_i}$. For the conditional (!) expectation under Q we have

$$E^Q\left(\frac{1}{\sigma(T_i)} \Delta X(T_i) \mid \mathcal{F}_{T_i}\right) = \frac{\mu(T_i)}{\sigma(T_i)} \Delta T_i + E^Q(\Delta W(T_i) \mid \mathcal{F}_{T_i})$$

Then

$$E^Q\left(\frac{1}{\sigma(T_i)} \Delta X(T_i) \mid \mathcal{F}_{T_i}\right) = 0$$

if $E^Q(\Delta W(T_i) \mid \mathcal{F}_{T_i}) = C(T_i) \Delta T_i$ where $C(T_i) = -\frac{\mu(T_i)}{\sigma(T_i)}$ (we correct for the drift). Given the considerations above we have to apply a change of measure

$$\left.\frac{dQ}{dP}\right|_{\Delta T_i} := \exp\left(C(T_i) \Delta W(T_i) - \frac{1}{2} C(T_i)^2 \Delta T_i\right)$$

for the time step ΔT_i (apply (2.11) with $Z = \Delta W(T_i)$, $c = C(T_i) \Delta T_i$ and $\sigma = \sqrt{\Delta T_i}$).

This is just the change of measure needed to make the increment $\Delta X(T_i)$ drift-free. Since the increments $\Delta W(T_i)$ are independent, we get the change of measure for the process X from T_0 to T_n by multiplying the Radon-Nikodým densities, i.e.,

$$\left.\frac{dQ}{dP}\right|_{T_n} := \prod_{i=0}^{n-1} \left.\frac{dQ}{dP}\right|_{\Delta T_i} = \prod_{i=0}^{n-1} \exp\left(C(T_i) \Delta W(T_i) - \frac{1}{2} C(T_i)^2 \Delta T_i\right)$$

$$= \exp\left(\sum_{i=0}^{n-1} C(T_i) \Delta W(T_i) - \frac{1}{2} \sum_{i=0}^{n-1} C(T_i)^2 \Delta T_i\right).$$

(2.12)

The change of measure $\left.\frac{dQ}{dP}\right|_{T_n}$ will make all increments $\Delta X(T_i)$ drift-free for $i = 0, \ldots, n-1$. Due to the independence of the increments $\Delta W(T_i)$, we obtain independence of the $\left.\frac{dQ}{dP}\right|_{\Delta T_i}$ and thus

$$E^Q(\Delta W(T_i)) = E^P\left(\Delta W(T_i) \prod_{j=0}^{n-1} \left.\frac{dQ}{dP}\right|_{\Delta T_j}\right) = \underbrace{\prod_{j \neq i} E^P\left(1 \left.\frac{dQ}{dP}\right|_{\Delta T_j}\right)}_{=1} E^P\left(\Delta W(T_i) \left.\frac{dQ}{dP}\right|_{\Delta T_i}\right).$$

The term (2.12) is a discrete version of (2.7). To some extent we have just proven a version of the change of measure theorem for time-discrete Itô processes.

The term $-C(t)^2 \, dt$ in (2.7) (or $-C(T_i)^2 \, \Delta T_i$ in (2.12)) may also be motivated as follows: The random variable

$$Z(t) := \left. \frac{dQ}{dP} \right|_{\mathcal{F}_t}$$

represents a density of the measure $Q|_{\mathcal{F}_t}$ with respect to the measure $P|_{\mathcal{F}_t}$. Since $Q|_{\mathcal{F}_t}$ should be a probability measure we must have

$$Q|_{\mathcal{F}_t}(\Omega) = E^{P|_{\mathcal{F}_t}}(Z(t)) = 1 \tag{2.13}$$

and with $Z(0) = 1$ this follows if Z is a martingale. Thus the *drift correction* $-C(t)^2 \, dt$ follows from Lemma 50 because Z is a lognormal process. ◁

2.10 Stochastic Integration

In the previous sections the following integrals were considered:

Maps:

- $\displaystyle\int_{t_1}^{t_2} f(t)\,dt$ Lebesgue or Riemann integral.
  Integral of a real valued function with respect to t.

Random Variables:

- $\displaystyle\int_{\Omega} Z(\omega)\,dP(\omega)$ Lebesgue integral.
  Integral of a random variable Z with respect to a measure P (cf. expectation).

Stochastic Processes:

- $\displaystyle\int_{\Omega} X(t_1,\omega)\,dP(\omega)$ Lebesgue integral.
  Integral of a random variable $X(t_1)$ with respect to a measure P; see Figure 2.10.

- $\displaystyle\int_{t_1}^{t_2} X(t)\,dt$ Lebesgue integral or Riemann integral.
  The (pathwise) integral of the stochastic process X with respect to t.

- $\displaystyle\int_{t_1}^{t_2} X(t)\,dW(t)$ Itô integral.
  The (pathwise) integral of the stochastic process X with respect to a Brownian motion W

The notion of a stochastic integral may be extended to more general integrands and/or more general integrators. For completeness we mention:

Definition 62 (Integral with Respect to a Semimartingale as Integrator):
Let Y denote a semimartingale (see Remark 63) of the form

$$Y(t) = A(t) + M(t),$$

where $A(t)$ is a process with locally bounded variation and $M(t)$ a local martingale. Let $X(t)$ denote a previsible process. Then we define

$$\int_{t_1}^{t_2} X(t)\,dY(t) \ := \ \int_{t_1}^{t_2} X(t)\,dA(t) \ + \ \int_{t_1}^{t_2} X(t)\,dM(t).$$

Remark 63 (Stochastic Integral): The class of processes (integrands) for which we may define a stochastic integral depends on the properties of the integrators (and vice versa). For continuous integrators (as the Brownian motion) the integrands merely

44

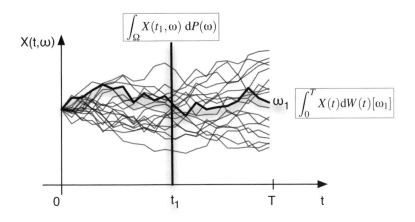

Figure 2.10. *Integration of stochastic processes.*

have to be adapted processes. To allow more general integrators one has to restrict to a smaller class of integrands, e.g., previsible processes. Compare Example 4.1 and Remark 4.4 in [13]. For more detailed discussion of the stochastic integral see [5], §5.5, and (especially for more general integrators) [13], §4, and [20], §3.

Further Reading: On stochastic processes: As introduction, see [27, 25]. For an in-depth discussion, see [20, 29, 31]. ◁|

2.11 Partial Differential Equations (PDEs)

We consider *partial differential equations* only marginally. In Section 7.2.2 we derive the Black-Scholes partial differential equation. The bridge from *stochastic differential equations* (SDE) to *partial differential equations* (PDE) is given through the Feynman-Kač theorem below.

2.11.1 Feynman-Kač Theorem

Theorem 64 (Feynman-Kač): Let X denote a d-dimensional Itô process, $X = (X_1, \ldots, X_d)$, following the stochastic differential equation (SDE):

$$dX_i(t) = \mu_i(t, X)\, dt + \sigma_i(t, X)\, dW_i^{\mathbb{Q}}(t) \qquad \text{on } [0, T] \text{ under } \mathbb{Q}.$$

Furthermore let V denote the solution of the parabolic partial differential equation (PDE):

$$\frac{\partial V}{\partial t} + \sum_{i=1}^{d} \mu_i(t, x)\frac{\partial V}{\partial x_i} + \frac{1}{2}\sum_{i,j=1}^{d} \gamma_{i,j}(t, x)\frac{\partial^2 V}{\partial x_i \partial x_j} \;=\; 0 \qquad \text{on } I \times \Omega$$

$$V(T, \cdot) \;=\; \phi(\cdot) \qquad \text{on } \mathbb{R}^d$$

with $\gamma_{i,j} = \sigma_i \sigma_j \rho_{i,j}$ and $dW_i(t)\, dW_j(t) = \rho_{i,j}\, dt$.

$$V(t, x) = E^{\mathbb{Q}}\left(\phi(X(T)) \mid X(t) = x\right) \qquad \text{for } (t, x) \in [0, T] \times \mathbb{R}^d \tag{2.14}$$

Remark 65 (Solving Backward in Time): Note that the PDE solves V backward in time. V is given at the final time T and the PDE described V for $t < T$. The meaning of this will become apparent in the following interpretation, however; to fully understand the interpretation in our context the knowledge of the next chapter is helpful.

 Interpretation: A stochastic differential equation decribes how a stochastic process X changes from $X(t)$ to $X(T)$. The change is the increment $\Delta X = \int_t^T dX$—a random variable. The increment describes how values change and give the probability for such a change.

If we now look at a stochastic process that is a function of X, say $V(t) = V(t, X(t))$, then Itô's lemma allows us to derive the stochastic differential equation for V, i.e., we have a formula for dV. The change from $V(t)$ to $V(T)$ is the increment $\Delta V = \int_t^T dV$, and again the increment describes how values change and give the probabilities for such a change. However, the probabilities do not change. If X moves from $X(t) = x$ to $X(T) = y$ with some (transition) probability density $\phi(t, x; T, y)$, then V moves from $V(t, x) =: u$ to $V(T, y) =: v$ with the same (transition) probability density $\phi(t, x; T, y)$.

So for V just the attained values change. The underlying transition probabilities are the ones of X. These are, of course, just the direct consequences of our assumption that V is a *function* of X. This assumption "splits" the definition of the stochastic process V into two parts: The transition probabilities are given by X. The values that are attained are given by $(t, x) \mapsto V(t, x)$.

Now, if we consider the function V to be the conditional expectation operator in (2.14), then it is not surprising that there is a rule of how to calculate $V(t)$ from $V(T)$ using the coefficients of the SDE of X, because these coefficients essentially contain the transition probabilities of X. This rule for calculating V is a partial differential equation.

The theorem makes two restrictive assumptions on the process V, namely:

- $V(t)$ is a function of some underlying state variables $X(t)$, and

- $V(t)$ is the conditional expectation of $V(T)$.

However, as we will learn in the next chapter, under suitable (and meaningful) assumptions, all the stochastic processes describing the prices of financial derivatives will fulfill these assumptions. Thus, the theorem allows us to derive the price of a financial derivative V as a function of some other quantity X through a PDE, given we know that function at some future time T. For financial derivatives, the time T function $V(T)$ is often known (e.g., for a call option on X we know that at time T its value is $\max(X(T) - K, 0)$). Solving the PDE gives the function $V(0)$ from $V(T)$. If today's value $x_0 := X(0)$ of X is known, then the function $V(0)$ gives today's value of V as $V(0, x_0)$. ◁|

 Further Reading: In [34] a short proof of the Feynman-Kač theorem is given. The instructive books of Wilmott, e.g., [40], give, besides an introduction to mathematical finance, an overview on PDE methods. The numerical methods for pricing derivatives by PDEs are discussed, e.g., in [10, 35, 40]. ◁|

2.12 List of Symbols

The following list of symbols summarizes the most important concepts from Chapter 2:

Symbol	Object	Interpretation
ω	element of Ω	State. In the context of stochastic processes: path.
Ω	set	State space.
X	random variable	Map which assigns an event/outcome (e.g., a number) to a state. Example: the payoff of a financial product (this may be interpreted as a snapshot of the financial product itself).
X	stochastic process	Sequence (in time) of random variables (e.g., the evolution of a financial product (could be its payoffs but also its value)).
$X(t)$, X_t	stochastic process evaluated at time t ($\equiv$ random variable)	See above.
$X(\omega)$	stochastic process evaluated in state ω	Path of X in state ω.
W	Brownian motion	Model for a continuous (random) movement of a particle with independent increments (position changes).
$\mathcal{F}$	σ-algebra (set of sets)	Set of information configurations (set of sets of states).
$\{\mathcal{F}_t \mid t \geq 0\}$	filtration	$\mathcal{F}_t$ is the information known at time t.

CHAPTER 3

Replication

> Nowadays people know the price of everything and the value of nothing.
>
> *Oscar Wilde*
> *The Picture of Dorian Gray [38]*

3.1 Replication Strategies

3.1.1 Introduction

We motivate the important principle of *replication* by considering the simplest financial derivative, the *forward contract*. Consider the following products:

A **(Forward Contract on Rain):** At time $T_1 > 0$ the amount of rain fallen $R(T_1)$ is measured (in millimeter) at a predefined place and the dollar amount

$$A(T_1) := (R(T_1) - X) \cdot \frac{\$}{\text{mm}}$$

is paid. Here, X denotes a constant reference amount of rain.

B **(Forward Contract on IBM Stock):** At time $T_1 > 0$ the value $S(T_1)$ of an IBM stock is fixed and the dollar amount

$$B(T_1) := (S(T_1) - X)$$

is paid. Here, X denotes a constant reference value.

These products may be interpreted as a guarantee or insurance.[1] The product A is a *weather derivative*, the product B is an *equity derivative*. What is a fair value for the product A and the product B? What do we expect to pay in T_0 (today) for such a guarantee?

Consider the product A: It appears that the determination of its fair value requires an exact assessment of the probability of rain at time T_1. So let $R_{T_1} : \Omega \to \mathbb{R}$ denote a random variable (rain quantity at time T_1, modeled over a suitable probability space—for $\omega \in \Omega$ the $R_{T_1}(\omega)$ denotes the quantity of rain that falls in state ω. Then, the random variable $A(T_1) := R_{T_1} - X$ defines the payoff of product A. We wish to determine the value $A(T_0)$ of this product at time T_0.

For the trivial case of a single-point distribution, i.e.,

$$R_{T_1}(\omega) = R^* = \text{const.} \qquad \forall\, \omega \in \Omega,$$

i.e., the amount of rain at time T_1 is R^* with probability 1; in other words: It is certain that in T_1 the amount of rain falling is R^*. Then product A pays at time T_1 the amount $A^* := R^* - X$ with probability 1. In this case, the product corresponds to a savings account. Let $P(T_0)$ denote the value that has to be invested at time T_0 (into a savings account) to receive in T_1 (including interest) the amount 1 (=: $P(T_1)$), then product A pays in T_1 the value A^* times the value of P. Since A is equivalent to A^* times P, we have

$$A(T_0) = A^*\, P(T_0)$$

(we assumed that the interest rate paid is independent of the amount invested, i.e., the interest is proportional to the amount invested). With $P(T_1) = 1$ this may be written as

$$\frac{A(T_0)}{P(T_0)} = \frac{A^*}{P(T_1)}.$$

For the (quite unrealistic) case of a one-point distribution (i.e., a deterministic payment) we may derive the value of the product A by comparing it to the value of another product with deterministic payoff (the savings account).

In the general case, where a probability distribution of $R(T_1)$ is known, it is questionable that we can replace the value $\frac{A^*}{P(T_1)}$ by, e.g., the expectation—

$$\frac{A(T_0)}{P(T_0)} \stackrel{?}{=} \mathrm{E}\left(\frac{A(T_1)}{P(T_1)}\right)$$

---

[1] The product B is a guarantee to buy the stock S at time T_1 for the amount X, since the product B pays the difference required to buy the stock at its value $S(T_1)$. The product A is an insurance against a change in rain quantity, which would be sensible for the operator of an irrigation system supplying water to farmers. If he suffers from a loss of earnings during a rainy season, he gets paid back a quantity proportional to the rain fallen.

—if so, this would imply that the risk (e.g., the variance of $A(T_1)$) does not influence the value. Since product A is an insurance against the risk of variations in rainfall, this appears to be nonsense.

Product B is very similar to product A. Instead of the amount of rain $R(T_1)$ the value of a stock $S(T_1)$ determines the payout. The considerations of the previous section apply accordingly. However, for B it is possible to determine its value in $T_0 = 0$ independent of the specific probability distribution of $S(T_1)$:

In time T_0 we take a loan with fixed interest rate such that the amount to be repaid at time T_1 is X. This loan pays us in T_0 the amount $X\,P(T_1; T_0)$.[2] In addition we acquire the stock at its current market price $S(T_0)$. In total we have to pay in T_0 the amount

$$V(T_0) = S(T_0) - XP(T_1; T_0). \tag{3.1}$$

At time T_1 this portfolio (stock + loan) (*replication portfolio*) will have the value

$$V(T_1) = S(T_1) - X,$$

i.e. it matches exactly the value of the product B at time T_1.

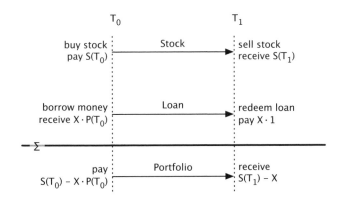

Figure 3.1. *Buy and hold replication strategy.*

Thus, we have found a strategy (*replication strategy*) to construct a portfolio for which its value in T_1 matches our product B exactly. This property is fulfilled in any state $\omega \in \Omega$ independent of the probability distribution of $S(T_1)$. Furthermore, the cost to acquire the portfolio in T_0, i.e., $V(T_0)$, is known. As a random variable $B(T_1)$

---

[2] We denote here by $P(T_1; T_0)$ the amount paid by a loan in T_0, which has to be repaid in T_1 by the amount $1 =: P(T_1; T_1)$ and such a loan is acquired X times. See Section 1.3.4 on the notation $P(T_1; T_0)$.

is indistinguishable from $V(T_1)$. Thus we have (valuing on a fair basis) that the price $B(T_0)$ of product B is equal to the cost of the portfolio $V(T_0)$. To determine $V(T_0)$ it is only required to know *today's* price for the stock and *today's* price for a loan (see Equation (3.1)).

The replication strategy used in this example is called *buy-and-hold* since all parts of the product required for replication are bought at time T_0. See Figure 3.1.

The essential difference between product A and product B is that for product B the quantity that carries the uncertainty (the stock) can be bought. In other words, it is possible to buy (and sell) "risk". We assume that we may buy and sell parts of stocks (and other traded products) in arbitrary (real-valued) quantities.

The key for the construction of the replication portfolio is that:

- It is possible to express today's value of a deterministic future payment, and thus there is a vehicle to transfer a deterministic future payment to an earlier time: $P(T_1; t)$ is a traded product.

- It is possible to buy or sell the underlying (the risk carrier) at any time in any quantity: $S(t)$ is a traded product.

It is a surprising consequence of the construction of a replication portfolio that:

- The real probabilities do not enter into the current value of the replication portfolio.

Our strategy is to construct a portfolio at time T_0 and wait until time T_1 (*buy-and-hold strategy*). Obviously this kind of strategy may be refined by restructuring the portfolio at other times. Dynamic (infinitesimal) restructuring will allow the replication of arbitrary (continuous) payouts, given that the underlying random variables (*the underlyings*) are traded products. It is this condition that prevents the replication of product A: We cannot buy or sell the random variable "rain".[3]

Consider again the equation

$$\frac{B(T_0)}{P(T_1; T_0)} = E^{\mathbb{P}}\left(\frac{B(T_1)}{P(T_1; T_1)}\right).$$

This equation would reduce the pricing (i.e., the calculation of $B(T_0)$) to the calculation of an expectation. The equation holds if

$$\frac{S(T_0)}{P(T_1; T_0)} = E^{\mathbb{P}}\left(\frac{S(T_1)}{P(T_1; T_1)}\right).$$

---

[3] Under certain conditions it would be possible to replicate product A: If there were a company whose stock value is perfectly correlated to the amount of rain falling, then the product may be replicated using stocks of that company. Of course, such a stock will only exist in some approximate sense, but then it might be possible to replicate product A in an approximate sense.

Obviously this equation cannot hold in general (or only the expectation of $S(T_1)$ would enter into the pricing and we would be ignorant of any risk). However, it is possible to change the measure such that the corresponding equation holds, i.e., there is a measure $\mathbb{Q}$ such that

$$\frac{S(T_0)}{P(T_1;T_0)} = E^{\mathbb{Q}}\left(\frac{S(T_1)}{P(T_1;T_1)}\right).$$

A change of measure is indeed an admissible tool when considering replication portfolios since—as we have seen—the real probabilities (the measure $\mathbb{P}$) do not enter into the calculation. To motivate this important concept let us consider the following (simple) example.

3.1.2 Replication in a Discrete Model

3.1.2.1 Example: Two Times (T_0, T_1), Two States (ω_1, ω_2), Two Assets (S, N)

Let S and N denote stochastic processes defined over a filtered probability space $(\Omega, \mathcal{F}, P, \mathcal{F}_t)$ with $\Omega = \{\omega_1, \omega_2\}$. Here $S(t, \omega)$ denotes the value of a financial product S and $N(t, \omega)$ denotes the value of a financial product N. We consider two points in time T_0 (present) and T_1 (future). We assume that S and N are traded at these times. Let the filtration be given by $F_{T_0} = \{\emptyset, \Omega\}$ and $F_{T_1} = \{\emptyset, \{\omega_1\}, \{\omega_2\}, \Omega\}$, i.e., in T_0 it is not possible to decide which of the states ω_1, ω_2 we are in, but in T_1 this information is known. Assume that the processes S and N are $\{F_t\}$-adapted, i.e., in T_0 we have $S(T_0, \omega_1) = S(T_0, \omega_2)$ and $N(T_0, \omega_1) = N(T_0, \omega_2)$. So, independent of the (unknown) state, the products have a defined value in T_0.

Given a derivative product (with the stochastic value process V) depending in T_1 on the attained state ω_i. We seek to determine the value of V in T_0. Our setup is illustrated in Figure 3.2.

To have the derivative product V replicated by a portfolio, we seek α, β such that

$$
\begin{aligned}
V(T_1, \omega_1) &= \alpha S(T_1, \omega_1) + \beta N(T_1, \omega_1), \\
V(T_1, \omega_2) &= \alpha S(T_1, \omega_2) + \beta N(T_1, \omega_2).
\end{aligned}
\tag{3.2}
$$

This system of equations has a solution (α, β) if

$$S(T_1, \omega_1)N(T_1, \omega_2) \neq S(T_1, \omega_2)N(T_1, \omega_1). \tag{3.3}$$

With this solution the value of the replication portfolio (and thus the cost of replication) is known in T_0, and thus the "fair" value of the derivative product V in T_0 as

$$V(T_0) = \alpha S(T_0) + \beta N(T_0). \tag{3.4}$$

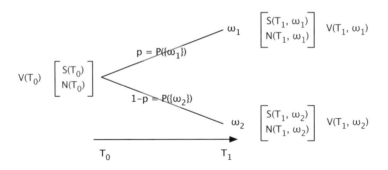

Figure 3.2. *Replication: The two-times two-states two-assets example.*

As before, the probabilities $P(\{\omega_1\})$, $P(\{\omega_2\})$ do not enter into the calculation of the cost of replication and thus $V(T_0)$. Let us investigate now whether $V(T_0)$ may be expressed in terms of an expectation. Assume that $N \neq 0$ and consider (3.2), (3.3), and (3.4) for N-relative prices. Equivalent to (3.2), (3.3), and (3.4) the portfolio satisfies in T_1:

$$\frac{V(T_1,\omega_1)}{N(T_1,\omega_1)} = \alpha\frac{S(T_1,\omega_1)}{N(T_1,\omega_1)} + \beta \qquad \frac{V(T_1,\omega_2)}{N(T_1,\omega_2)} = \alpha\frac{S(T_1,\omega_2)}{N(T_1,\omega_2)} + \beta \qquad (3.5)$$

with the solvability condition being

$$\frac{S(T_1,\omega_1)}{N(T_1,\omega_1)} \neq \frac{S(T_1,\omega_2)}{N(T_1,\omega_2)} \qquad (3.6)$$

and in T_0 we have

$$\frac{V(T_0)}{N(T_0)} = \alpha\frac{S(T_0)}{N(T_0)} + \beta. \qquad (3.7)$$

Obviously we have

$$\frac{V(T_0)}{N(T_0)} = E\left(\frac{V(T_1)}{N(T_1)}\Big|\mathcal{F}_{T_0}\right) \quad \Leftrightarrow \quad \frac{S(T_0)}{N(T_0)} = E\left(\frac{S(T_1)}{N(T_1)}\Big|\mathcal{F}_{T_0}\right).$$

Now let $q \in \mathbb{R}$ such that

$$\frac{S(T_0)}{N(T_0)} = q\frac{S(T_1,\omega_1)}{N(T_1,\omega_1)} + (1-q)\frac{S(T_1,\omega_2)}{N(T_1,\omega_2)},$$

i.e.,

$$q = \frac{\dfrac{S(T_0)}{N(T_0)} - \dfrac{S(T_1,\omega_2)}{N(T_1,\omega_2)}}{\dfrac{S(T_1,\omega_1)}{N(T_1,\omega_1)} - \dfrac{S(T_1,\omega_2)}{N(T_1,\omega_2)}}$$

(cf. (3.6)). If in addition to (3.6) we have

$$0 \le q \le 1, \tag{3.8}$$

then $\mathbb{Q}^N(\{\omega_1\}) := q$, $\mathbb{Q}^N(\{\omega_2\}) := 1 - q$ defines a probability measure, and under $\mathbb{Q}^N$ we have

$$\frac{S(T_0)}{N(T_0)} = E^{\mathbb{Q}^N}\left(\frac{S(T_1)}{N(T_1)}\Big|\mathcal{F}_{T_0}\right), \quad \text{i.e.,} \quad \frac{V(T_0)}{N(T_0)} = E^{\mathbb{Q}^N}\left(\frac{V(T_1)}{N(T_1)}\Big|\mathcal{F}_{T_0}\right). \tag{3.9}$$

So instead of calculating the parameters (α, β) for the replication portfolio we may alternatively calculate the measure $\mathbb{Q}^N$, i.e., the parameter q. At first it appears to be equally complex to calculate $\mathbb{Q}^N$ as it is to calculate the replication strategy. However, determining $\mathbb{Q}^N$ has a striking advantage: The calculation of $\mathbb{Q}^N$ is independent of the derivative V, but the pricing formula (3.9) is valid for all derivatives V. If $\mathbb{Q}^N$ has been determined once, all derivatives V may be priced as a $\mathbb{Q}^N$-expectation.

In Equations (3.5) to (3.9) we have considered N-relative prices, i.e., the value of any product V was expressed in fractions of N, i.e., by $\frac{V}{N}$. As long as $S \ne 0$, we may repeat these considerations with S-relative prices, i.e., we have

$$\frac{V(T_1,\omega_1)}{S(T_1,\omega_1)} = \alpha + \beta\frac{N(T_1,\omega_1)}{S(T_1,\omega_1)} \qquad \frac{V(T_1,\omega_2)}{S(T_1,\omega_2)} = \alpha + \beta\frac{N(T_1,\omega_2)}{S(T_1,\omega_2)} \tag{3.10}$$

with the same (α, β) and we obtain the same value for the replication portfolio $V(T_0)$, namely

$$\frac{V(T_0)}{S(T_0)} = \alpha + \beta\frac{N(T_0)}{S(T_0)}. \tag{3.11}$$

If we determine the measure $\mathbb{Q}^S$, such that

$$\frac{N(T_0)}{S(T_0)} = E^{\mathbb{Q}^S}\left(\frac{N(T_1)}{S(T_1)}\Big|\mathcal{F}_{T_0}\right), \quad \text{i.e.,} \quad \frac{V(T_0)}{S(T_0)} = E^{\mathbb{Q}^S}\left(\frac{V(T_1)}{S(T_1)}\Big|\mathcal{F}_{T_0}\right), \tag{3.12}$$

then the measure $\mathbb{Q}^S$ is different from $\mathbb{Q}^N$ - we have for example

$$\mathbb{Q}^S(\{\omega_1\}) = \frac{\dfrac{N(T_0)}{S(T_0)} - \dfrac{N(T_1,\omega_2)}{S(T_1,\omega_2)}}{\dfrac{N(T_1,\omega_1)}{S(T_1,\omega_1)} - \dfrac{N(T_1,\omega_2)}{S(T_1,\omega_2)}}.$$

$P(\{\omega_1\}) = p_{11} \cdot p_{21} \cdot p_{31}$
$P(\{\omega_2\}) = p_{11} \cdot p_{21} \cdot (1-p_{31})$
etc.

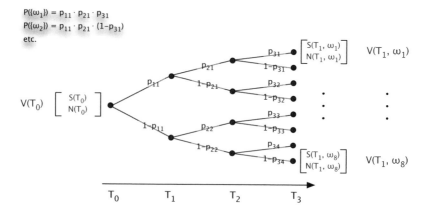

Figure 3.3. *Replication: Generalization to multiple states.*

However, the measure $\mathbb{Q}^S$ also allows us to calculate the value of all derivatives V as a $\mathbb{Q}^S$-expectation via (3.12).

We conclude this section with some remarks:

- Under the measure $\mathbb{Q}^N$ the relative price $\frac{S}{N}$ is a martingale: This is the defining property for the measure $\mathbb{Q}^N$.

- Under the measure $\mathbb{Q}^N$ the relative price $\frac{V}{N}$ of any replication portfolio V is a martingale:[4] This allows the calculation of the price of V as $\mathbb{Q}^N$-expectation of N-relative payout.

- The choice of the product that functions as reference (numéraire) is arbitrary (as long as it is nonzero). The measure $\mathbb{Q}$, under which any numéraire-relative replication portfolio becomes a martingale, depends on the chosen numéraire. This makes it possible to change the numéraire measure pair $(N, \mathbb{Q}^N)$, e.g., if this simplifies the calculation of the expectation.

- It is necessary to consider relative prices, such that
  - $\mathbb{Q}^N$ is independent of V,
  - $\mathbb{Q}^N$ is a probability measure, i.e., $\mathbb{Q}^N(\Omega) = 1$.

---

[4] This follows since the replication portfolio is a linear combination of martingales and the expectation is linear.

Exercise: Change of Measure

1. Reconsider the above under the numéraire S, calculate $\mathbb{Q}^S$ and show, that $\mathbb{Q}^N \neq \mathbb{Q}^S$.

2. Instead of relative prices consider absolute prices, i.e., choose as numéraire 1 and determine the measure $\mathbb{Q}^1$ for which

$$V(T_0) = E^{\mathbb{Q}^1}(V(T_1)|\mathcal{F}_{T_0}).$$

   Show that $\mathbb{Q}^1$ depends on the specific $V(T_1)$ and is thus not universal for all replication portfolios.

In case we wish to replicate a payoff $X(T)$, which depends on multiple states $\omega_1, \ldots, \omega_n$, then the above may be extended either by considering multiple time steps $T_0, T_1, \ldots, T_{n-1} = T$ (dynamic replication) or multiple assets $N, S_1, \ldots, S_{n-1}$—see Figure 3.3.

3.2 Foundations: Equivalent Martingale Measure

3.2.1 Challenge and Solution Outline

 Motivation: According to the previous example, the evaluation of a product through the value of a corresponding replication strategy may be given as the $\mathbb{Q}^N$-expectation of the N-relative price (where N denotes the chosen reference asset—the *numéraire*).

How can we determine the measure $\mathbb{Q}^N$, given that we know the price processes under the real measure $\mathbb{P}$? Note that under the measure $\mathbb{Q}^N$, the N-relative prices are martingales, i.e., as Itô processes they are drift-free (Lemma 53)—this is the defining property of $\mathbb{Q}^N$—, and a change of measure $\mathbb{P} \to \mathbb{Q}^N$ implies a change of drift for Itô processes (Theorem 59).

Thus, if we know the price processes under the real measure $\mathbb{P}$ (i.e., given a "model"), first we can derive the N-relative price processes under the real measure $\mathbb{P}$ by using the quotient rule. Then we can derive the equivalent martingale measure $\mathbb{Q}^N$ from the change of drift by using Girsanov's theorem (Theorem 59) (see Figure3.4).

Surprisingly, in our applications we *never* need to calculate the equivalent martingale measure: Since we know the processes under $\mathbb{Q}^N$ (we know their drift under $\mathbb{Q}^N$ and only the drift changes under a change of measure), we know the conditional probability densities under $\mathbb{Q}^N$, and these are enough to calculate expectations (see Definition 10).

What remains is to clarify under which conditions a given payoff function may be replicated and under which conditions an equivalent martingale measure exists. In this chapter we give a short overview of the corresponding mathematical foundations. In our later applications we will not discuss the existence of the equivalent martingale measure. ◁|

Problem Description

Given: $\mathcal{M} = \{X_1, \ldots, X_n\}$, where X_i denotes price processes under the (real) measure $\mathbb{P}$, and a contingent claim (payoff profile) $V^{(T)}$, where $V^{(T)}$ is a $\mathcal{F}_T$ measurable random variable.

Wanted: Price indication, i.e., the value $V(t)$ of $V^{(T)}$ at time $t < T$, especially $V(0)$, where V is a $\{\mathcal{F}_t\}$-adapted stochastic process.

Solution (Sketched)

- *Choice of numéraire*: Let $N \in \mathcal{M}$ denote a price process that may function as reference asset (*numéraire*). Without loss of generality let $N = X_1$.

- *Existence of a martingale measure for N-relative prices*: By Theorem 74 there exists a measure $\mathbb{Q}^N$, such that $\frac{X_i}{N}$ is a martingale with respect to $\mathcal{F}_t$, $\forall\, i = 1, \dots, n$.

- *Definition of the (candidate) of a value process of the replication portfolio*: Define

$$V(t) := N(t)\, \mathrm{E}^{\mathbb{Q}^N}\left(\frac{V^{(T)}}{N(T)} | \mathcal{F}_t\right).$$

Then $\frac{V(t)}{N(t)}$ is a $\mathbb{Q}^N$-martingale with respect to $\mathcal{F}_t$ (*tower law*).

- *Martingale representation theorem gives trading strategy*: Since the processes $\frac{X}{N} = (\frac{X_1}{N}, \dots, \frac{X_n}{N})$ and $\frac{V}{N}$ are martingales under $\mathbb{Q}^N$, there exists a $\phi = (\phi_1, \dots, \phi_n)$ such that

$$\frac{V(t)}{N(t)} = \frac{V(0)}{N(0)} + \int_0^t \phi(s) \cdot \mathrm{d}\left(\frac{X(s)}{N(s)}\right) \qquad (3.13)$$

(*Martingale Representation Theorem*).

- *The portfolio process ϕ may be chosen to be self-financing* by setting

$$\phi_1(t) := \frac{V(0)}{N(0)} + \sum_{i=2}^n \int_0^t \phi_i(s)\, \mathrm{d}\left(\frac{X_i(s)}{N(s)}\right) - \sum_{i=2}^n \phi_i(t)\frac{X_i(t)}{N(t)}.$$

Note: $\mathrm{d}\left(\frac{X_1}{N}\right) = \mathrm{d}\left(\frac{N}{N}\right) = 0$, i.e., (3.13) holds unchanged.

- *The portfolio process ϕ describes a replication portfolio for $V^{(T)}$*: We have

$$V^{(T)} = \sum_{i=1}^n \phi_i(T)X_i(T) = \sum_{i=1}^n \phi_i(0)X_i(0) + \int_0^T \phi(s) \cdot \mathrm{d}X(s).$$

- *The evaluation does not require explicit determination of the replication portfolio*: $V(t)$ is the value of the replication portfolio at time t and we have

$$\frac{V(t)}{N(t)} = \mathrm{E}^{\mathbb{Q}^N}\left(\frac{V^{(T)}}{N(T)} | \mathcal{F}_t\right). \qquad (3.14)$$

Real Measure   Martingale Measure

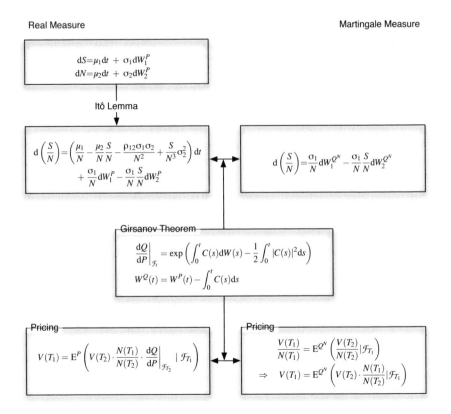

Figure 3.4. *Real measure versus martingale measure.*

3.2.2 Steps toward the Universal Pricing Theorem

We will now list the central building blocks of risk-neutral pricing, towards the *universal pricing theorem* (3.14):

- *Trading strategies* and characterization of *self-financing* of a trading strategy— in the sequel in Numbers 66 to 72.

- *Equivalent martingale measure*, to obtain a trading strategy from the martingale representation theorem, as a candidate for the replication strategy—in the sequel in Numbers 73 to 74.

- *Replication of given payoff functions* and *universal pricing theorem* followed by definition of a martingale process $t \mapsto E^{Q^N}(\frac{V^{(T)}}{N(T)} \mid \mathcal{F}_t)$ for the given payoff function V^T—in the sequel in Numbers 76 to 79.

This program is found in this order in most of the literature, some being less technical ([3]), some being more technical ([5, 27]). We will usually sketch the theory without technical proofs but include references to the literature.

Basic Assumptions (Part 1 of 3)

Let $\mathcal{M} = \{X_1, \ldots, X_n\}$ denote a family of (Itô) stochastic processes, defined over the filtered probability space $(\Omega, \mathcal{F}, \mathbb{P}, \{\mathcal{F}_t\})$

$$dX_i = \mu_i^{\mathbb{P}} dt + \sum_{j=1}^{m} \sigma_{i,j} dW_j,$$

where $\{\mathcal{F}_t\}$ is (the augmentation of) the filtration generated by the (independent) Brownian motions W_j and the coefficients $\mu^{\mathbb{P}}$ and σ fulfill the integrability conditions

$$\int_0^T |\mu_i^{\mathbb{P}}(s)| \, ds \; < \; \infty \qquad \int_0^T \sum_{j=1}^{m} |\sigma_{i,j}(s)|^2 \, ds \; < \; \infty \qquad \mathbb{P}\text{-almost surely.}$$

The elements of $\mathcal{M}$ are price processes of traded assets ($\mathcal{M}$ represents the market). We consider these only up to a finite time horizon T and thus furthermore assume $\mathcal{F} = \mathcal{F}_T$.

3.2.2.1 Self-Financing Trading Strategy

Definition 66 (Portfolio, Trading Strategy, Self-Financing): ⌐

1. An n-dimensional $\{\mathcal{F}_t\}$-progressively measurable process $\phi = (\phi_1, \ldots, \phi_n)$ with

$$\int_0^T |\phi_i \mu_i^{\mathbb{P}}(s)| \, ds \; < \; \infty \qquad \int_0^T \sum_{j=1}^m |\phi_i \sigma_{i,j}(s)|^2 \, ds \; < \; \infty \qquad \mathbb{P}\text{-almost surely}$$

   is called *portfolio process* or *trading strategy*.

2. The *value of the portfolio* ϕ at time t is given by the scalar product

$$V_\phi(t) := \phi(t) \cdot X(t) = \sum_{i=1}^n \phi_i(t) X_i(t).$$

   The process V_ϕ is called *wealth process* of the portfolio ϕ.

3. The *gain process* of the portfolio ϕ is defined by

$$G_\phi(t) := \int_0^t \phi(\tau) \cdot dX(\tau) = \sum_{i=1}^n \int_0^t \phi_i(\tau) \cdot dX_i(\tau).$$

4. The portfolio process ϕ is called *self-financing*, if

$$V_\phi(t) = V_\phi(0) + G_\phi(t) \; \forall \; t \in [0, T] \qquad \mathbb{P}\text{-almost surely}, \qquad (3.15)$$

   i.e.,

$$dV_\phi = \phi \cdot dX. \qquad (3.16)$$

   ⌙

Remark 67: The integrability condition in 1 of Definition 66 ensures that the Itô integral $\int_0^t \phi(s) \cdot dX(s)$ exists.

Interpretation: We interpret $\{X_i \mid i = 1, \ldots, n\}$ as a family of stock price processes and $\phi(t) := (\phi_1(t), \ldots, \phi_n(t))$ as a stock portfolio, i.e., ϕ_i denotes the number of stocks X_i in the portfolio.

The relation (3.16) may be interpreted as follows: A change in the portfolio value V_ϕ comes only from changes in the stocks X, *as if* the portfolio remained unchanged, i.e., we hold a portfolio of ϕ stocks and gain over dt the amount $\phi \cdot dX$.

The interpretation of condition (3.16) becomes clear if we consider the time-discrete variant of a self-financing strategy:

- At time T_i there exists a $\phi(T_i)$ of the products $X(T_i)$. Its value is $V_\phi(T_i) = \phi(T_i) \cdot X(T_i)$.

- At time T_i the products X are traded at prices $X(T_i)$ and the portfolio is re-arranged in a *self-financing* manner. The portfolio changes by $\Delta\phi(T_i) = \phi(T_{i+1}) - \phi(T_i)$. This change does not imply a change in value:

$$\Delta\phi(T_i) \cdot X(T_i) = 0. \tag{3.17}$$

- Over the interval $\Delta T_i := T_{i+1} - T_i$ the value of the products X changes by $\Delta X(T_i) := X(T_{i+1}) - X(T_i)$. The value of the portfolio thus changes by $\phi(T_{i+1}) \cdot \Delta X(T_i)$, i.e. the value then is $\phi(T_{i+1}) \cdot X(T_{i+1})$.

- For the value change of the portfolio we thus find

$$
\begin{aligned}
\Delta V_\phi(T_i) &= \phi(T_{i+1}) \cdot X(T_{i+1}) - \phi(T_i) \cdot X(T_i) \\
&= \phi(T_{i+1}) \cdot \Delta X(T_i) + \underbrace{\Delta\phi(T_i) \cdot X(T_i)}_{= 0 \text{ by } (3.17)}.
\end{aligned}
$$

This corresponds to the continuous case:

$$\mathrm{d}V_\phi = \phi \cdot \mathrm{d}X.$$

Note that this interpretation is consistent with the definition of the elementary Itô integral, see also Remark 38.

- (and so on.)

Remark: We will discuss this time-discrete variant of a trading strategy (which does not result in a complete replication) in Chapter 7. ◁|

3.2.2.2 Relative Prices

Definition 68 (Numéraire): ⌐
A price process $N \in \mathcal{M}$ is called *numéraire* (on $[0, T]$) if

$$\mathbb{P}(\{N(t) > 0 \mid \forall\, t \leq T\}) = 1.$$

⌐

Basic Assumptions (Part 2 of 3)

For the remainder of this chapter we assume that X_1 is a numéraire and we will use the symbol $N := X_1$. Furthermore we require that the chosen numéraire is such that the integrability condition formulated in Definition 66 is equivalent to the corresponding integrability condition under the normalized system of the relative price processes $\tilde{X} = (\frac{X_2}{N}, \ldots, \frac{X_n}{N})$, i.e., with

$$d\tilde{X} = \tilde{\mu} \, dt + \tilde{\sigma} \, dW$$

we require that for all $\{\mathcal{F}_t\}$-progressively measurable $\tilde{\phi} = (\phi_2, \ldots, \phi_n)$, $\phi = (0, \phi_2, \ldots, \phi_n)$

$$\int_0^T |\phi(s) \cdot \mu(s)| + \|\phi(s) \cdot \sigma(s)\|_2^2 \, ds < \infty \qquad \mathbb{P}\text{-almost surely,}$$

$$\Leftrightarrow \quad \int_0^T |\tilde{\phi}(s) \cdot \tilde{\mu}(s)| + \|\tilde{\phi}(s) \cdot \tilde{\sigma}(s)\|_2^2 \, ds < \infty \qquad \mathbb{P}\text{-almost surely.}$$

Remark 69 (Assumptions on the Numéraire): We pose an additional assumption on the numéraire, namely that the integrability condition can be equivalently formulated with respect to the normalized system $\frac{X}{N}$, i.e., the relative price processes. In many cases this follows from a more specific choice for the numéraire, e.g., for a locally riskless numéraire $dN = r(t)N(t) \, dt$ or for $dN = r(t)N(t) \, dt + \sum_{j=1}^m \sigma_{1,j}(t)N(t) \, dW_j$ with bounded $\sigma_{1,j}$. Thus, in many works the requirement on the normalized system does not appear in this form since it is implied by the specific choice of the numéraires.

Lemma 70 (Condition of Self-Financing Is Invariant under a Move to Relative Prices): Let $\phi = (\phi_1, \ldots, \phi_n)$ denote a portfolio. Then we have

$$dV_\phi = \sum_{i=1}^n \phi_i \, dX_i \Leftrightarrow d\frac{V_\phi}{N} = \sum_{i=1}^n \phi_i \, d\frac{X_i}{N}.$$

In other words, ϕ is self-financing if and only if

$$d\frac{V_\phi}{N} = \sum_{i=1}^n \phi_i \, d\frac{X_i}{N}.$$

Proof: Let

$$dV_\phi = \sum_{i=1}^n \phi_i \, dX_i. \tag{3.18}$$

Then we find from

$$dX_i \left(\frac{1}{N} + d\left(\frac{1}{N} \right) \right) + X_i\, d\left(\frac{1}{N} \right) = d\frac{X_i}{N}$$

$$
\begin{aligned}
d\frac{V_\phi}{N} &= dV_\phi \cdot \left(\frac{1}{N} + d\left(\frac{1}{N} \right) \right) + V_\phi\, d\left(\frac{1}{N} \right) \\
&\overset{(3.18)}{=} \sum_{i=1}^{n} \phi_i\, dX_i \left(\frac{1}{N} + d\left(\frac{1}{N} \right) \right) + \sum_{i=1}^{n} \phi_i X_i\, d\left(\frac{1}{N} \right) \\
&= \sum_{i=1}^{n} \phi_i\, d\frac{X_i}{N} = \sum_{i=2}^{n} \phi_i\, d\frac{X_i}{N}.
\end{aligned}
$$

Conversely we have from

$$d\frac{V_\phi}{N} = \sum_{i=1}^{n} \phi_i\, d\frac{X_i}{N} \tag{3.19}$$

with $d\frac{X_i}{N}\,(N + dN) + \frac{X_i}{N}\, dN = dX_i$, that

$$
\begin{aligned}
dV_\phi &= d\left(\frac{V_\phi}{N} N \right) = d\frac{V_\phi}{N}\,(N + dN) + \frac{V_\phi}{N}\, dN \\
&\overset{(3.19)}{=} \sum_{i=1}^{n} \phi_i\, d\frac{X_i}{N}(N + dN) + \sum_{i=1}^{n} \phi_i \frac{X_i}{N}\, dN \\
&= \sum_{i=1}^{n} \phi_i\, dX_i.
\end{aligned}
$$

$\square$|

Remark 71: Note that due to the choice of the numéraire, ϕ_1 does not enter into the sum over $d\frac{X_i}{N}$. We will use this in the following lemma to construct a self-financing replication portfolio.

The move to relative prices makes it possible to construct a self-financing portfolio from a partial portfolio $\phi_2, \ldots, \phi_n$ which fulfills for a given process V the relation

$$d\frac{V}{N} = \sum_{i=2}^{n} \phi_i\, d\frac{X_i}{N}. \tag{3.20}$$

Note that in (3.20) V stands for an arbitrary process, not limited to V_ϕ (the value process of the portfolio). This process becomes the value process of a self-financing portfolio by the following choice of ϕ_1 (replication):

Lemma 72 (Self-Financing Strategy for Given Partial Portfolio and Given Initial Value): Let $\tilde{\phi} = (\phi_2, \ldots, \phi_n)$ be $\{\mathcal{F}_t\}$-progressively measurable and such that

$$\int_0^t |\tilde{\phi}(s)\tilde{\mu}(s)| + \|\tilde{\phi}(s) \cdot \tilde{\sigma}(s)\|_2^2 \, ds \; < \infty \qquad \mathbb{P}\text{-almost surely.}$$

Then we have that $\phi = (\phi_1, \ldots, \phi_n)$ where

$$\phi_1(t) := \frac{V_0}{N(0)} + \sum_{i=2}^n \int_0^t \phi_i(s) \, d\left(\frac{X_i(s)}{N(s)}\right) - \sum_{i=2}^n \phi_i(t) \frac{X_i(t)}{N(t)} \qquad (3.21)$$

defines a self-financing strategy with $V_\phi(0) = V_0$.

Proof: In order to show that ϕ is self-financing it is sufficient to show that (Lemma 70)

$$d\frac{V_\phi}{N} = \sum_{i=1}^n \phi_i \, d\frac{X_i}{N}. \qquad (3.22)$$

From Equation (3.21) it follows (note that $\frac{X_1}{N} = 1$ and $d\frac{X_1}{N} = 0$) that

$$\sum_{i=1}^n \phi_i(t) \frac{X_i(t)}{N(t)} = \frac{V_0}{N(0)} + \sum_{i=1}^n \int_0^t \phi_i(s) d\left(\frac{X_i(s)}{N(s)}\right)$$

and thus (3.22). The initial condition $V_\phi(0) = V_0$ follows from setting $t = 0$. $\qquad \square$

The relation (3.20) would follow from the martingale representation theorem if the corresponding processes were martingales. It thus becomes natural to ask for a measure under which the relative prices $\frac{X_i}{N}$ become martingales.

3.2.2.3 Equivalent Martingale Measure

Definition 73 (Equivalent Martingale Measure):
Let N denote a numéraire. A probability measure $\mathbb{Q}^N$ defined on $(\Omega, \mathcal{F})$ is called *equivalent martingale measure with respect to N (equivalent N-martingale measure)* (on $\mathcal{M}$), if

1. $\mathbb{Q}^N$ and $\mathbb{P}$ are equivalent, and

2. the N relative price processes $\frac{X_i}{N}$ ($i = 1, \ldots, n$) are $\mathcal{F}_t$-martingales with respect to $\mathbb{Q}^N$.

Theorem 74 (Equivalent Martingale Measure, Existence and Uniqueness): Let $\tilde{X} = (\frac{X_2}{N}, \ldots, \frac{X_n}{N})$ with

$$d\tilde{X} = \tilde{\mu} \, dt + \tilde{\sigma} \cdot dW(t).$$

Suppose that there exists a progressively measurable process $C : [0, T] \times \Omega \to \mathbb{R}^{n-1}$ such that $\int_0^T \|C(s)\|^2 \, ds < \infty$ and

$$\tilde{\mu} = \tilde{\sigma} \cdot C \qquad \lambda \times \mathbb{P}\text{-almost surely on } [0, T] \times \Omega.$$

Let $Z(t) := \exp(\int_0^t C(s) \cdot dW(s) - \frac{1}{2} \int_0^t \|C(s)\|^2 \, ds)$.

1. (Existence) If $\mathbb{E}^\mathbb{P}(Z(T)) = 1$ then the measure $\mathbb{Q}^N$ defined through

$$\left.\frac{d\mathbb{Q}^N}{d\mathbb{P}}\right|_{\mathcal{F}_t} = Z(t) \qquad (3.23)$$

   is an equivalent martingale measure.

2. (Uniqueness) If further the process C defined in (74) is unique, then $\mathbb{Q}^N$ is the unique equivalent martingale measure.

Remark 75 (State Price Deflator): If $\mathbb{E}^\mathbb{P}(Z(T)) < 1$, then (3.23) does not define a Radon-Nikodým process of a probability measure. Then it is not possible to define an equivalent martingale (probability) measure through (3.23). However, it is still possible to define a universal pricing theorem through

$$\frac{V(t)}{N(t)} = \mathbb{E}^\mathbb{P}\left(\frac{V^{(T)}}{N(T)}Z(T) \,\big|\, \mathcal{F}_t\right).$$

The process $\frac{Z}{N}$ is called *state price deflator*.

3.2.2.4 Payoff Replication

Given the existence of an equivalent martingale measure, we can define a self-financing trading strategy replicating a given contingent claim V^T.

Definition 76 (Admissible Trading Strategy): ⌐

A self-financing trading strategy ϕ is called *admissible* if

$$V_\phi \geq -K$$

for some finite K. ⌋

Definition 77 (Attainable Contingent Claim, Complete): ⌐

Let $T > 0$. A $\mathcal{F}_T$-measurable random variable $V^{(T)}$ is called *attainable contingent claim* (or *replicable payoff*) if there exists (at least one) admissible trading strategy ϕ such that

$$V_\phi(T) = V^{(T)}.$$

The market $\mathcal{M}$ is called *complete* if any contingent claim (payoff function) is attainable (replicable). ⌋

Basic Assumptions (Part 3 of 3)

Using the martingale representation theorem on $E^{\mathbb{Q}^N}(\frac{V^{(T)}}{N(T)} \mid \mathcal{F}_t)$ we find a representation $\tilde{\theta} \cdot dW^{\mathbb{Q}^N}$. In order to find the replication portfolio $\tilde{\phi}$ such that $\tilde{\theta} \cdot dW^{\mathbb{Q}^N} = \tilde{\phi} \cdot d\tilde{X} = \tilde{\phi} \cdot \tilde{\sigma} \cdot dW^{\mathbb{Q}^N}$ we need to solve $\tilde{\phi} \cdot \tilde{\sigma} = \tilde{\theta}$. We assume that for any $\tilde{\theta}$, $\int_0^T \|\tilde{\theta}(s)\|_2^2 \, ds < \infty$, there exists a $\tilde{\phi}$, $\int_0^T \|\tilde{\theta}(s)\|_2^2 \, ds < \infty$ such that

$$\tilde{\phi} \cdot \tilde{\sigma} = \tilde{\theta}.$$

What follows is

Theorem 78 (Self-Financing Replication Portfolio of a Given Payoff Function): Let $\mathbb{Q}^N$ denote an equivalent martingale measure and $V^{(T)}$ a given payoff function (contingent claim) with $E^{\mathbb{Q}^N}\left(\left|\frac{V^{(T)}}{N(T)}\right|^2\right) < \infty$. Furthermore let

$$V(t) := N(t) \, E^{\mathbb{Q}^N}\left(\frac{V^{(T)}}{N(T)} \mid \mathcal{F}_t\right).$$

From that it follows that $\frac{V}{N}$ is a $\mathbb{Q}^N$-martingale. Furthermore let $\phi_2, \dots, \phi_n$ be as in the martingale representation theorem, i.e.,

$$d\frac{V}{N} = \sum_{i=2}^{n} \phi_i \, d\frac{X_i}{N}.$$

For the portfolio $\phi = (\phi_1, \dots, \phi_n)$ let ϕ_1 be chosen as in Lemma 72 with $V_\phi(0) := V(0)$. Then ϕ is a self-financing replication portfolio of V; i.e., we have

$$V(t) = V_\phi = \sum_{i=1}^{n} \phi_i X_i.$$

In other words, $V^{(T)}$ is an attainable contingent claim.

Proof: That $\frac{V}{N}$ is a $\mathbb{Q}^N$-martingale follows immediately from the definition and the *tower law*[5]. From the definition of the martingale measure, $\frac{X_i}{N}$ are $\mathbb{Q}^N$-martingales and martingale representation theorem gives the existence of $\phi_2, \dots, \phi_n$ with

$$d\frac{V}{N} = \sum_{i=2}^{n} \phi_i \, d\frac{X_i}{N}.$$

Choosing ϕ_1 as in Lemma 72 will make ϕ self-financing.

---

[5] See Exercise 2 on page 479

For the value process V_ϕ we find from Lemma 70 that

$$d\frac{V_\phi}{N} = \sum_{i=2}^{n} \phi_i \, d\frac{X_i}{N}$$

and thus $d\frac{V_\phi}{N} = d\frac{V}{N}$, so together with $V_\phi(0) = V(0)$

$$V(t) = V_\phi(t) = \sum_{i=1}^{n} \phi_i \, dX_i.$$

$\Box|$

Since the definition of the martingale measure $\mathbb{Q}^N$ was totally independent of V we have

Theorem 79 (Risk-Neutral Valuation Formula, Universal Pricing Theorem):
Let ϕ denote an admissible self-financing trading strategy and $\mathbb{Q}^N$ an equivalent martingale measure with respect to the numéraire N with $E^{\mathbb{Q}^N}\left(\left|\frac{V_\phi(T)}{N(T)}\right|^2\right) < \infty$. Then

$$\frac{V_\phi(t)}{N(t)} = E^{\mathbb{Q}^N}\left(\frac{V_\phi(T)}{N(T)} \,\Big|\, \mathcal{F}_t\right). \tag{3.24}$$

3.3 Excursus: Relative Prices and Risk-Neutral Measures

3.3.1 Why relative prices?

In our simple example from Section 3.1.2, using time-discrete processes with discrete states, it became apparent that the useful martingale measure may be chosen independently of the payoff function only if we consider relative prices. However, looking at the continuous time theory from Section 3.2.2, the consideration of relative prices seems to be less motivated. To motivate the *relative prices* we repeat steps of the continuous time theory by considering absolute prices:

Let

$$dX_i = \mu_i \, dt + \sigma_i \, dW_i^{\mathbb{P}}, \quad i = 1, \ldots, n$$

denote the price processes of n traded assets. We assume that $\sigma_i > 0$, $\forall \, i$. The processes are given under the real measure $\mathbb{P}$. Since $\sigma_i > 0$, we have from Girsanov's theorem that there exists a measure $\mathbb{Q}^1$ such that the processes are $\mathbb{Q}^1$-martingales, i.e.,

$$dX_i = \sigma_i \, dW_i^{\mathbb{Q}^1}, \quad i = 1, \ldots, n.$$

If V^T denotes a given payoff function, then

$$V(t) := \mathrm{E}^{\mathbb{Q}^1}(V^{(T)} \mid \mathcal{F}_t) \tag{3.25}$$

is a $\mathbb{Q}^1$-martingale, and from the martingale representation theorem we get the existence of a portfolio process $(\phi_1, \ldots, \phi_n)$ with

$$dV(t) = \sum_{i=1}^{n} \phi_i \, dX_i.$$

At this point we have to ask ourselves whether $(\phi_1, \ldots, \phi_n)$ is a self-financing replication portfolio, i.e., whether

$$V(t) = \sum_{i=1}^{n} \phi_i \, X_i.$$

Without this identity $V(t) - V(0)$ is just the gain process of the trading strategy $(\phi_1, \ldots, \phi_n)$.[6]

The key in the construction of the self-financing replication portfolio was the ability to use one asset as storage, here, e.g., X_1. We divided by this asset and applied the

---

[6] Recall the definition of self-financing: The required condition is $d(\sum_{i=1}^{n} \phi_i X_i) = \sum_{i=1}^{n} \phi_i \, dX_i$.

representation theorem only to the remaining $n - 1$ processes $\frac{X_2}{X_1}, \ldots, \frac{X_n}{X_1}$. After that we could use ϕ_1 to ensure self-financing since $\frac{X_1}{X_1}$ ($= 1$) did not carry any risk (see Lemma 70). So:

- If we use the representation theorem on $(X_1, \ldots, X_n)$, we find a representation of the gain process, but it is not clear if the portfolio $(\phi_1, \ldots, \phi_n)$ is self-financing and replicating.

- If we use the representation theorem on $(X_2, \ldots, X_n)$ and set ϕ_1 through Lemma 70, then the portfolio $(\phi_1, \ldots, \phi_n)$ is self-financing, but it is not clear if the portfolio replicates $V^{(T)}$.

In order to construct a self-financing replicating trading strategy there has to be an asset that (locally) does not carry any risk. The n processes $X_1, \ldots, X_n$ have to be driven by $n - 1$ Brownian motions, i.e., the $n - 1$ processes $\frac{X_2}{X_1}, \ldots, \frac{X_n}{X_1}$ have to be driven by $n - 1$ Brownian motions.

We consider price processes as before, however now let $\sigma_1 = 0$ and, e.g., $\mu_1 = r\, X_1$ with $r > 0$ and $X_1(0) > 0$. In this case X_1 corresponds to the value of a riskless ($\sigma_1 = 0$) asset with continuously compounded interest rate r:

$$dX_1 = r\, X_1\, dt, \quad \text{i.e.,} \quad X_1(t) = X_1(0)\, \exp(r\, t).$$

Obviously, there exists no measure under which X_1 would be a martingale. The product is riskless, i.e., X_1 is not stochastic and the expectation is thus independent of the probability measure. To make all processes martingales via a change of measure we have to move to relative prices. Dividing all processes by $N := X_1$ we have $\frac{X_1}{N} = 1$, which trivially is a martingale (under any measure). The remaining processes $\frac{X_i}{N}$ are either risky (i.e., with nonzero volatility) or martingales (i.e., if they are riskless they are constant), since the presence of a riskless N-relative process with drift different from 1 would imply arbitrage.[7]

It is not necessary to chose the riskless asset as numéraire. We could have divided by any asset. Either $\frac{X_1}{N}$ is stochastic (i.e., has nonzero volatility) and can be turned into a martingale by a change of measure, or $\frac{X_1}{N}$ is drift-free.[8]

---

[7] If there were two riskless assets with different local rates of return, then we could construct a portfolio with zero initial cost having a positive payout with probability 1. This portfolio is constructed by buying the high yield asset and financing this by (short) selling the low yield asset.

[8] This statement does not hold globally, it holds locally.

Interpretation: The transformation to relative prices corresponds to the transformation of the riskless rate of return (drift) to 0. This corresponds to the change of the frame of reference in physics. Since there is only *one* riskless asset, there is only one measure under which *all* processes become martingales. With this measure we can use the martingale representation theorem to derive the *universal pricing theorem*. ◁|

3.3.2 Risk-Neutral Measure

For each numéraire N there usually exists a different equivalent martingale measure $\mathbb{Q}^N$. As *risk-neutral measure* we denote the equivalent martingale measure $\mathbb{Q}^B$ corresponding to the (locally) riskless asset B with

$$\mathrm{d}B = r\, B\, \mathrm{d}t.$$

The name *risk-neutral measure* stems from the following considerations: Under the measure $\mathbb{Q}^B$ we have for any financial asset S that $\frac{S}{B}$ is a martingale. Thus, relative to B, S does not have any additional drift. So under $\mathbb{Q}^B$ the asset S has the same local rate of return as the riskless asset B, i.e., we have (written as a lognormal process)

$$\mathrm{d}S = r\, S\, \mathrm{d}t + \sigma\, S\, \mathrm{d}W^{\mathbb{Q}^B}.$$

If $\mathbb{Q}^B$ had been the real measure, then this would imply that the real local rate of return of S would be the same as B. In such a market all assets would have the same local rate of return, namely r, independent of their risk σ. In other words: On average, the market participants are *neutral with respect to risk*; they are neither risk-affine nor risk-averse.

In general one would expect that under the real measure all risky assets ($\sigma > 0$) have a local rate of return, i.e., drift, $> r$, since investors like to get rewarded for the risk they take. There are, however, counterexamples: For a lottery the expected payoff is usually much less than the initial investment, since the lottery pays out only a part of the total investments.

Part II

First Applications: Black-Scholes Model, Hedging, and Greeks

CHAPTER 4

Pricing of a European Stock Option under the Black-Scholes Model

We assume here that interest rates are deterministic and that a bank account exists which may be used to invest or draw cash at a *continuously compounded* rate $r(t)$, i.e., the investment of $B(0) = 1$ in $t = 0$ evolves as

$$dB(t) = r(t) \, B(t) \, dt,$$

i.e.,

$$B(t) \;=\; \exp\left(\int_0^t r(\tau) \, d\tau \right) \tag{4.1}$$

(here r denotes a known real valued function of time, not a random variable). Furthermore we assume that the stock $S(t)$ follows

$$dS(t) \;=\; \mu^{\mathbb{P}}(t) \, S(t) \, dt \;+\; \sigma(t) \, S(t) \, dW^{\mathbb{P}}(t) \qquad \text{under the real measure } \mathbb{P}, \tag{4.2}$$

(if σ is a constant, then σ is called Black-Scholes volatility). The measure $\mathbb{P}$ denotes the *real measure*. Let $K \in \mathbb{R}$. We wish to derive the value $V(0)$ of the contract paying

$$V(T) \;=\; \max(S(T) - K, 0)$$

in $t = T$.

We use the techniques from Chapter 3 (the following steps will be repeated similarly in other applications): As a numéraire $N(t)$, i.e., as reference quantity, we choose the bank account $N(t) := B(t)$. From Theorem 74 we have the existence of a measure $\mathbb{Q}^N$, equivalent to $\mathbb{P}$, such that $\frac{S(t)}{N(t)}$ and $\frac{V(t)}{N(t)}$ are both martingales.[1] From Theorem 59 S is

---

[1] By $V(t)$ we denote the value of the replication portfolio consisting of $S(t)$, $B(t)$ such that $V(T) = \max(S(T) - K, 0)$.

given $\mathbb{Q}^N$ with a changed drift, i.e.,

$$dS(t) = \mu^{\mathbb{Q}^N}(t)\, S(t)\, dt + \sigma(t)\, S(t)\, dW^{\mathbb{Q}^N}(t) \qquad \text{under } \mathbb{Q}^N,$$

(where $W^{\mathbb{Q}^N}$ denotes a $\mathbb{Q}^N$-Brownian motion).[2] From the quotient rule 49 we find

$$
\begin{aligned}
d\!\left(\frac{S(t)}{B(t)}\right) &= \frac{S(t)}{B(t)}[\mu^{\mathbb{Q}^N}(t)\, dt + \sigma(t)\, dW^{\mathbb{Q}^N}(t)] - \frac{S(t)}{B(t)}[r(t)\, dt] \\
&\quad - \frac{S(t)}{B(t)}[\mu^{\mathbb{Q}^N}(t)\, dt + \sigma(t)\, dW^{\mathbb{Q}^N}(t)][r(t)\, dt] \\
&\quad - \frac{S(t)}{B(t)}[r(t)\, dt][r(t)\, dt]
\end{aligned}
$$

and by formal expansion, see Theorem 46:

$$= \frac{S(t)}{B(t)}[(\mu^{\mathbb{Q}^N}(t) - r(t))\, dt + \sigma(t)\, dW^{\mathbb{Q}^N}(t)] \quad \text{under } \mathbb{Q}^N.$$

From Lemma 53 it follows that $\mu^{\mathbb{Q}^N}(t) = r(t)$. For the process $Y := \log(S)$ we then have by Lemma 50

$$d(\log(S(t))) = (r(t) - \tfrac{1}{2}\sigma^2(t))\, dt + \sigma(t)\, dW^{\mathbb{Q}^N}(t) \quad \text{under } \mathbb{Q}^N,$$

i.e., $\log(S(T))$ has normal distribution with mean $\bar{\mu} := \log(S(0)) + \bar{r}T - \tfrac{1}{2}\bar{\sigma}^2 T$ and standard deviation $\bar{\sigma}\sqrt{T}$, where we define $\bar{r}$ and $\bar{\sigma}$ as[3]

$$\bar{r} = \frac{1}{T}\left(\int_0^T r(t)\, dt\right) \qquad\qquad \bar{\sigma} = \left(\frac{1}{T}\int_0^T \sigma^2(t)\, dt\right)^{1/2}.$$

That the distribution of $\log(S(T))$ is normal follows from the definition of the Itô process: By

$$d(\log(S(t))) = (r(t) - \tfrac{1}{2}\sigma^2(t))dt + \sigma(t)dW^{\mathbb{Q}^N}(t)$$

---

[2] Since B has no stochastic component, i.e., does not depend on the path parameter $\omega \in \Omega$, (4.1) holds under $\mathbb{Q}^N$, too.

[3] Often the model is considered with a constant (time independent) rate r and a constant *volatility* σ. In this case we have $\bar{r} = r$ and $\bar{\sigma} = \sigma$. This and the shortening of notation are the reasons that we introduce the averaged quantities $\bar{r}$ and $\bar{\sigma}$.

we have just stated that

$$
\begin{aligned}
\log(S(T)) &= \log(S(0)) + \int_0^T r(t) - \frac{1}{2}\sigma^2(t)\, dt + \int_0^T \sigma(t)\, dW^{\mathbb{Q}^N}(t) \\
&= \log(S(0)) + \bar{r}T - \frac{1}{2}\bar{\sigma}^2 T + \int_0^T \sigma(t)dW^{\mathbb{Q}^N}(t) \\
&= \log(S(0)) + \bar{r}T - \frac{1}{2}\bar{\sigma}^2 T + \bar{\sigma}W^{\mathbb{Q}^N}(T).
\end{aligned}
$$

We thus know the dynamics of $S(t)$ and $B(t)$ and more importantly the distribution of $S(T)$ under the measure $\mathbb{Q}^N$. Furthermore we know that the value of $V(0)$ satisfies

$$
\frac{V(0)}{B(0)} = E^{\mathbb{Q}^N}\left(\frac{V(T)}{B(T)}\right),
$$

thus (with $B(0) = 1$, $B(T) = \exp(\bar{r}T)$)

$$
\begin{aligned}
V(0) = B(0)\, E^{\mathbb{Q}^N}\left(\frac{V(T)}{B(T)}\right) &= 1\, E^{\mathbb{Q}^N}\left(\frac{\max(S(T) - K, 0)}{\exp(\bar{r}T)}\right) \\
&= \exp(-\bar{r}T)\, E^{\mathbb{Q}^N}\left(\max(S(T) - K, 0)\right) \\
&= \exp(-\bar{r}T)\, E^{\mathbb{Q}^N}\left(\max(\exp(\log(S(T))) - K, 0)\right)
\end{aligned}
$$

and since $\log(S(T))$ is normally distributed with mean $\bar{\mu}$ and standard deviation $\bar{\sigma}\sqrt{T}$

$$
= \exp(-\bar{r}T) \int_{-\infty}^{\infty} \max(\exp(y) - K, 0)\, \frac{1}{\bar{\sigma}\sqrt{T}}\phi\left(\frac{y - \bar{\mu}}{\bar{\sigma}\sqrt{T}}\right) dy,
$$

where $\phi(x) := \frac{1}{\sqrt{2\pi}}\exp(-x^2/2)$ denotes the density of the standard normal distribution. The above integral may be represented as

$$
V(0) = S(0)\Phi(d_+) - \exp(-\bar{r}T)K\Phi(d_-), \tag{4.3}
$$

where $\Phi(x) := \frac{1}{\sqrt{2\pi}}\int_{-\infty}^{x}\exp(-\frac{y^2}{2})\, dy$ and $d_\pm = \frac{1}{\bar{\sigma}\sqrt{T}}\left[\log(\frac{S(0)}{K}) + \bar{r}T \pm \frac{\bar{\sigma}^2 T}{2}\right]$.[4]

---

[4] We denote the cumulative normal distribution function by Φ. In some books, especially in connection with the Black-Scholes model, it is denoted by N (which we usually use for the numéraire). Instead of d_+, d_- one often uses the symbols d_1, d_2.

To derive this formula we consider for given $K, \mu, \sigma \in \mathbb{R}$ $(\sigma > 0)$

$$\int_{-\infty}^{\infty} \max(\exp(y) - K, 0) \, \frac{1}{\sigma} \phi\left(\frac{y - \mu}{\sigma}\right) dy$$

$$= \int_{\log(K)}^{\infty} (\exp(y) - K) \, \frac{1}{\sqrt{2\pi}\sigma} \exp\left(-\frac{(y - \mu)^2}{2\sigma^2}\right) dy$$

$$= \frac{1}{\sqrt{2\pi}\sigma} \int_{\log(K)}^{\infty} \exp(y) \, \exp\left(-\frac{(y - \mu)^2}{2\sigma^2}\right) dy$$

$$- \frac{1}{\sqrt{2\pi}\sigma} K \int_{\log(K)}^{\infty} \exp\left(-\frac{(y - \mu)^2}{2\sigma^2}\right) dy.$$

It is

$$\frac{1}{\sqrt{2\pi}\sigma} \int_{\log(K)}^{\infty} \exp(y) \, \exp\left(-\frac{(y - \mu)^2}{2\sigma^2}\right) dy$$

$$= \frac{1}{\sqrt{2\pi}\sigma} \int_{\log(K)}^{\infty} \exp\left(-\frac{(y - \mu)^2 - 2\sigma^2 y}{2\sigma^2}\right) dy$$

with $(y - \mu)^2 - 2\sigma^2 y = (y - (\mu + \sigma^2))^2 - 2\mu\sigma^2 - \sigma^4$

$$= \frac{1}{\sqrt{2\pi}\sigma} \int_{\log(K)}^{\infty} \exp\left(-\frac{(y - (\mu + \sigma^2))^2}{2\sigma^2} + \mu + \frac{1}{2}\sigma^2\right) dy$$

$$= \frac{1}{\sqrt{2\pi}\sigma} \exp\left(\mu + \frac{1}{2}\sigma^2\right) \int_{\log(K)}^{\infty} \exp\left(-\frac{(y - (\mu + \sigma^2))^2}{2\sigma^2}\right) dy$$

and with the substitution $\frac{y - (\mu + \sigma^2)}{\sigma} \mapsto y$

$$= \frac{1}{\sqrt{2\pi}} \exp\left(\mu + \frac{1}{2}\sigma^2\right) \int_{\frac{\log(K) - (\mu + \sigma^2)}{\sigma}}^{\infty} \exp\left(-\frac{y^2}{2}\right) dy.$$

Similarly we have with the substitution $\frac{y - \mu}{\sigma} \mapsto y$

$$\frac{1}{\sqrt{2\pi}\sigma} K \int_{\log(K)}^{\infty} \exp\left(-\frac{(y - \mu)^2}{2\sigma^2}\right) dy = \frac{1}{\sqrt{2\pi}} K \int_{\frac{\log(K) - \mu}{\sigma}}^{\infty} \exp\left(-\frac{y^2}{2}\right) dy.$$

Defining

$$\Phi(x) := \frac{1}{\sqrt{2\pi}} \int_{-\infty}^{x} \exp\left(-\frac{y^2}{2}\right) dy$$

we thus get from

$$\frac{1}{\sqrt{2\pi}} \int_x^\infty \exp\left(-\frac{y^2}{2}\right) dy = \Phi(-x)$$

that

$$\int_{-\infty}^\infty \max(\exp(y) - K, 0) \frac{1}{\sigma} \phi\left(\frac{y-\mu}{\sigma}\right) dy$$

$$= \exp\left(\mu + \frac{1}{2}\sigma^2\right) \Phi\left(\frac{\mu + \sigma^2 - \log(K)}{\sigma}\right) - K \Phi\left(\frac{\mu - \log(K)}{\sigma}\right).$$

Remark 80 (Implied Black-Scholes Volatility): From Equation (4.3) and the definition of $d_\pm$ we see that under the model (4.2) the price of an option depends on $S(0)$, $\bar{r}$, and $\bar{\sigma}$ only (apart from product parameters like T and K). The parameter

$$\bar{\sigma} = \left(\frac{1}{T} \int_0^T \sigma^2(t)\, dt\right)^{1/2}$$

is called the Black-Scholes volatility, while $\sigma(t)$ is called the instantaneous volatility of the model (4.2). The Black-Scholes volatility is the square root of the average instantaneous variance (which is the square of the instantaneous volatility)[5].

For fixed $S(0)$, $\bar{r}$ the pricing formula (4.3) is a bijection

$$\bar{\sigma} \mapsto V(0)$$
$$[0, \infty) \to [\max(S(0) - \exp(-\bar{r}T)K, 0)\,,\ S(0)]$$

and thus we may calculate the Black-Scholes volatility corresponding to a given price $V(0)$. This volatility is called the implied Black-Scholes volatility.

---

[5] Note: The variance of the sum of two independent normally distributed random variables is the sum of their respective variances.

CHAPTER 5

Excursus: The Density of the Underlying of a European Call Option

Lemma 81 (Probability Density of the Underlying of a European Call Option):
Let S denote a price process and N a numéraire on the probability space $(\mathbb{R}, \mathcal{B}(\mathbb{R}), \mathbb{Q}^N)$. Let $K \in \mathbb{R}$ and $T_0 < T_1 \leq T_2$, where T_1 denotes a fixing date and T_2 a payment date of a European option with payout

$$V(K; T_2) := \max(S(T_1) - K, 0).$$

Let us further assume that the numéraire N has been chosen such that $N(T_2) = 1$. Then we have for the (risk-neutral) probability density[1] $\phi_{S(T_1)}(s)$ of $S(T_1)$ under $\mathbb{Q}^N$

$$\phi_{S(T_1)}(s) = \frac{1}{N(T_0)} \left. \frac{\partial^2 V(K, T_0)}{\partial K^2} \right|_{K=s},$$

i.e.,

$$\frac{\partial^2 V(K, T_0)}{\partial K^2} = N(T_0)\, \phi_{S(T_1)}(K) = \frac{d\mathbb{Q}^N}{dS}(\{\omega : S(\omega) = K\}).$$

Remark 82 (Application): We may apply Lemma 81 to European stock options ($T_1 = T_2$ and $N(t) = \exp(r\,(T_1 - t))$), see Chapter 4, or to caplets, see Chapter 10.

If a model is given (e.g., as a "black box" through some pricing software) and if the model allows the pricing of arbitrary European options, then we can use this lemma to

---

[1] To be precise, $\phi_{S(T_1)}(s)$ is the conditional probability density, conditioned on $\mathcal{F}_{T_0}$. For simplicity we assume that $\mathcal{F}_{T_0} = \{\emptyset, \Omega\}$ and write $E^{\mathbb{Q}^N}(\cdot)$ for $E^{\mathbb{Q}^N}(\cdot \,|\mathcal{F}_{T_0})$.

infer the probability distribution of $S(T)$ created by this model. This allows for some simple model tests: If the probability density is 0 in any region, then the model should be discarded if these regions play any role in the pricing. If the model exhibits regions with negative probability density, then the model is not arbitrage-free. It is easy to set up interpolatory models that fail when tested by this lemma. See Chapter 6.

Proof: Under the measure $\mathbb{Q}^N$, N-relative prices are martingales. Thus we have

$$\frac{V(K;T_0)}{N(T_0)} = \mathrm{E}^{\mathbb{Q}^N}\left(\frac{V(K;T_2)}{N(T_2)}\right)$$

i.e.,

$$V(K;T_0) = N(T_0)\,\mathrm{E}^{\mathbb{Q}^N}\left(\frac{V(K;T_2)}{N(T_2)}\right)$$

$$= N(T_0)\int_\Omega \max(S(T_1;\omega) - K, 0)\,\mathrm{d}\mathbb{Q}^N(\omega)$$

$$= N(T_0)\int_{-\infty}^{\infty} \max(s - K, 0)\,\phi_{S(T_1)}(s)\,\mathrm{d}s,$$

and thus

$$\frac{\partial}{\partial K}V(K;T_0) = N(T_0)\int_{-\infty}^{\infty} -1_{[0,\infty)}(s - K)\phi_{S(T_1)}(s)\,\mathrm{d}s$$

$$= N(T_0)\int_{K}^{\infty} -\phi_{S(T_1)}(s)\,\mathrm{d}s,$$

i.e.,

$$\frac{\partial^2}{\partial K^2}V(K;T_0) = N(T_0)\,\phi_{S(T_1)}(K).$$

□|

As a direct consequence we have

Lemma 83 (Prices of European Options Are Convex): Prices of European options are a convex function of the strike.

 Further Reading: The original article containing Lemma 81 is [52]. It will play role in the interpolation of option prices, i.e., for the implied volatility surface, and in the definition of a Markov functional model; see Chapters 6 and 27.   ◁|

CHAPTER 6

Excursus: Interpolation of European Option Prices

6.1 No-Arbitrage Conditions for Interpolated Prices

We consider the price $V(K;0)$ of a European option with payoff $\max(S(T) - K, 0)$ as a function of the strike K. We assume $S(T) \geq 0$. Let $P(T;0)$ denote the value of the zero-coupon bond with notional 1.[1] From Lemma 83, the mapping

$$K \mapsto V(K)$$

(we write shortly $V(K)$ for $V(K;0)$), which maps a strike K to the corresponding option price $V(K)$ (for fixed, given maturity T and notional 1), has to satisfy the following conditions:

- $V(0) = P(T;0)\, S(0)$ (since $S(T) \geq 0$),

- $V(K) = V(0) - P(T;0)\, K$ for $K < 0$ (since $S(T) \geq 0$),
  i.e., V is linear for $K < 0$,

- $\lim\limits_{K \to \infty} V(K) = 0$,

---

[1] See Definition 97 on the definition and Section 1.3.4 on the notation $P(T;0)$.

- $\dfrac{\mathrm{d}^2 V}{\mathrm{d}K^2} \geq 0$ (from Lemma 81), i.e., $V(K)$ is convex.

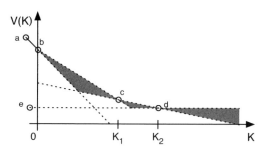

Figure 6.1. *Arbitrage-free option prices for European options (payoff* $\max(S(T) - K, 0)$*) with strike* $K < K_1$ *or* $K_1 < K < K_2$ *or* $K > K_2$*; assuming that two option prices* $V(K_1) > V(K_2)$ *(for* $K_1 < K_2$*) are given.*

Given strikes $K_1 < K_2$ and corresponding option prices $V(K_1)$ and $V(K_2)$, respectively, we find geometric conditions for the possible (arbitrage-free) strike-price points $(K, V(K))$. The price points $c = (K_1, V(K_1))$ and $d = (K_2, V(K_2))$ as well as $a = (-1, V(0) - P(T; 0))$, $b = (0, V(0))$, and $e = (0, V(K_2))$ define lines $\overline{ab}$, $\overline{bc}$, $\overline{cd}$, and $\overline{de}$ which are boundaries of the set of admissible strike-price points. As depicted in Figure 6.1 we have:

- For $K < K_1 < K_2$:

$$V(K) \geq V(0) - P(T; 0)\,K \qquad\qquad ((K, V(K)) \geq \overline{ab}).$$
$$V(K) \geq V(K_1)\frac{K_2 - K}{K_2 - K_1} + V(K_2)\frac{K - K_1}{K_2 - K_1} \qquad ((K, V(K)) \geq \overline{cd}).$$
$$V(K) \leq V(0) + (V(K_1) - V(0))\frac{K}{K_1} \qquad ((K, V(K)) \leq \overline{bc}).$$

- For $K_1 < K < K_2$:

$$V(K) \geq V(K_0) + (V(K_1) - V(0))\frac{K}{K_1} \qquad ((K, V(K)) \geq \overline{bc}).$$
$$V(K) \geq V(K_2) \qquad\qquad ((K, V(K)) \geq \overline{de}).$$
$$V(K) \leq V(K_1)\frac{K_2 - K}{K_2 - K_1} + V(K_2)\frac{K - K_1}{K_2 - K_1} \qquad ((K, V(K)) \leq \overline{cd}).$$

- For $K_1 < K_2 < K$:

$$V(K) \geq V(K_1)\frac{K_2 - K}{K_2 - K_1} + V(K_2)\frac{K - K_1}{K_2 - K_1} \qquad ((K, V(K)) \geq \overline{cd}).$$
$$V(K) \leq V(K_2) \qquad\qquad ((K, V(K)) \leq \overline{de}).$$

6.2 Arbitrage Violations through Interpolation

In this section we give some examples to show that simple interpolation of prices or implied volatilities may lead to arbitrage.

6.2.1 Example 1: Interpolation of Four Prices

Consider a stock S with a current value of $S(0) = 1.0$ and a European option on S having maturity $T = 1.0$ and strike K, i.e., we consider the payoff

$$\max(S(1.0) - K, 0).$$

Assume that for strikes $K = 0.5, 0.75, 1.25$, and 2.0 the following prices are given:

i	Strike K_i	Price $V(K_i)$	Implied Volatility $\sigma(K_i)$
1	0.50	0.5277	0.4
2	0.75	0.3237	0.4
3	1.25	0.1739	0.6
4	2.00	0.0563	0.6

The given implied volatility is the σ of a Black-Scholes model with assumed interest rate (short rate) of $r = 0.05$. Note, however, that we start with given prices which are model-independent.

We will now discuss several "obvious" interpolation methods applied to this data. One approach consists of the interpolation of prices, the other approach consists of the interpolation of implied volatilities and calculating interpolated prices from the interpolated volatilities. These interpolations are not model-independent. The interpolation method itself constitutes a model. This is obvious with the second approach, where a model is involved in the calculation but also applies to the first approach. From Lemma 81 we have that the interpolation method constitutes a model for the underlying's probability density. It does not model the underlying's dynamics.

6.2.1.1 Linear Interpolation of Prices

In Figure 6.2 we show the linear interpolation of the given option prices. The prices do not allow arbitrage, which is obvious from the convexity of their linear interpolation.

However, the linear interpolation of prices has severe disadvantages: The linear interpolation of option prices implies a model under which the underlying may not attain values $K \neq K_i$; the corresponding probability density is zero for these values. For values corresponding to the given strikes K_i a point measure is assigned: see Figure 6.2, right. In addition the corresponding implied volatilities look strange.

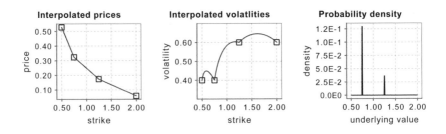

Figure 6.2. *Linear interpolation of option prices.*

6.2.1.2 Linear Interpolation of Implied Volatilities

In Figure 6.3 we show the linear interpolation of the corresponding implied volatilities (center), while the corresponding prices are recalculated via the Black-Scholes formula (left). The linear interpolation of implied (Black-Scholes) volatilities may lead to

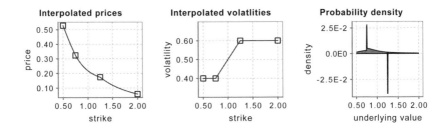

Figure 6.3. *Linear interpolation of implied volatilities.*

prices allowing for arbitrage: At the edges $K = 0.75$ and $K = 1.25$ we have a point measure, the measure for $S(T) = 1.25$ being negative. Thus the implied measure is not a probability measure; see Figure 6.3, right. From Lemma 81 we have: There is no arbitrage-free pricing model that generates this interpolated price curve.

On the other hand, linear interpolation of implied volatilities also has a nice property: If the given prices correspond to prices from a Black-Scholes model, i.e., if the implied volatilities are constant, then, trivially, the interpolated prices correspond to the same Black-Scholes model.

6.2.1.3 Spline Interpolation of Prices and Implied Volatilities

To remove the disadvantage of degenerated densities (i.e., the formation of point measures) as it appears for a linear interpolation, we move to a smooth (differentiable) interpolation method, e.g., spline interpolation. Figure 6.4 shows the spline interpolation of the given data. The upper row shows the spline interpolation of the option prices. The lower row shows the spline interpolation of the implied volatilities. Using (cubic) spline interpolation the probability densities are continuous functions.

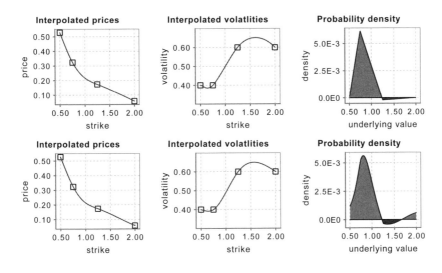

Figure 6.4. *Spline interpolation of option prices (upper row) and implied volatilities (lower row).*

However, it shows large regions with negative densities (Figure 6.4, right), thus the density is not a probability density and arbitrage possibilities exist. The strike-price curve is not convex (Figure 6.4, left).

A spline interpolation of prices generates negative densities, because the spline interpolation of convex sample points is not convex.

6.2.2 Example 2: Interpolation of Two Prices

The example from Section 6.2.1 of a linear interpolation of implied volatilities may suggest that the problem arises at the joins of the linear interpolation, i.e., the behavior at $K_2 = 0.75$ and $K_3 = 1.25$. One could hope that a local smoothing would solve this problem. Instead we will give an example showing that a linear growth of implied volatility may be inadmissible alone, although the interpolated prices are admissible.

We consider two prices $V(K_2) > V(K_3)$. The monotony ensures that these two prices alone do not allow for an arbitrage.

6.2.2.1 Linear Interpolation for Decreasing Implied Volatilities

Consider the following prices:

i	Strike K_i	Price $V(K_i)$	Implied Volatility $\sigma(K_i)$
2	0.75	0.4599	0.9
3	1.25	0.0018	0.1

The implied volatility decreases with the strike K. Figure 6.5 shows the linear interpolation of the implied volatilities (center). The density (see Figure 6.5, right)

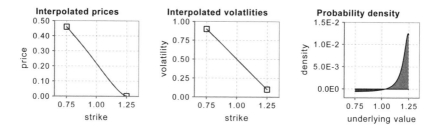

Figure 6.5. *Linear interpolation for decreasing implied volatility.*

shows a large region with negative values: thus it is not a probability density. The reason is a decay of the implied volatility too fast.

Conclusion: An arbitrarily fast decrease of implied volatility is not possible.

6.2.2.2 Linear Interpolation for Increasing Implied Volatilities

For the example of increasing implied volatility we consider the following prices:

i	Strike K_i	Price $V(K_i)$	Implied Volatility $\sigma(K_i)$
2	0.75	0.2897	0.2
3	1.25	0.2532	0.8

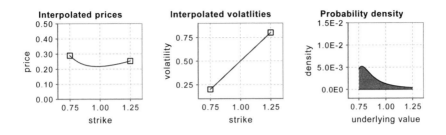

Figure 6.6. *Linear interpolation for increasing implied volatility.*

Figure 6.6 shows the linear interpolation of the increasing implied volatilities.

At first sight the density implied by the linear interpolation of the implied volatilities exhibits no flaw (positive, no point measure). However, it is not a probability density. The integral below the segment is larger than 1. That this is inevitable is obvious from the price curve. The price curve is convex (and thus the density positive), but not monotone. The arbitrage is obvious, since the option with strike 1.0 is cheaper than the option with strike 1.25. If the prices should converge to 0 for increasing strike, then it is inevitable that the convexity will be violated. Thus it is inevitable that the density will exhibit a region with negative values beyond the interpolation region (this, however, will make the integral of the density to 1).

Conclusion: An arbitrarily fast increase of implied volatility is not possible.

6.3 Arbitrage-Free Interpolation of European Option Prices

The examples of the previous sections bring up the question for arbitrage-free interpolation methods.

Every arbitrage-free pricing model defines an arbitrage-free interpolation method, if it is able to reproduce the given prices. However, this insight is almost useless, since:

- Most models are not able to reproduce arbitrarily given prices exactly.

  - The Black-Scholes model (Chapter 4) and (in its simplest form) the LIBOR market model (Chapter 19) allow, by the choice of their volatility, the perfect fit for only one European option per maturity.

– Extended models (e.g., models with stochastic volatility) allow for a calibration to more than one option price per maturity. They often do this in an approximative way, i.e., the residual error of the fitting is minimized, but not necessarily 0. However, this may also be a desired effect, since such a fitting is more robust against errors in the input data.

- Extended models that fit more than one option price per maturity usually require great effort to find the corresponding model parameters. These models are intended for the pricing of complex derivatives (which justifies the effort), but not primarily as an interpolation method to price European options. Examples of such model extensions are *stochastic volatility* or *jump-diffusion* extensions of the LIBOR market model, [23].

- Some models even require a continuum of European option prices $K \mapsto V(T, K)$ as input. Then they require an interpolation. An example for such a model is the Markov functional model; see Chapter 27.

Thus, special methods and models have been developed to specifically achieve the interpolation of given European option prices. Often they reproduce given prices only approximately in the sense of a "best fit"—they are not an interpolation in the original sense. The construction of an (interpolating or approximating) price function $(T, K) \mapsto V(T, K)$ of European option prices with maturity T and strike K is called *modeling of the volatility surface*.

Definition 84 (Volatility Surface):   ⌐
Given a continuum $(T, K) \mapsto V(T, K)$ of prices of European options with maturity T and strike K. Let $\sigma(T, K)$ denote the *implied volatility* of $V(T, K)$, i.e., the volatility for which a risk-neutral pricing with a lognormal model $dS = rS\,dt + \sigma(T, K)S\,dW$, $S(0) = S_0$ (Black-Scholes model) will reproduce the prices $V(T, K)$. Then $(T, K) \mapsto \sigma(T, K)$ is called *volatility surface*.   ⌋

The modeling of a volatility surface may be viewed as important postprocessing of market data. On the other hand it may be viewed as an integral part of the pricing model. A detailed discussion of volatility surface modeling is beyond the scope of this little excursus and we restrict ourself to citing a few methods:

 Further Reading: An introduction to the volatility surface is given by Gatheral [15].

The *mixture of lognormal* approach [54] uses the fact that if the probability density $\phi_S(T)$ of the underlying $S(T)$ is given as convex combination

$$\phi_{S(T)} = \sum_{i=1}^{n} \lambda_i \phi_{\sigma_n}, \qquad \sum_{i=1}^{n} \lambda_i = 1, \quad \lambda_i \geq 0$$

of lognormal densities ϕ_{σ_i}, then the price of a European option is given as a convex combination (using the same weights λ_i's) of corresponding Black-Scholes formulas.[2] In some cases this allows for a choice of the ϕ_{σ_i}'s and λ_i's that exactly reproduces given option prices. However, the method is not able to reproduce arbitrary arbitrage-free prices.

Fengler [64] applies a cubic spline interpolation with an additional shape constraint to given option prices. This leads to an arbitrage-free smoothing algorithm for the implied volatility.

The *SABR* model [75] is a four-parameter model (σ, α, β, ρ) for European interest rate options (the underlying is an interest rate; see Chapter 8). It reproduces given prices only approximately, but models a more realistic behavior of the dependency of the volatility surface from the spot value of the underlying. ◁|

 Experiment: At http://www.christian-fries.de/finmath/ applets/OptionPriceInterpolation.html several interpolation methods may be applied to a user-defined configuration of European option prices. The figures in this chapter were produced with this software.

◁|

---

[2] This follows directly from the linearity of the integral with respect to the integrator $dQ = \phi(S)\,dS$.

CHAPTER 7

Hedging in Continuous and Discrete Time and the *Greeks*

7.1 Introduction

The evaluation theory we have presented so far relies on the possibility of replicating a financial derivative by a portfolio of traded assets (replication portfolio) and a self-financing strategy.

If we assume a perfect replication, we can evaluate a derivative product without explicitly deriving the replicating strategy. This is made possible by the introduction of the equivalent martingale measure and the universal pricing theorem. However, in practice, the price calculated by this risk-neutral valuation theory makes sense only if the corresponding replication is conducted. Thus, it is necessary to explicitly determine the replication portfolio in order to do the replication.

In the first part of this chapter we will show that the replication strategy may be derived from the risk-neutral valuation surprisingly simply.

Depending on the state space modeled, the following basic requirement has to hold in order to replicate arbitrary payoff functions:

- To replicate arbitrary payoff functions modeled over a continuous state space, it is in general necessary to trade continuously and in infinitesimal amounts of the underlyings. The existence of such a replicating trading strategy is ensured by the martingale representation theorem, see Theorem 54.

- To replicate arbitrary payoff functions modeled over a finite, discrete state space, it is only necessary to trade at discrete times. However, the number of traded assets (underlyings) has to match the number of possible state transitions over one time step, see Section 3.1.2.

Both modelings (continuous or discrete replication) do not represent reality exactly: on the one hand it is not possible to trade continuously in infinitesimal amounts; on the other hand the number of possible states is usually much larger than the number of traded assets.[1]

Apart from the question of how to derive the replication strategy for time-continuous trading, we are especially interested in the replication when trading takes place at discrete points in time only. In this case, which is closer to reality, a complete replication is not possible. A residual risk remains.

Definition 85 (Hedge):
A replication portfolio (almost) replicating a derivative product and thus, if considered together with the derivative product neutralizing (reducing) the total risk, is called a (partial or incomplete) *hedge*. The corresponding trading strategy is called *hedging*.

First we will derive the delta hedge, delta-gamma hedge, and vega hedge from the risk-neutral evaluation theory (Section 7.2). We will then transfer these hedges to the case of trading in discrete time, these being incomplete hedges, and analyze the residual error (Section 7.4).

The Bouchaud-Sornette method (Section 7.5) determines the replication portfolio that minimizes the residual risk. The residual risk has to be measured in the real measure $\mathbb{P}$. There is a real risk of a loss due to incomplete hedging. The real measure is no longer irrelevant since a complete replication is no longer possible.

7.2 Deriving the Replications Strategy from Pricing Theory

Let $\mathcal{M} := \{S_0, S_1, \ldots, S_n\}$ denote a set of Itô price processes, representing a complete market of traded assets. Let the process $N = S_0$ be a numéraire and $\mathbb{Q}^N$ the corresponding martingale measure. Let $t \mapsto V(t)$ denote the price process of a derivative product, i.e.,

$$V(t) = N(t) \, \mathrm{E}^{\mathbb{Q}^N}\left(\frac{V(T)}{N(T)} | \mathcal{F}_t\right).$$

We assume that the value of the derivative product V can be written as a function of $(t, S_0(t), S_1(t), \ldots, S_n(t))$, i.e., we have

$$V(t) = V(t, S_0(t), S_1(t), \ldots, S_n(t)). \qquad (7.1)$$

---

[1] At a stock exchange the prices that can be attained by a stock are in fact discrete.

If $V(t)$ is given by an evaluation theory—i.e., an analytic formula or pricing algorithm—as a function of the underlyings, then the self-financing replication portfolio

$$\Pi(t) := \phi_0(t)S_0(t) + \phi_1(t)S_1(t) + \cdots + \phi_n(t)S_n(t) \qquad (7.2)$$

with $\Pi(t) = V(t)$ (replication) may be determined as follows:

The property that Π is self-financing may be written as (see Definition 66)

$$d\Pi(t) = \phi_0(t)\, dS_0 + \phi_1(t)\, dS_1(t) + \cdots + \phi_n(t)\, dS_n(t) = \sum_{i=0}^{n} \phi_i\, dS_i.$$

Furthermore we find from Itô's lemma (Theorem 47) applied to (7.1)

$$dV(t) = \frac{\partial V(t)}{\partial t}\, dt + \sum_{i=0}^{n} \frac{\partial V(t)}{\partial S_i}\, dS_i + \frac{1}{2} \sum_{i,j=0}^{n} \underbrace{\frac{\partial^2 V(t)}{\partial S_i \partial S_j}\, dS_i\, dS_j}_{=(\ldots)\, dt}.$$

From $d\Pi(t) = dV(t)$ (replication), we find $\phi_0, \ldots, \phi_n$ by comparing coefficients. It is

$$\phi_0(t) = \frac{\partial V(t)}{\partial S_0}, \quad \phi_1(t) = \frac{\partial V(t)}{\partial S_1}, \quad \ldots, \quad \phi_n(t) = \frac{\partial V(t)}{\partial S_n} \qquad (7.3)$$

$$\frac{\partial V(t)}{\partial t}\, dt + \frac{1}{2} \sum_{i,j=0}^{n} \frac{\partial^2 V(t)}{\partial S_i \partial S_j}\, \underbrace{dS_i\, dS_j}_{=(\ldots)\, dt} = 0. \qquad (7.4)$$

Conclusion: Under a model, which permits the evaluation of V through a martingale measure and universal pricing theorem, i.e. without explicit construction of the replication portfolio, the replication portfolio may be calculated *a posteriori* from the partial derivative of the price after the underlyings.

 Interpretation: Comparing coefficients of dV and $d\Pi$ we find Equations (7.3) for the replication portfolio process ϕ. The *partial differential equation* (7.4) is just a reformulation that

$$V(t) = V(t, S_0(t), S_1(t), \ldots, S_n(t))$$

is the value process of a self-financing portfolio in the *coordinates* $(S_0(t), S_1(t), \ldots, S_n(t))$. Equation (7.4) may be used together with the *final condition*

$$V(T, S_0(T), S_1(T), \ldots, S_n(T)) = V^T$$

to determine $V(t)$ for $t < T$ from the given payoff profile V^T. This is the entry point to the pricing of derivatives by partial differential equations; see Section 2.11. ◁|

7.2.1 Deriving the Replication Strategy under the Assumption of a Locally Riskless Product

The ansatz $V(t) = V(t, S_0(t), S_1(t), \ldots, S_n(t))$ is just one possible way of interpreting V as a function of the underlyings. It is a specific choice of the coordinate system leading to very natural equations for the portfolio parameters and the partial differential equations for V. If $N = S_0$ is locally risk-free, we may easily give V in the coordinates $(t, S_1(t), \ldots, S_n(t))$ alone. A special case which we consider now: Let $N = S_0$ denote a locally risk-free numéraire, i.e., assume that the process N (i.e., S_0) is of the form

$$dN = \frac{\partial N}{\partial t} dt, \tag{7.5}$$

i.e., we assume that dN does not have a dW-component.[2] For $N > 0$ (N is a numéraire) we have $r := \frac{\partial \log(N)}{\partial t}$

$$dN = \frac{\partial N}{\partial t} dt = \frac{\partial \log(N)}{\partial t} N \, dt = rN \, dt. \tag{7.6}$$

We assume that, for a given N, $V(t)$ is given as a function in the coordinates $(t, S_1(t), \ldots, S_n(t))$ and denote this function by V again, i.e.,

$$V(t) = V(t, S_1(t), \ldots, S_n(t)). \tag{7.7}$$

In other words: N is no longer a modeled free quantity.[3]

If $V(T)$ is given in this from, we may derive the self-financing replication portfolio (7.2) with $\Pi(t) = V(t)$ as follows: From Itô's lemma (Theorem 47), applied to $V(t, S_1(t), \ldots, S_n(t))$ we have

$$dV(t) = \frac{\partial V(t)}{\partial t} dt + \sum_{i=1}^{n} \frac{\partial V(t)}{\partial S_i} dS_i + \frac{1}{2} \sum_{i,j=1}^{n} \frac{\partial^2 V(t)}{\partial S_i \, \partial S_j} \underbrace{dS_i \, dS_j}_{=(\ldots) \, dt}.$$

With $d\Pi(t) = \sum_{i=0}^{n} \phi_i \, dS_i = dV(t)$ (self-financing of Π, replication of V) we find $\phi_0, \ldots, \phi_n$ by comparing the coefficients of dt and dS_i, $i \geq 1$ as

$$\phi_1(t) = \frac{\partial V(t)}{\partial S_1}, \quad \ldots, \quad \phi_n(t) = \frac{\partial V(t)}{\partial S_n}$$

$$\phi_0(t) \frac{\partial N}{\partial t} dt = \frac{\partial V(t)}{\partial t} dt + \frac{1}{2} \sum_{i,j=1}^{n} \frac{\partial^2 V(t)}{\partial S_i \, \partial S_j} \underbrace{dS_i \, dS_j}_{=(\ldots) \, dt}. \tag{7.8}$$

---

[2] The property is local, since $\frac{\partial N}{\partial t}$ may very well be stochastic. An example would be $dN = rN \, dt$, as it was considered for the Black-Scholes model in Chapter 4.

[3] An example is the Black-Scholes model, where $N(t) = \exp(r\,t)$ and r is a fixed model parameter.

The equation for ϕ_0 seems unnaturally complex when compared to the equation for $\phi_1, \ldots, \phi_n$. This is a consequence of representation (7.7), which we had assumed for V; see Section 7.2.3. However, since the portfolio is self-financing, it is not necessary to calculate ϕ_0 from (7.8); ϕ_0 follows from the condition of self-financing portfolios.

7.2.2 Black-Scholes Differential Equation

Writing $\frac{\partial N}{\partial t} = \frac{\partial \log(N)}{\partial t} N =: rN$ the equation for ϕ_0 in (7.8) becomes

$$\phi_0(t)r(t)N(t) \, dt = \frac{\partial V(t)}{\partial t} \, dt + \frac{1}{2} \sum_{i,j=1}^{n} \frac{\partial^2 V(t)}{\partial S_i \, \partial S_j} \underbrace{dS_i \, dS_j}_{=(\ldots) \, dt}.$$

Since

$$V = \phi_0 N + \sum_{i=1}^{n} \phi_i S_i \overset{(7.8)}{=} \phi_0 N + \sum_{i=1}^{n} \frac{\partial V(t)}{\partial S_i} S_i$$

we also have

$$\phi_0 N = V - \sum_{i=1}^{n} \frac{\partial V(t)}{\partial S_i} S_i$$

and denoting the instantaneous correlations by $\rho_{i,j}$, i.e., $dS_i \, dS_j = \sigma_i \sigma_j \rho_{i,j} \, dt$ we arrive at

$$\frac{\partial V(t)}{\partial t} + r(t) \sum_{i=1}^{n} \frac{\partial V(t)}{\partial S_i} S_i + \frac{1}{2} \sum_{i,j=1}^{n} \frac{\partial^2 V(t)}{\partial S_i \, \partial S_j} \sigma_i \sigma_j \rho_{i,j} = r(t)V(t).$$

This is a variant of the Black-Scholes partial differential equation in the coordinates $(t, S_1, \ldots, S_n)$.

7.2.3 Derivative $V(t)$ as a Function of Its Underlyings $S_i(t)$

The representation of the portfolio process components ϕ_i as a partial derivative of V as well as the partial differential equation for V depend of course on the chosen coordinate system of the underlyings. Instead of comparing coefficients of dS_i, $i \geq 1$ and dt (as above), in general, it is possible to compare the coefficients of dt and the differentials of the driving Brownian motions dW_i.

To illustrate that we have just considered a change (or rather substitution) of coordinates, let us reconsider V written in the two different coordinate systems above, i.e,

$$V(t) = V^{n+1}(t, S_0(t), S_1(t), \ldots, S_n(t))$$
$$V(t) = V^n(t, S_1(t), \ldots, S_n(t)).$$

With $S_0 = N$ and $dN(t) = r(t)N(t)\,dt$ (this assumption makes it possible to replace the dependence on S_0 by a dependence on t) it is

$$V^n(t, S_1, \ldots, S_n) = V^{n+1}(t, N(t), S_1, \ldots, S_n),$$

thus

$$\frac{\partial V^n}{\partial t} = \frac{\partial V^{n+1}}{\partial t} + \frac{\partial V^{n+1}}{\partial N}\frac{\partial N}{\partial t} = \frac{\partial V^{n+1}}{\partial t} + \phi_0 \frac{\partial N}{\partial t},$$

i.e.,

$$\frac{\partial V^n}{\partial t} - \phi_0 \frac{\partial N}{\partial t} = \frac{\partial V^{n+1}}{\partial t}.$$

Together with $dS_i\,dS_j = 0$ if $i = 0$ or $j = 0$ in (7.4) this gives

$$\frac{\partial V(t)^{n+1}}{\partial t}\,dt + \frac{1}{2}\sum_{i,j=0}^{n}\frac{\partial^2 V^{n+1}(t)}{\partial S_i\,\partial S_j}\underbrace{dS_i\,dS_j}_{=(\ldots)\,dt} = 0$$

$$\Leftrightarrow \qquad \frac{\partial V^n(t)}{\partial t}\,dt - \phi_0 \frac{\partial N}{\partial t}\,dt + \frac{1}{2}\sum_{i,j=1}^{n}\frac{\partial^2 V^n(t)}{\partial S_i\,\partial S_j}\underbrace{dS_i\,dS_j}_{=(\ldots)\,dt} = 0$$

$$\Leftrightarrow \qquad \frac{\partial V^n(t)}{\partial t} + r(t)\sum_{i=1}^{n}\frac{\partial V^n(t)}{\partial S_i}S_i + \frac{1}{2}\sum_{i,j=1}^{n}\frac{\partial^2 V^n(t)}{\partial S_i\,\partial S_j}\sigma_i\sigma_j\rho_{i,j} = r(t)V^n(t). \qquad (7.9)$$

7.2.3.1 Path-Dependent Options

Our assumption that $V(t)$ is a function of $S_i(t)$ excludes path-dependent options. However, our considerations may be easily extended to the case of path-dependent options by writing V as a function of the corresponding path quantities. For example, considering a path-dependent option of the form

$$V(t) = V\left(t, S(t), \int_0^t S(\tau)\,d\tau\right)$$

with

$$I(t) = \int_0^t S(\tau)\,d\tau$$

we derive

$$dV(t) = \frac{\partial V(t)}{\partial t}\,dt + \frac{\partial V(t)}{\partial S}\,dS + \frac{\partial V(t)}{\partial I}\,dI$$

and substitute $dI(t) = S(t)\,dt$. For the replication portfolio we thus get

$$\phi_1 = \frac{\partial V(t, S, I)}{\partial S}.$$

7.2.4 Example: Replication Portfolio and Partial Differential Equation of a European Option under a Black-Scholes Model

As an illustration we reconsider these calculations for a simple model, the Black-Scholes model, with price processes N for the savings account and S for the stock:

$$dN(t) = rN(t)\,dt, \qquad\qquad N(0) = 1,$$
$$dS(t) = \mu S(t)\,dt + \sigma S(t)\,dW(t), \qquad\qquad S(0) = S_0.$$

For a European option with strike K and maturity T, i.e. *Payout* $\max(S(T) - K, 0)) \overset{!}{=} V(T)$, the Black-Scholes-Merton formula (4.3) for the value V at time t is

$$V(t, S(t)) = S(t)\Phi(d_+) - \exp(-\bar{r}(T - t))K\Phi(d_-),$$

where

$$d_{\pm} = \frac{1}{\bar{\sigma}\sqrt{(T - t)}}\left[\log\left(\frac{S(t)}{K}\right) + \bar{r}(T - t) \pm \frac{\bar{\sigma}^2(T - t)}{2}\right].$$

See (4.3) in Chapter 4.

This gives V as a function of t and $S(t)$, since r is seen as constant and thus $N(t)$ is known as a deterministic function of t. With $N(t) = \exp(r\,t)$ we may write V as a function of $(t, N(t), S(t))$, namely

$$V(t, N(t), S(t)) = S(t)\Phi(d_+) - N(t)\frac{K}{N(T)}\Phi(d_-), \qquad (7.10)$$

where

$$d_{\pm} = \frac{1}{\bar{\sigma}\sqrt{(T - t)}}\left[\log\left(\frac{S(t)}{N(t)}\frac{N(T)}{K}\right) \pm \frac{\bar{\sigma}^2(T - t)}{2}\right].$$

It is

$$
\begin{aligned}
\frac{\partial V}{\partial S(t)} &= \Phi(d_+) + S(t)\phi(d_+)\frac{1}{S(t)}\frac{1}{\sigma\sqrt{T - t}} - N(t)\frac{K}{N(T)}\phi(d_-)\frac{1}{S(t)}\frac{1}{\sigma\sqrt{T - t}} \\
&= \Phi(d_+) + \frac{1}{S(t)}\frac{1}{\sigma\sqrt{T - t}}\left(S(t)\phi(d_+) - N(t)\frac{K}{N(T)}\phi(d_-)\right) \\
&= \Phi(d_+) + \frac{1}{S(t)}\frac{1}{\sigma\sqrt{T - t}}\left(S(t)\phi(d_+) - N(t)\frac{K}{N(T)}\phi(d_-)\right).
\end{aligned}
$$

With

$$\phi(d_\pm) = C \exp(-\frac{1}{2}d_\pm^2)$$

$$= C \exp\left[-\frac{\log\left(\frac{S(t)}{N(t)}\frac{N(T)}{K}\right)^2 \pm 2\log\left(\frac{S(t)}{N(t)}\frac{N(T)}{K}\right)\frac{\bar{\sigma}^2(T-t)}{2} + \frac{\bar{\sigma}^4(T-t)^2}{4}}{2\bar{\sigma}^2(T-t)}\right]$$

$$= C \exp\left[\pm\frac{1}{2}\log\left(\frac{N(t)}{S(t)}\frac{K}{N(T)}\right) - \frac{\log\left(\frac{S(t)}{N(t)}\frac{N(T)}{K}\right)^2 + \frac{\bar{\sigma}^4(T-t)^2}{4}}{2\bar{\sigma}^2(T-t)}\right]$$

$$= \left(\frac{N(t)}{S(t)}\frac{K}{N(T)}\right)^{\pm 1/2} C \exp\left[-\frac{\log\left(\frac{S(t)}{N(t)}\frac{N(T)}{K}\right)^2 + \frac{\bar{\sigma}^4(T-t)^2}{4}}{2\bar{\sigma}^2(T-t)}\right]$$

we get

$$\phi(d_-) = \frac{S(t)}{N(t)}\frac{N(T)}{K}\phi(d_+)$$

and thus

$$\frac{\partial V}{\partial S(t)} = \Phi(d_+) + \frac{1}{S(t)}\frac{1}{\sigma\sqrt{T-t}}\underbrace{\left(S(t)\phi(d_+) - N(t)\frac{K}{N(T)}\phi(d_-)\right)}_{=0} = \Phi(d_+).$$

Similarly, we can derive

$$\frac{\partial V}{\partial N(t)} = -\frac{K}{N(T)}\Phi(d_-).$$

To sum up, we have $V(t, N(t), S(t)) = \phi_0(t)N(t) + \phi_1(t)S(t)$ with

$$\phi_0(t) = -\frac{K}{N(T)}\Phi(d_-), \qquad \phi_1(t) = \Phi(d_+).$$

Indeed, we might simply have read off the representation of the replication portfolio from (7.10).

For the function

$$V(t, n, s) = s\Phi(d_+) - n\frac{K}{N(T)}\Phi(d_-)$$

with

$$d_\pm = \frac{1}{\bar{\sigma}\sqrt{(T-t)}}\left[\log\left(\frac{s}{n}\frac{N(T)}{K}\right) \pm \frac{\bar{\sigma}^2(T-t)}{2}\right].$$

it is

$$\frac{\partial V(t)}{\partial t}dt + \frac{1}{2}\frac{\partial^2 V(t)}{\partial n\,\partial n}dN\,dN + \frac{\partial^2 V}{\partial n\,\partial s}dN\,dS + \frac{1}{2}\frac{\partial^2 V(t)}{\partial s\,\partial s}dS\,dS = 0$$

$dN\,dN = 0$
$dN\,dS = 0$

$\Longleftrightarrow$

$$\frac{\partial V(t)}{\partial t}dt + \frac{1}{2}\frac{\partial^2 V(t)}{\partial s\,\partial s}\sigma^2 = 0.$$

See Exercise 10. Furthermore, the final condition

$$V(T, N(T), s) = \max(s - K, 0)$$

holds.

7.2.4.1 Interpretation of V as a Function in (t, S)

For the Black-Scholes model, the classical way is to write V as a function of t and $S(t)$ only. In this case we get (see (7.8))

$$\phi_1 = \frac{\partial V(t)}{\partial S} = \Phi(d_+)$$

$$\phi_0 r N = \frac{\partial V(t)}{\partial t} + \frac{1}{2}\sigma^2 S^2 \frac{\partial^2 V(t)}{\partial S^2},$$

In addition we have from $V = \phi_0 N + \phi_1 S$

$$\phi_0 N = V - \phi_1 S,$$

which brings us once again to the Black-Scholes partial differential equation

$$rV = r\phi_0(t)N(t) + r\phi_1 S(t)$$

$$= \frac{\partial V(t)}{\partial t} + \frac{1}{2}\sigma^2 S^2 \frac{\partial^2 V(t)}{\partial S^2} + r\frac{\partial V(t)}{\partial S}S$$

and thus

$$\frac{\partial V(t)}{\partial t} + r\frac{\partial V(t)}{\partial S}S + \frac{1}{2}\sigma^2 S^2 \frac{\partial^2 V(t)}{\partial S^2} = rV.$$

Definition 86 (Delta Hedge):
This choice of the replication portfolio given in Equations (7.2) and (7.3) is called
Delta hedge.

Interpretation: Considering the derivative product V as a function of the underlyings S, then the portfolio Π given by the delta hedge is the local linear approximation of V; see Figure 7.2. For a fixed path ω the map $t \mapsto S(t, \omega)$ is almost surely continuous (a property of Itô processes) and on that path we have $V(T) = \int_0^T dV = \int_0^T d\Pi$. This corresponds to the main theorem of ordinary calculus: $f(T) = \int_0^T df = \int_0^T f'(t)\, dt$ (e.g. for differentiable f). This illustrates why the portfolio Π is only required to coincide with V upon their first derivative by S, i.e., the so-called *delta*: $\frac{\partial V}{\partial S} = \frac{\partial \Pi}{\partial S}$. ◁|

7.3 Greeks

The partial derivatives of the value $V(t)$ of a derivative product upon product- or model-parameters are important quantities. They describe how the product reacts to infinitesimal changes and are essential for the construction of the replication portfolios. These quantities are usually denoted by Greek letters and therefore called *Greeks*. Here we give definitions of the most important Greeks:

Definition 87 (Delta):
The first-order partial derivative of the price of a derivative product with respect to the underlyings is called *delta*. It is the first-order sensitivity of the derivative product to price changes of the market-traded assets (or market quotes).

Definition 88 (Gamma):
The second-order partial derivative of the price of a derivative product with respect to the underlyings is called *gamma*. It is the second-order sensitivity of the derivative product to price changes of the market-traded assets (or market quotes).

Definition 89 (Vega[4]):
The first-order partial derivative of the price of a derivative product with respect to the underlyings log-volatility is called *vega*. It is the first-order sensitivity of the derivative product to log-volatility changes of the market-traded European options. ⌟

Definition 90 (Theta):
The first-order partial derivative of the price of a derivative product with respect to time is called *theta*. It is the first-order sensitivity of the derivative product to time. ⌟

Definition 91 (Rho):
The first-order partial derivative of the price of a derivative product with respect to the interest rate (r) is called *rho*. It is the first-order sensitivity of the derivative product to a change in interest rate.

---

[4] Vega is not a Greek letter.

The definition of *rho* is only used for models with nonstochastic interest rates. For interest rate derivatives (evaluated under a model of stochastic interest rates), as will be considered in the later chapters, the interest rates are underlyings and *rho* is just a *delta*. ⌋

7.3.1 Greeks of a European Call-Option under the Black-Scholes Model

Given the Black-Scholes model with price processes for the savings account N and the stock S:

$$
\begin{aligned}
dN(t) &= rN(t)\,dt, & N(0) &= 1, \\
dS(t) &= \mu S(t)\,dt + \sigma S(t)\,dW(t), & S(0) &= S_0,
\end{aligned}
$$

consider a european stock option with *payout* $\max(S(T) - K, 0)) \overset{!}{=} V(T)$. From Chapter 4 the price $V(0)$ of the option under this model is given by the Black-Scholes-Merton formula (4.3):

$$
V(0) = S_0 \Phi(d_+) - \exp(-rT)K\Phi(d_-),
$$

where

$$
\Phi(x) := \frac{1}{\sqrt{2\pi}} \int_{-\infty}^{x} \exp(-\frac{y^2}{2})\,dy
$$

and

$$
d_\pm = \frac{1}{\sigma\sqrt{T}}\left[\log\left(\frac{S(0)}{K}\right) + rT \pm \frac{\sigma^2 T}{2}\right].
$$

Thus, the price is a function of S_0, r, K, σ, and T.

From the Black-Scholes-Merton formula we may derive the Greeks analytically. They are given in Table 7.1.

7.4 Hedging in Discrete Time: Delta and Delta-Gamma Hedging

If a delta hedge is applied continuously, then it achieves a perfect replication. At any time, the portfolio $V - \Pi$ is neutral with respect to infinitesimal changes in the underlyings S_i. However, if the delta hedge is applied only at discrete times t_k, i.e., the portfolio process $(\phi_0, \phi_1, \dots, \phi_n)$ is kept constant on time intervals $[t_k, t_{k+1})$, then

Greek	Definition	within the Black-Scholes model
Delta (Δ)	$\dfrac{\partial V}{\partial S_0}$	$= \Phi(d_+)$
Gamma (Γ)	$\dfrac{\partial^2 V}{\partial S_0^2}$	$= \dfrac{\Phi'(d_+)}{S_0 \sigma \sqrt{T}}$
Vega	$\dfrac{\partial V}{\partial \sigma}$	$= S_0 \sqrt{T} \Phi'(d_+)$
Theta (Θ)	$-\dfrac{\partial V}{\partial T}$	$= -\dfrac{S_0 \Phi'(d_+) \sigma}{2\sqrt{T}} - rK \exp(-rT) \Phi(d_-)$
Rho (ρ)	$\dfrac{\partial V}{\partial r}$	$= KT \exp(-rT) \Phi(d_-)$

Table 7.1. *Greeks within the Black-Scholes model.*

the replication is no longer exact.[5] While in continuous time

$$dV(t) = \frac{\partial V(t)}{\partial t}\,dt + \sum_{i=0}^{n} \frac{\partial V(t)}{\partial S_i}\,dS_i + \frac{1}{2}\sum_{i,j=0}^{n} \frac{\partial^2 V(t)}{\partial S_i\,\partial S_j}\,dS_i\,dS_j,$$

$$= \sum_{i=0}^{n} \frac{\partial V(t)}{\partial S_i}\,dS_i + \underbrace{\frac{\partial V(t)}{\partial t}\,dt + \frac{1}{2}\sum_{i,j=0}^{n} \frac{\partial^2 V(t)}{\partial S_i\,\partial S_j}\,dS_i\,dS_j}_{=0},$$

is exact by Itô's lemma (and (7.4)), we have for the time-discrete case ($\Delta V = V(t + \Delta t) - V(t)$)

$$\Delta V(t) = \frac{\partial V(t)}{\partial t}\,\Delta t + \sum_{i=0}^{n} \frac{\partial V(t)}{\partial S_i}\,\Delta S_i + \frac{1}{2}\sum_{i,j=0}^{n} \frac{\partial^2 V(t)}{\partial S_i\,\partial S_j}\,\underbrace{\Delta S_i\,\Delta S_j}_{\neq(\dots)\,\Delta t} \qquad (7.11)$$

$$= \sum_{i=0}^{n} \frac{\partial V(t)}{\partial S_i}\,\Delta S_i + \frac{1}{2}\sum_{i,j=0}^{n} \frac{\partial^2 V(t)}{\partial S_i\,\partial S_j}\,\underbrace{(\Delta S_i\,\Delta S_j - \gamma_{i,j}\,\Delta t)}_{\neq 0} + \text{h.o.t.}, \qquad (7.12)$$

---

[5] See also the discussion of *self-financing* (Definition 66).

where $\gamma_{i,j}$ is given by $dS_i\, dS_j = \gamma_{i,j}\, dt$ and h.o.t. $= O(|\Delta t|^2, |\Delta t\, \Delta S_i|, |\Delta S_i|^3)$. For comparison, for the replication portfolio we have in the time-discrete case

$$\Delta\Pi(t) = \sum_{i=0}^{n} \phi_i\, \Delta S_i. \tag{7.13}$$

Assuming we have an exact replication at time t ($V(t) = \Pi(t)$), then there is no choice for the replication portfolio ϕ such that $\Delta V(t) = \Delta\Pi(t)$ in general. An exact replication at time $t + \Delta t$ is not guaranteed.

To ensure that the replication portfolio is self-financing, we have to reformulate the requirement in the time-discrete setting. Trading at discrete times t_k the self-financing property is

$$\sum_{i=0}^{n} (\phi_i(t_k) - \phi_i(t_{k-1}))S_i(t_k) = 0.$$

This may be ensured, e.g., by the choice of ϕ_0 according to

$$\phi_0(t_k) = \phi_0(t_{k-1}) - \frac{1}{S_0(t_k)}\left(\sum_{i=1}^{n} (\phi_i(t_k) - \phi_i(t_{k-1}))S_i(t_k) \right) \tag{7.14}$$

(we have $S_0 \neq 0$ since we assumed that S_0 is a numéraire). While in continuous time V may be replicated exactly with a self-financing portfolio, there is a choice whether ϕ_0 is chosen to reduce the replication error or ϕ_0 is chosen to keep the portfolio self-financing.

The restriction to a self-financing portfolio leads to an error propagation of the replication error via (7.14); see Section 7.4.2.

7.4.1 Delta Hedging

A *delta hedge* determines the replication portfolio such that the delta of the replication portfolio Π matches the delta of V. In other words, the delta of the replication error $V - \Pi$ is zero. The replication error is first-order insensitive to movements of the underlyings. If we consider the requirement of self-financing to be ensured by the choice of ϕ_0, then we chose

$$\phi_i(t_k) = \frac{\partial V(t_k)}{\partial S_i}$$

$$\phi_0(t_k) = \phi_0(t_{k-1}) - \frac{1}{S_0(t_k)}\left(\sum_{i=1}^{n} (\phi_i(t_k) - \phi_i(t_{k-1}))S_i(t_k) \right).$$

For the error $\Delta V(t) - \Delta\Pi(t)$ we get from (7.12) and (7.13)

$$\Delta V(t) - \Delta\Pi(t) = \sum_{i=0}^{n} \left(\frac{\partial V(t)}{\partial S_i} - \phi_i \right) \Delta S_i + \frac{1}{2} \sum_{i,j=0}^{n} \frac{\partial^2 V(t)}{\partial S_i \, \partial S_j} \underbrace{(\Delta S_i \, \Delta S_j - \gamma_{i,j} \, \Delta t)}_{\neq 0} + \text{h.o.t.}$$

and with $\phi_i = \dfrac{\partial V(t)}{\partial S_i}$ for $i \neq 0$

$$= \left(\frac{\partial V(t)}{\partial S_0} - \phi_0 \right) \Delta S_0 + \frac{1}{2} \sum_{i,j=0}^{n} \frac{\partial^2 V(t)}{\partial S_i \, \partial S_j} \underbrace{(\Delta S_i \, \Delta S_j - \gamma_{i,j} \, \Delta t)}_{\neq 0} + \text{h.o.t.,}$$

where h.o.t. $= O(|\Delta t|^2, |\Delta t \, \Delta S_i|, |\Delta S_i|^3)$.

7.4.2 Error Propagation

The choice of ϕ_0 at time $t = 0$ is determined by the initial condition $\Pi(0) = V(0)$ and at later times $t = t_k$, $k = 1, 2, \ldots$ by the requirement for a self-financing trading strategy:

$$\sum_{i=0}^{n} \phi_i(t_k) S_i(t_k) = \sum_{i=0}^{n} \phi_i(t_{k-1}) S_i(t_k). \tag{7.15}$$

This results in an error propagation. Assuming that at time t_k the replication error due to the previous choice of $\phi_i(t_{k-1})$ is $V(t_k) - \sum_{i=0}^{n} \phi_i(t_{k-1}) S_i(t_k) =: e(t_k)$, then we have with

$$V(t_k) = \sum_{i=0}^{n} \frac{\partial V(t_k)}{\partial S_i} S_i(t_k), \qquad \Pi(t_k) = \sum_{i=0}^{n} \phi_i(t_k) S_i(t_k) \tag{7.16}$$

and $\phi_i(t_k) = \dfrac{\partial V(t_k)}{\partial S_i}$ for $k \neq 0$ (delta hedge), such that

$$e(t_k) = V(t_k) - \sum_{i=0}^{n} \phi_i(t_{k-1}) S_i(t_k) \overset{(7.15)}{=} V(t_k) - \Pi(t_k) \overset{(7.16)}{=} \left(\frac{\partial V(t_k)}{\partial S_0} - \phi_0(t_k) \right) S_0.$$

Thus the requirement for a self-financing portfolio implies

$$\phi_0(t) = \frac{\partial V(t)}{\partial S_0} - \frac{1}{S_0}(V(t) - \Pi(t)).$$

This shows how the replication error at time t_k is propagated over the time step Δt by the requirement for a self-financing strategy through ϕ_0.

Summarizing, we have for the delta hedge

$$\Delta V(t) - \Delta\Pi(t) = (V(t) - \Pi(t))\frac{\Delta S_0}{S_0} + \frac{1}{2}\sum_{i,j=0}^{n}\frac{\partial^2 V(t)}{\partial S_i\,\partial S_j}\underbrace{(\Delta S_i\,\Delta S_j - \gamma_{i,j}\,\Delta t)}_{\neq 0} + \text{h.o.t.},$$

where h.o.t. $= O(|\Delta t|^2, |\Delta t\,\Delta S_i|, |\Delta S_i|^3)$.

7.4.2.1 Example: Time-Discrete Delta Hedge under a Black-Scholes Model

We consider a Black-Scholes model with the notation as above. For the replication portfolio $\phi_0(t)N(t) + \phi_1(t)S(t)$ we have

$$\phi_1(t) = \frac{\partial V(t)}{\partial S(t)} = \Phi\left(\frac{1}{\sigma\sqrt{T-t}}\left[\log\left(\frac{S(t)}{K}\right) + r(T-t) + \frac{\sigma^2}{2}\right]\right).$$

We use this choice to trade within the replication portfolio at discrete times t_k, $k = 0, 1, \ldots$. The size of the "cash position" is chosen such that the portfolio remains self-financing, i.e.

$$\phi_1(t_k) = \frac{\partial V(t_k)}{\partial S(t_k)} = \Phi\left(\frac{1}{\sigma\sqrt{T-t_k}}\left[\log\left(\frac{S(t_k)}{K}\right) + r(T-t_k) + \frac{\sigma^2}{2}\right]\right)$$

$$\phi_0(t_k) = \frac{1}{N(t_k)}(\phi(t_{k-1})N(t_k) + \phi_1(t_{k-1})S(t_k) - \phi_1(t_k)S(t_k))$$

At time $t_0 = 0$ the portfolio is set up according to the option value known from the evaluation formula:[6]

$$\phi_0(t_0) = \frac{1}{N(t_0)}(V(t_0) - \phi_1(t_0)S(t_0)).$$

Figure 7.1 shows the result of a weekly delta hedge.

---

[6] Eventually, the initial investment in the replication portfolio is not sufficient, see Figure 7.1, right.

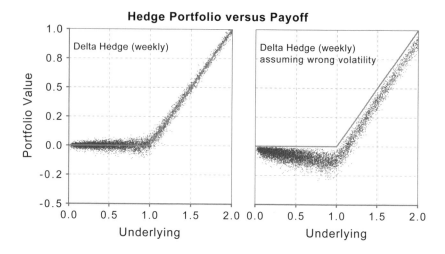

Figure 7.1. *Samples of the value of the replication portfolio using weekly delta hedging. The market (N, S) follows a Black-Scholes model $dN(t) = rN(t) \, dt$, $dS(t) = \mu S(t) \, dt + \sigma S(t) \, dW(t)$ with $(N(0), S(0)) = (1.0, 1.0)$, $r = \mu = 0.05$ and $\sigma(t) = 0.5$. We aim to replicate the payout $X(T) = \max(S(T) - K, 0))$ with $K = 1.0$ and $T = 6.0$ (line). The replication portfolio $\Pi = \phi_0 N + \phi_1 S$ is chosen according to a delta hedge at times $t_k = k \times \frac{1}{52}$. Shown are the realized values $\Pi(T)$ (dots) over $S(T)$. In the figure on the left the delta hedge uses the correct model parameters. In the figure on the right the portfolio is constructed assuming a volatility of 0.4 (instead of 0.5). As a result the option value is underestimated and the replication portfolio is set up without enough initial value, such that at maturity $T = 6.0$ the mean of the replication portfolio is systematically below the option payout. In addition, the variance of the replication portfolio is increased, since the delta hedge used is wrong.*

7.4.3 Delta-Gamma Hedging

Motivation (The Delta Hedge Again): A delta hedge (Figure 7.2) applied in time-discrete steps $\Delta t_j = t_{j+1} - t_j$ is not exact since the option value is, in general, a nonlinear function of the changes of the underlyings $\Delta S_i = S_i(t_{j+1}) - S_i(t_j)$. However, the replication portfolio is linear in the underlyings.

For motivation reconsider a market consisting of a stock S and some other numéraire asset N (as in the Black-Scholes model) and a derivative product V as a function of (t, S): $V = V(t, S)$. At fixed time t_j, V will be a nonlinear function $S \mapsto V(t_j, S)$ of S in general. However, the replication portfolio $\Pi = \phi_0 N + \phi_1 S$ is linear in S. Locally, for a fixed point $S(t_j, \omega)$ the replication portfolio may match the

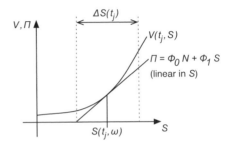

Figure 7.2. *Delta hedge.*

value and the first derivative in S (*the delta*) of V: For infinitesimal changes dS the infinitesimal changes of derivative (dV) and replication portfolio $d\Pi$ agree. If the movements of S are within in a larger (noninfinitesimal) range $\Delta S(t_j)$, then derivative and replication portfolio may deviate significantly. ◁

The error given by the only linear approximation of the derivative product may be reduced by adding products to the replication portfolio which are themselves non-linear, e.g., options. This is possible if such products are traded and may be used for replication: Some standardized options are traded in sufficient quantities and may be used to replicate derivative products that are not traded in liquid quantities. An example of such a case would be options traded at standardized maturities, which could be used to build a replication portfolio for an option with a nonstandard maturity (and/or strike).

We consider a portfolio consisting of $S_0, S_1, \ldots, S_n$ and additional (derivative) products $C_1, \ldots, C_m$

$$\Pi(t) = \phi_0(t)S_0(t) + \cdots + \phi_n(t)S_n(t) + \psi_1(t)C_1(t) + \cdots + \psi_m(t)C_m(t)$$

Assuming that, like $V(T)$, the $C_k(t)$'s may be written as functions $C_k(t, S_0, \ldots, S_n)$ of the S_i's, we have for the replication portfolio $\Pi(t)$ instead of (7.13)

$$
\begin{aligned}
\Delta\Pi(t) &= \sum_{i=0}^{n} \phi_i \, \Delta S_i + \sum_{k=1}^{m} \psi_k \, \Delta C_k \\
&= \sum_{i=0}^{n} \left(\phi_i + \sum_{k=1}^{m} \psi_k \frac{\partial C_k}{\partial S_i} \right) \Delta S_i + \sum_{k=1}^{m} \psi_k \frac{\partial C_k}{\partial t} \, \Delta t \\
&\quad + \sum_{i,j=1}^{n} \sum_{k=1}^{m} \psi_k \frac{\partial^2 C_k}{\partial S_i \, \partial S_j} \, \Delta S_i \, \Delta S_j + \text{h.o.t.} \\
&= \sum_{i=0}^{n} \left(\phi_i + \sum_{k=1}^{m} \psi_k \frac{\partial C_k}{\partial S_i} \right) \Delta S_i + \sum_{i,j=1}^{n} \sum_{k=1}^{m} \psi_k \frac{\partial^2 C_k}{\partial S_i \, \partial S_j} (\Delta S_i \, \Delta S_j - \gamma_{i,j} \, \Delta t) + \text{h.o.t.}
\end{aligned}
$$

where again h.o.t. $= O(|\Delta t|^2, |\Delta t \, \Delta S_i|, |\Delta S_i|^3)$. Let us compare this with the expansion of V from (7.12):

$$
\Delta V(t) = \sum_{i=0}^{n} \frac{\partial V(t)}{\partial S_i} \Delta S_i + \frac{1}{2} \sum_{i,j=0}^{n} \frac{\partial^2 V(t)}{\partial S_i \, \partial S_j} \underbrace{(\Delta S_i \, \Delta S_j - \gamma_{i,j} \, \Delta t)}_{\neq 0} + \text{h.o.t.}
$$

If the portfolio process $(\phi_0, \ldots, \phi_n, \psi_1, \ldots, \psi_m)$ solves the equations

$$
\sum_{k=1}^{n} \frac{\partial^2 C_k}{\partial S_i \, \partial S_j} \psi_k = \frac{\partial^2 V}{\partial S_i \, \partial S_j} \qquad i \le j = 1, \ldots, n \qquad (7.17)
$$

$$
\phi_i + \sum_{k=1}^{n} \frac{\partial C_k}{\partial S_i} \psi_k = \frac{\partial V}{\partial S_i} \qquad i = 1, \ldots, n, \qquad (7.18)
$$

then the residual risk is $\Delta V(t) - \Delta\Pi(t) = O(|\Delta t|^2, |\Delta t \, \Delta S_i|, |\Delta S_i|^3)$. To neutralize the gamma of a derivative product V it requires at most as many hedge derivatives C_k as partial derivatives $\frac{\partial^2 V}{\partial S_i \, \partial S_j}$, $i \le j = 0, \ldots, n$ (gamma) are nonzero. It requires at most $m = n(n-1)/2$ additional derivative products.

Remark 92 (Linear Product, Static Hedge): If a derivative product V is a linear function of the underlyings, then the delta hedge replicates the product globally. In this case, not only is a gamma hedge unnecessary (gamma and all higher order derivatives are zero), but the dynamic adjustment of the replication portfolio is not required. In this case the hedge is called *static* and the derivative product is called *linear product*.

7.4.3.1 Example: Time-Discrete Delta-Gamma Hedge under a Black-Scholes Model

We consider a Black-Scholes model with the notation as above. Let C_1 denote an option with maturity T^* and payoff profile $\max(S(T^*) - K^*, 0)$. We aim to replicate an option V with maturity $T < T^*$ and payoff profile $\max(S(T) - K, 0)$. We allow trading at discrete times $0 = t_0 < t_1 < \ldots$.

For the option V to be replicated we have

$$\Delta V(t_k) = \frac{\partial V(t_k)}{\partial t} \, \Delta t_k + \frac{\partial V(t_k)}{\partial S} \, \Delta S(t_k) + \frac{1}{2} \frac{\partial^2 V(t_k)}{\partial S^2} (\Delta S(t_k))^2$$
$$+ O(|\Delta t_k|^2, |\Delta t_k \, \Delta S(t_k)|, |\Delta S(t_k)|^3).$$

For the replication portfolio Π we have

$$\Delta \Pi(t_k) = \phi_0(t) \, \Delta N(t_k) + \phi_1(t_k) \, \Delta S(t) + \psi_1(t_k) \, \Delta C(t_k)$$
$$= \phi_0(t) \, \Delta N(t_k) + \phi_1(t_k) \, \Delta S(t) + \psi_1(t_k) \frac{\partial C(t_k)}{\partial t} \, \Delta t_k$$
$$+ \psi_1(t_k) \frac{\partial C(t_k)}{\partial S} \, \Delta S(t_k) + \psi_1(t_k) \frac{1}{2} \frac{\partial^2 C(t_k)}{\partial S^2} (\Delta S(t_k))^2$$
$$+ O(|\Delta t_k|^2, |\Delta t_k \, \Delta S(t_k)|, |\Delta S(t_k)|^3).$$

For the replication portfolio $\Pi(t) = \phi_0(t)N(t) + \phi_1(t)S(t) + \psi_1(t)C(t)$ we find

$$\phi_1(t) = \frac{\partial V(t)}{\partial S(t)} = \Phi\left(\frac{1}{\sigma\sqrt{T-t}}\left[\log\left(\frac{S(t)}{K}\right) + r(T-t) + \frac{\sigma^2}{2}\right]\right).$$

We chose this to trade within the replication portfolio at discrete times t_k. The size of the "cash position" is chosen such that the portfolio remains self-financing, i.e.,

$$\phi_1(t_k) = \frac{\partial V(t_1)}{\partial S(t_1)} = \Phi\left(\frac{1}{\sigma\sqrt{T-t_k}}\left[\log\left(\frac{S(t_k)}{K}\right) + r(T-t_k) + \frac{\sigma^2}{2}\right]\right)$$

$$\phi_0(t_k) = \frac{1}{N(t_k)}(\phi_0(t_{k-1})N(t_k) + \phi_1(t_{k-1})S(t_k) - \phi_1(t_k)S(t_k))$$

At time $t_0 = 0$ the portfolio is set up according to the option value known from the evaluation formula.[7]

$$\phi_0(t_0) = \frac{1}{N(t_k)}(V(t_0) - \phi_1(t_0)S(t_0))$$

---

[7] Eventually the initial investment into the replication portfolio is not sufficient; see Figure 7.1, right.

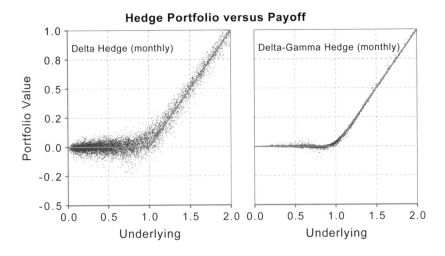

Figure 7.3. *Samples of the value of the replication portfolio using monthly hedging: delta versus delta-gamma hedge.*

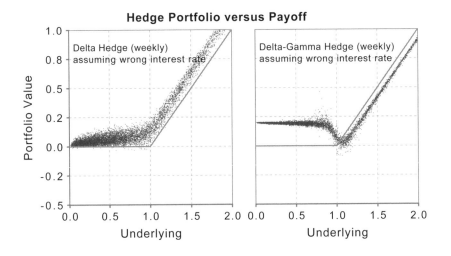

Figure 7.4. *Samples of the value of the replication portfolio using weekly hedging with wrong interest rate.*

Interpretation (Role of the Hold Period Δt): The delta hedge neutralizes the first-order error in S; the delta-gamma hedge neutralizes the first- and second-order errors in S (Figure 7.3). The residual error of the delta-gamma hedge is of the order

$$O(|\Delta t|^2, |\Delta t \, \Delta S_i|, |\Delta S_i|^3).$$

A question arises:

- Why is the error in Δt not considered?

First, the answer is given in the interpretation of the hedge strategies: They aim to minimize the risk in the underlyings S_i, i.e., the dependence of $V(t) - \Pi(t)$ on $S_i(t)$. The interpretation is that the length of the hold period Δt is not known a priori. In our considerations we thus assume $\Delta t \to 0$. Then the residual error is $O(\Delta S_i^2)$ for the delta hedge and $O(\Delta S_i^3)$ for the delta-gamma hedge.

Furthermore: An error in ΔS_j is stochastic, i.e., a risk. An error in Δt is deterministic. If the hold period Δt is known a priori, then the error in Δt may be compensated for if the portfolio is not required to match the derivative value initially, i.e., $\Pi(0) = V(0)$. The derivative value $V(0)$ corresponds to the replication portfolio for infinitesimal hold periods $\Delta t \to 0$. We will consider a known hold period $\Delta t > 0$ in Section 7.5. ◁|

7.4.4 Vega Hedging

As in the delta-gamma hedging, where in addition to a hedge of the option's delta a hedge of the option's gamma is considered, we may consider further dependencies of the pricing function. If the dependence on the log-volatility of the underlyings, i.e. $\frac{\partial V}{\partial \sigma_i}$ for $dS_i = \mu_i S_i \, dt + \sigma_i S_i \, dW_i$, is neutralized, then this called *vega hedging*.

7.5 Hedging in Discrete Time: Minimizing the Residual Error (Bouchaud-Sornette Method)

The delta hedge transfers the optimal trading strategy for trading in continuous time to the time-discrete case, for which the strategy is not necessarily optimal. A more adequate calculation of a risk minimizing trading strategy may be derived from the residual risk: We look for a trading strategy that minimizes[8] the residual risk. This is the idea of the method of Bouchaud-Sornette [49]. Since for the time-discrete case the replication portfolio does not give an exact replication, we have yet to clarify in which sense a replication portfolio is optimal, i.e., in which norm the hedge error

---

[8] In contrast to neutralizes.

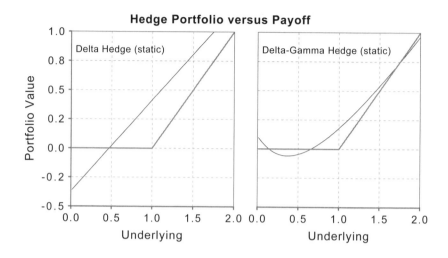

Figure 7.5. *Value of the replication portfolio without rehedging. The replication portfolio is set up in t = 0.0 corresponding to a delta- or delta-gamma hedge and kept fixed until option maturity T (model and option parameter are as in Figure 7.1).*

is measured. The optimal replication portfolio derived from this criterion is not necessarily identical with the replication portfolio of the delta hedge and depends on the norm that is used to measure the residual risk.

At time T a payoff profile $V(T)$ has to be replicated as closely as possible by a replication portfolio $t \mapsto \Pi(t)|_{t=T}$. We consider the mean squared error, i.e., the variance of $V(T) - \Pi(T)$:

$$\text{Var}(V(T) - \Pi(T)).$$

We also wish to minimize the variance in a conditional sense and repeat the required definitions and a lemma:

Definition 93 (Conditional Variance): ⌐

Let $C \subset \mathcal{F}$ denote two σ-algebra and X an $\mathcal{F}$-measurable random variable over a probability space $(\Omega, \mathbb{P}, \mathcal{F})$. Then

$$\text{Var}(X|C) := \text{E}(X^2|C) - (\text{E}(X|C))^2$$

is the *conditional variance* of X under C. ⌐

Remark 94 (Conditional Variance): The conditional variance is a C-measurable random variable.

Lemma 95 (Conditional Variance): Let $C \subset \mathcal{F}$ denote two σ-algebras, X an $\mathcal{F}$-measurable random variable and Y a C-measurable random variable over the probability space $(\Omega, \mathbb{P}, \mathcal{F})$. Then

$$\text{Var}(X + Y|C) = \text{Var}(X|C).$$

Proof: The proof is elementary. From $\text{E}(X \cdot Y|C) = Y \cdot \text{E}(X|C)$ and $\text{E}(Y^2|C) = Y^2$ we get

$$
\begin{aligned}
\text{Var}(X + Y \mid C) \overset{\text{Def.}}{=} \ & \text{E}((X + Y)^2 \mid C) - (\text{E}(X + Y \mid C))^2 \\
= \ & \text{E}(X^2 \mid C) + 2\text{E}(X\,Y \mid C) + \text{E}(Y^2 \mid C) \\
& - \text{E}(X \mid C)^2 - 2\text{E}(X \mid C)\,\text{E}(Y \mid C) - \text{E}(Y \mid C)^2 \\
= \ & \text{E}(X^2 \mid C) + \cancel{2Y\,\text{E}(X \mid C)} + \cancel{Y^2} - \text{E}(X \mid C)^2 - \cancel{2Y\,\text{E}(X \mid C)} - \cancel{Y^2} \\
= \ & \text{E}(X^2 \mid C) - \text{E}(X \mid C)^2 = \text{Var}(X \mid C).
\end{aligned}
$$

$\square|$

Let $0 = t_0 < t_1 < \cdots < t_m = T$ denote a time discretization and $\phi(t) = (\phi_0, \ldots, \phi_n)$ a portfolio process $\Pi(t) = \phi(t) \cdot S(t) = \sum_{i=0}^{n} \phi_i(t) S_i(t)$, with piecewise constant ϕ_i, i.e., we assume that trading takes place only at discrete times ($\phi_i(t) = \phi_i(t_j)$ for $t_j \le t < t_{j+1}$).

7.5.1 Minimizing the Residual Error at Maturity T

Trading Strategy: At time t_k the self-financing portfolio $\phi(t_k)$ is chosen such that the residual risk is minimized, i.e.,

$$\text{Var}(V(T) - \Pi(T) \mid \mathcal{F}_{t_k}) \rightarrow \min,$$

where $\{\mathcal{F}_t\}$ denotes the filtration generated by S.

To simplify notation let $S_0 := P(T)$ denote the *bond with maturity* T, i.e., the financial product that guarantees a payment of 1 at time T, $P(T) : t \mapsto P(T; t)$ with $P(T; T) \equiv 1$. This product will later lay the foundation for the theory of interest rates; see Definition 97 in Chapter 8. It is not required that one of the products is the *Bond with maturity* T, but it simplifies the notation since

$$\text{Var}(V(T) - \Pi(T)) = \text{Var}\left(\frac{V(T)}{S_0(T)} - \frac{\Pi(T)}{S_0(T)}\right)$$

and instead

$$\Pi(T) = \Pi(0) + \sum_{l=0}^{m-1} \phi(t_l)\,\Delta S(t_l).$$

115

We may equivalently write

$$\Pi(T) = \frac{\Pi(T)}{S_0(T)} = \frac{\Pi(0)}{S_0(0)} + \sum_{l=0}^{m-1} \sum_{i=0}^{n} \phi_i(t_l) \, \Delta \frac{S_i(t_l)}{S_0(t_l)}$$

$$= \frac{\Pi(0)}{S_0(0)} + \sum_{l=0}^{m-1} \sum_{i=1}^{n} \phi_i(t_l) \, \Delta \frac{S_i(t_l)}{S_0(t_l)}$$

—note that ϕ_0 does not enter in the last sum. Thus ϕ_0 may be chosen such that the condition of self-financing is fulfilled and ϕ_i, $i \geq 1$ determines the "optimal hedge".

With the notation $\tilde{S} = \dfrac{S}{S_0}$ we have

$$V(T) - \Pi(T) = V(T) - \Pi(0) - \sum_{l=0}^{m-1} \phi(t_l) \cdot \Delta \tilde{S}(t_l)$$

$$= V(T) - \Pi(0) - \sum_{l=0}^{k-1} \phi(t_l) \cdot \Delta \tilde{S}(t_l)$$

$$- \sum_{l=k+1}^{m-1} \phi(t_l) \cdot \Delta \tilde{S}(t_l) - \phi(t_k) \cdot \Delta \tilde{S}(t_k)$$

and with Lemma 95

$$\mathrm{Var}(V(T) - \Pi(T) \mid \mathcal{F}_{t_k})$$

$$= \mathrm{Var}\left(V(T) - \sum_{l=k+1}^{m-1} \phi(t_l) \cdot \Delta \tilde{S}(t_l) - \phi(t_k) \cdot \Delta \tilde{S}(t_k) \,\Big|\, \mathcal{F}_{t_k} \right)$$

$$= \mathrm{Var}\left(V(T) - \sum_{l=k+1}^{m-1} \phi(t_l) \cdot \Delta \tilde{S}(t_l) \,\Big|\, \mathcal{F}_{t_k} \right)$$

$$- 2 \cdot \mathrm{Cov}\left(V(T) - \sum_{l=k+1}^{m-1} \phi(t_l) \cdot \Delta \tilde{S}(t_l) \,,\, \phi(t_k) \cdot \Delta \tilde{S}(t_k) \,\Big|\, \mathcal{F}_{t_k} \right)$$

$$+ \mathrm{Var}\left(\phi(t_k) \cdot \Delta \tilde{S}(t_k) \mid \mathcal{F}_{t_k} \right).$$

Finally we get from

$$\frac{1}{2} \frac{\partial}{\partial \phi(t_k)} \mathrm{Var}(V(T) - \Pi(T) \mid \mathcal{F}_{t_k}) = 0$$

the equation for $\phi(t_k)$

$$\phi(t_k) \cdot \mathrm{Cov}(\Delta \tilde{S}(t_k), \Delta \tilde{S}(t_k) \mid \mathcal{F}_{t_k}) \cdot \phi(t_k)^\top$$
$$- \mathrm{Cov}(V(T) - \sum_{l=k+1}^{m-1} \phi(t_l) \cdot \Delta \tilde{S}(t_l), \Delta \tilde{S}(t_k) \mid \mathcal{F}_{t_k}) = 0. \qquad (7.19)$$

7.5.2 Minimizing the Residual Error in Each Time Step

The strategy described in Section 7.5.1 focuses at each time t_k on the minimization of the residual risk at maturity T. An alternative trading strategy is to require at each time t_k that the residual risk at time t_{k+1} is minimized by the choice of the portfolio $\phi(t_k)$.

$$\mathrm{Var}(V(t_{k+1}) - \Pi(t_{k+1}) \mid \mathcal{F}_{t_k}) \to \min.$$

Obviously (consider the last time step $k = m - 1$ in (7.19)) the equation for $\phi(t_k)$ then follows as

$$\phi(t_k) \cdot \mathrm{Cov}(\Delta \tilde{S}(t_k), \Delta \tilde{S}(t_k) \mid \mathcal{F}_{t_k}) \cdot \phi(t_k)^\top - \mathrm{Cov}(V(t_{k+1}), \Delta \tilde{S}(t_k) \mid \mathcal{F}_{t_k}) = 0.$$

Remark 96 (Measure): The measure under which the minimization, and thus the covariance in (7.19), has to be considered is the real measure $\mathbb{P}$. A consideration under a martingale measure essentially corresponds to a measuring of the risk under a different norm.

 Experiment: At http://www.christian-fries.de/finmath/ applets/HedgeSimulator.html a *hedge simulator* can be found. There, in a Monte Carlo simulation, paths $t \mapsto (N(t, \omega), S(t, \omega))$ are generated corresponding to a given model. Along each path a self-financing trading strategy is applied to a replication portfolio. At maturity T the value of the replication portfolio is compared to the option value $V(T)$. It is possible to choose the model parameters assumed in the construction of the replication portfolio independent from the model parameters, which determine the evolution of the market. A mismatch in the parameters results in an increased residual risk (see Figure 7.1). ◁|

Part III

Interest Rate Structures, Interest Rate Products, and Analytic Pricing Formulas

Motivation and Overview - Part III

Part III will consider the interest rate structures and the analytical pricing of interest rate derivatives. The methods for pricing financial products may roughly be classified into three groups. These are

- **Model-independent analytic pricing**: The financial product can be separated into a portfolio of traded assets. The value of the derivative product is then given by the value of the portfolio. The portfolio represents a replication portfolio, i.e., a static hedge.

- **Model-dependent analytic pricing**: The replication of the financial product requires dynamic hedging, i.e., a continuous adjustment of the portfolio, and thus requires a model for the underlying stochastics. The value of the replication portfolio depends on the model. However, the product and the chosen model are simple enough to derive an analytic pricing formula.

- **Model-dependent numeric pricing**: The replication of the financial product requires dynamic hedging and thus requires a model for the underlying stochastics. The value of the replication portfolio depends on the model. The complexity of the product or of the model requires a numerical calculation of the price.

In Chapter 8 we will define some elementary products and interest rates. In Section 9.1 simple interest rate products will be presented which allow model-independent analytic pricing. Here, the corresponding pricing is discussed right after the definition of the product. In Section 9.2 we define some simple options. In Chapter 10 and 11 we will show how to derive some analytic pricing formulas using some simple standard models. In Chapter 12 more exotic options will be presented, for which pricing requires generally numerical methods.

Numerical pricing methods and complex pricing models will be considered in Parts IV and V.

CHAPTER 8

Interest Rate Structures

8.1 Introduction

In previous sections we have considered a one-dimensional stochastic process as the underlying, representing a stock for example. We will now turn to the modeling of interest rates and later the pricing of interest rate derivatives. Interest rates potentially offer a richer structure than a single stock, since at each time t we have to consider an interest rate *curve* $F(T;t)$, $T \geq t$ instead of a scalar stock price $S(t)$.

Let $F(T;t)$ denote the interest earned on an investment made over the period $[t, T]$, if the contract is written at the beginning t of the period, i.e., we assume that $N(1 + F(T;t)(T - t))$ is the amount one receives at time T, if one invests the amount N at time t, and the contract is fixed at the beginning of the period, i.e., in t.[1] For different times $T_1 < T_2$ two such interest rates $F(T_1;t)$ and $F(T_2;t)$ may be different, the reason being the different interest earned on the two subperiods $[t, T_1]$ and $[T_1, T_2]$, e.g., if interest rates are expected to rise or fall in future. Thus, if one decomposes the time axis into small interest rate periods $[T_i, T_{i+1}]$ and defines

$$1 + F(T_k;t)(T_k - t) =: (1 + L(T_0, T_1;t)(T_1 - T_0)) \cdot \ldots \cdot (1 + L(T_{k-1}, T_k;t)(T_k - T_{k-1})), \quad (8.1)$$

then it is rational to represent each interest rate $L(T_i, T_{i+1};t)$ by its own stochastic process. Here we assume $T_0 = t$ and $L(T_i, T_{i+1};t)$ denotes the interest rate earned over the period $[T_i, T_{i+1}]$ as given by a contract which is fixed in t.[2] See Figure 8.1. It is common to have interest rate derivatives that depend on more than one interest rate and in this sense depend on the movement of the whole curve.

---

[1] We will comment below on the fact that the contract may be fixed at an earlier time than the start of the interest rate period.

[2] Equation (8.1) represents the relationship between interest earned over the period $[t, T_k]$ and interest earned from smaller subperiods, including compounding. Since all rates are fixed in time t, the difference is only interpretation. Thus the two sides in (8.1) must agree.

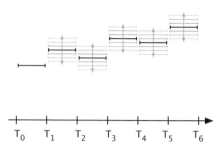

Figure 8.1. *Modeling an interest rate curve by a family of stochastic processes.*

8.1.1 Fixing Times and Tenor Times

If the interest rates $F(T_i; t)$ for periods $[t, T_i]$ for $i = 1, \ldots, k$ are known, then we can derive from Equation (8.1) the rates for the subperiods $[T_{i-1}, T_i]$. It is

$$L(T_{i-1}, T_i; t) = \frac{1}{T_i - T_{i-1}} \left(\frac{1 + F(T_i; t)(T_i - t)}{1 + F(T_{i-1}; t)(T_{i-1} - t)} - 1 \right). \tag{8.2}$$

The rate $L(T_{i-1}, T_i; t)$ is the interest rate for the *future* period T_{i-1}, T_i, that has been derived from the interest rate curve as of time t. The interest rate has been fixed at time t, i.e., the stochastic process is evaluated in t. The time t is the simulation time of the stochastic process; it thus determines that the random variable $L(T_{i-1}, T_i; t)$ has to be (at least) $\mathcal{F}_t$-measurable. The times T_i mark the start and the end of the periods. The period structure $\{T_0, T_1, \ldots, T_k\}$ is also called the *tenor structure*.

8.2 Definitions

All of the following random variables and stochastic processes are assumed to be defined over the same probability space $(\Omega, \mathcal{F}, \mathbb{P})$. As the building block of all interest rates we define the *bond*:

Definition 97 (Bond[3]):
Assume that a guaranteed payment of a unit currency[4] 1 in time T_2 is a traded product at any earlier time $t < T_2$ and its value in state $\omega \in \Omega$ is uniquely determined by (t, ω).

---

[3] The term defined is the *zero-coupon bond* (there are no intermediate payments (*coupons*) until maturity). We give the trivial extension to a *coupon bond* in Definition 108.
[4] See Remark 134.

The value of this product is called the price of the T_2 *bond as seen in* (t, ω) and will be denoted by $P(T_2; t, \omega)$. It defines a stochastic process on $[0, T_2]$ which we denote by $P(T_2)$:

$$P(T_2) : [0, T_2] \times \Omega \mapsto \mathbb{R}.$$

 Interpretation: The value of the bond $P(T_2; t, \omega)$ is the amount we have to invest at a given time t and a given state ω to receive 1 in T_2. Thus for a (riskless) investment of 1 at time t we get a guaranteed payment of $1/P(T_2; t, \omega)$ at time T_2, because we simply buy $1/P(T_2; t, \omega)$ times the bond. Note that we assume that all products may be traded in arbitrary fractions and that the price is unique, the same for selling and buying. Thus $P(T_2; t, \omega)$ not only implies the interest earned on an investment, it also implies the interest to be paid for a corresponding loan. ◁

Definition 98 (Forward Bond):
Let $0 \leq T_1 \leq T_2 < \infty$. As *forward bond* we denote the stochastic process

$$P(T_1, T_2) := \frac{P(T_2)}{P(T_1)},$$

defined on $[0, T_1]$.

 Interpretation: The *forward bond* $P(T_1, T_2; t)$ corresponds to the amount that has to be invested at time T_1 to receive a guaranteed payment of 1 in T_2, if that contract is finalized at time t in state ω. To receive 1 in T_2 one has to invest $P(T_2; t)$ in t. This investment is financed by borrowing until T_1, i.e., we borrow $P(T_2; t)\, 1/P(T_1; t)$ times the value of a $P(T_1)$ bond (compare Figure 8.2).

It is important to understand that $P(T_1, T_2; t)$ is not the value of a forward bond at time t. The value of that contract is simply 0. What is denoted by $P(T_1, T_2; t)$ is the amount that has to be paid in T_1 as part of a contract written in t. The subject of the contract is a bond which lies *forward* in time.

Obviously we have $P(T_2; t) = P(t, T_2; t)$. ◁

Often the bond (or forward bond) will be represented by a rate (or forward rate). However, the interest rate that corresponds to a bond depends on how we think of interest earned, e.g., if compounding, i.e., interest paid on interest received, is

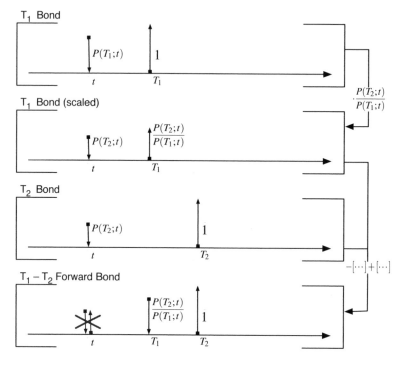

Figure 8.2. *Cash flow for a forward bond.*

considered. Thus, there are many different definitions of interest rates. For that reason we see an interest rate as a derived quantity. We define some interest rates.

Definition 99 (Forward Rate ((Forward) LIBOR)):
Let $T_1 < T_2$. The (forward) LIBOR[5] $L(T_1, T_2)$ is defined by

$$\frac{P(T_1)}{P(T_2)} = \frac{1}{P(T_1, T_2)} =: 1 + L(T_1, T_2)\,(T_2 - T_1),$$

i.e.,

$$L(T_1, T_2) := \frac{1}{T_2 - T_1}\,\frac{P(T_1) - P(T_2)}{P(T_2)}.$$

---

[5] LIBOR = London Inter Bank Offer Rate. The acronym LIBOR is often used for a rate following Definition 99, because the interbank rate follows this convention.

Definition 100 ((Continuously Compounded) Yield):
Let $T_1 < T_2$. The (continuously compounded) (forward) yield $r(T_1, T_2)$ is defined by

$$\frac{P(T_1)}{P(T_2)} = \frac{1}{P(T_1, T_2)} =: \exp(r(T_1, T_2)\,(T_2 - T_1)),$$

i.e.,

$$r(T_1, T_2) := \frac{-\log(P(T_1, T_2))}{T_2 - T_1} = -\frac{\log(P(T_2)) - \log(P(T_1))}{T_2 - T_1}.$$

Definition 101 (Instantaneous Forward Rate):
The *instantaneous forward rate* is defined by

$$f(t, T) := \lim_{T_2 \searrow T} L(T, T_2; t).$$

It may be interpreted as the interest rate for the infinitesimal period $[T, T + dT]$.

Remark 102 (Instantaneous Forward Rate): We have

$$f(t, T) = \lim_{T_2 \searrow T} L(T, T_2; t) = \lim_{T_2 \searrow T} \frac{1}{P(T_2; t)} \frac{P(T; t) - P(T_2; t)}{T_2 - T}$$
$$= -\frac{1}{P(T; t)} \frac{\partial P(T; t)}{\partial T} = -\frac{\partial \log(P(T; t))}{\partial T}.$$

Definition 103 (Short Rate):
Let $t \geq 0$. The short rate $r(t)$ is defined by

$$r(t) := \lim_{T \searrow t} r(t, T; t) = \frac{\partial}{\partial T} - \log(P(T; t))\Big|_{T=t}$$
$$= \frac{-\frac{\partial P(T; t)}{\partial T}}{P(T; t)}\Big|_{T=t} = -\frac{\partial P(T; t)}{\partial T}\Big|_{T=t}.$$

Remark 104 (Short Rate): Note that in Definitions 99, 100, and 101 we define families of stochastic processes, while the short rate from Definition 103 is a single, scalar stochastic process.

The *short rate* is a limit of the spot *forward rate* (LIBOR) and the *yield*. We have

$$r(t) = \lim_{T \searrow t} r(t, T; t) = \lim_{T \searrow t} L(t, T; t).$$

This *short rate* is the interest rate for the infinitesimal period $[t, t + dt]$, as seen in t.

Remark 105 (Forward Forward): The term "forward" is used ambiguously: In Definitions 99 and 100 we define stochastic processes for interest rates, i.e., for a given observation time t we define $L(T_1, T_2; t)$ and $r(T_1, T_2; t)$ as rates for the period $[T_1, T_2]$ as seen in t. If $T_1 > t$, then these rates are called "*forward*" (i.e. *forward LIBOR, forward yield*), since the rate is associated with a future period, lying forward in time. If $T_1 = t$ we would use the attribute "*spot*" instead.

However, on the other hand, the rate $L(T_1, T_2; T_1)$ (note $t = T_1$) is often denoted as *forward rate*, since it is an interest rate for an interval up to T_2. This is in contrast to the *short rate*, which is defined for an infinitesimal period. Being precise, terms like "forward forward rate" should be used.

There is a similar ambiguity for volatilities. The term *forward volatility* may be interpreted as the volatility of a forward rate or as a volatility of some process considered at a future period in time.

Remark 106 (Interest Rate Models): The various definitions of interest rates (as stochastic processes) are the starting point for the corresponding interest rate modeling. The corresponding models are listed in Table 8.1.

Interest Rate	Model
Forward rate L	LIBOR market model $\to$ Chapter 19
Instantaneous forward rate f	HJM framework $\to$ Chapter 22
Short rate r	Short rate models $\to$ Chapter 23

Table 8.1. *Interest Rate Models.*

Although the models above do not model the bond prices directly, we view zero-bond prices as the basic building blocks.

Remark 107 (Bond Prices as a Function of Interest Rates): The bond prices may be calculated from the interest rates. We have

$$P(T; t) = \exp(-r(t, T; t)(T - t))$$

$$P(T; t) = \exp\left(-\int_t^T f(t, \tau) \, d\tau\right)$$

$$P(T; t) = \prod_{i=0}^{n-1} (1 + L(T_i, T_{i+1}; t)(T_{i+1} - T_i))^{-1} \quad \text{for } t = T_0 < T_1 < \cdots < T_n = T$$

.

The short rate is an exception. Here the reconstruction of bond prices is possible only if the short rate process is known under the equivalent martingale measure $\mathbb{Q}^N$ corresponding to the numéraire $N(t) := \exp\left(\int_0^t r(\tau)\, d\tau\right)$. Then we have

$$P(T;t) = \mathrm{E}^{\mathbb{Q}^N}\left(\exp\left(-\int_t^T r(\tau)\, d\tau\right) \mid \mathcal{F}_t\right)$$

 Tip (Discount Factors as a Basic Market Data Object): We consider the price of zero bonds as given and view interest rates as derived quantities. It is natural to take this view in the implementation as well. If we want to provide information on market interest rates through a class[6], then we store internally a discretization of the bond price curve $j \mapsto P(T_j; 0)$ (also called *discount factors*). The class then provides the various interest rates under various conventions through methods. This design reduces the errors of misinterpretation of the stored data (especially if more than one developer works on the class), since the data stored is free of market conventions and the convention used to calculate the rates is explicit in the implementation of the corresponding method. This also reduces the documentation overhead for the data model.

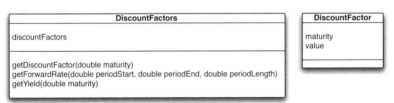

Figure 8.3. *UML diagram: The class* `DiscountFactors` *internally stores discount factors and provides various interest rates through corresponding methods.*

A problem of this design is that discount factors are usually not the quantities that are observable in the interest rate market. The market quotes the prices of various different interest rate products (e.g., futures or swaps), from which discount factors have to be calculated. This calculation is called *bootstrapping*. Under some conditions, it might be useful to store the original market data. One such application is the numerical calculation of partial derivatives with respect to a change in these input prices.[7]   ◁|

---

[6] See Chapter 30.

[7] The importance of the partial derivative with respect to the price of an *underlying* has been discussed in Chapter 7.

8.3 Interest Rate Curve Bootstrapping

Since bootstrapping of discount factors from market prices involves the inversion of a pricing formula, we have to discuss the interest rate products and their pricing. We will do so in Chapter 9 but give an anticipatory abstract description of the bootstrap algorithm here:

Let $0 = T_0 < T_1 < T_2 < \ldots$ denote a time discretization. For given discount factors $df_j = P(T_j; 0)$, $j = 0, 1, \ldots, i$, we assume the existence of an interpolation function $df(df_0, \ldots, df_i; t)$, having the property that an additional sample point beyond T_i will not change the interpolation in $t \le T_i$, i.e.,

$$df(df_0, \ldots, df_i; t) = df(df_0, \ldots, df_i, df_{i+1}; t) \qquad \forall\, t \le T_i \quad \forall\, i = 1, 2, 3, \ldots .$$

Let V_i^{market} denote given market prices of interest rate products for which the price may be expressed as a function V_i of the discount factors in $t \le T_i$, i.e.,

$$V_i = V_i(\, \{df(t) \mid t \le T_i\} \,).$$

We further assume that the discount factor $df(T_i)$ enters into the pricing, i.e., let $\dfrac{\partial}{\partial\, df(T_i)} V_i \neq 0$. Then the bootstrap algorithm is given by:

Induction Start (T_0):

- $df_0 := df(T_0) = P(T_0; 0) = P(0; 0) = 1.0.$

Induction Step ($T_{i-1} \to T_i$):

- Calculate $df_i := df(T_i)$ such that, using the discount factor interpolation, we have
$$V_i(\, \{df(df_0, \ldots, df_i; t) \mid t \le T_i\} \,) \stackrel{!}{=} V_i^{\text{market}}. \qquad (8.3)$$

In some cases Equation (8.3) may be directly solved for df_i, especially if it does not depend on the interpolation method used. Normally, a numerical solution is possible (see Appendix B.4).[8]

8.4 Interpolation of Interest Rate Curves

We consider, as before, a family of bond prices $T \mapsto P(T; 0)$, i.e., the discount factor curve, as the basic representation of the interest rate curve. If the prices $df_i := P(T_i; 0)$

---

[8] If interest rates are positive, a simple interval bisection like the Golden Section Search works, since $df_i \in [0, df_{i-1}]$.

are known for times $0 = T_0 < T_1 < T_2 < \ldots$, we seek a meaningful interpolation method to calculate interest rates for subperiods. The interpolation method should fulfill two basic requirements:

- The interpolation method should be sufficiently smooth, at least continuously differentiable. This is desirable because the calculation of an interest rate corresponds to a finite difference, i.e., converges to a derivative for decreasing period lengths.

- The interpolation method should preserve the monotonicity of discount factors, i.e., if we have monotone decreasing sample points, then the interpolation should be a monotone decreasing curve.

The following additional requirement is also desirable:

- If the sample points correspond to a set of constant rates, then their interpolation should give constant rates. In other words, the interpolation of sample points from a flat interest rate curve should be flat (where flat means flat with respect to a rate).

The linear interpolation of bond prices fulfills the second, but neither the first, nor the third requirement. The linear interpolation of forward rates fulfills the first and third requirement, but not necessarily the second. A simple interpolation method, which is also popular in practice, is the linear interpolation of the logarithm of the discount factors, i.e., the linear interpolation of $r(0, T_j) \, T_j$:

$$df(t) = \exp\left(\frac{T_{j+1} - t}{T_{j+1} - T_j} \log(df(T_j)) + \frac{t - T_j}{T_{j+1} - T_j} \log(df(T_{j+1})) \right).$$

This interpolation fulfills the second and the third requirement.

A more complete discussion of various interpolation methods for interest rate curves may be found in [76].

8.5 Implementation

We extend the design of the DiscountFactors class of Figure 8.3 by an interpolation algorithm and a bootstrap algorithm, see Figure 8.4.

If the interpolation method is realized as part of the getDiscountFactor() method, and if the methods which calculate interest rates from discount factors (like getForwardRate() or getYield()) only use getDiscountFactor() (and not the internal data model), then the interpolation method is available in all derived interest rates once it has been implemented in getDiscountFactor().[9]

---

[9] This is one reason for encapsulation of the internal data model, which should only be accessible to a small set of methods (even within the same class!).

131

The bootstrapper is then realized through one additional method `appendDiscountFactor(ProductSpecification productSpec, double marketPrice)`, where `productSpec` contains the description of the financial product for which an additional discount factor has to be calculated from the given market price `marketPrice`.

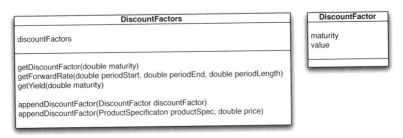

Figure 8.4. *UML Diagram: The class* `DiscountFactors` *internally stores discount factors and provides various interest rates through corresponding methods. The method* `appendDiscountFactor(ProductSpecification productSpec, double marketPrice)` *implements one induction step of a bootstrap algorithm.*

CHAPTER 9

Simple Interest Rate Products

So far we have defined a single interest rate product, the zero-coupon bond $P(T)$. In the following, we give the definitions of some basic interest rate products. Many definitions use Definition 99 of the forward rate (which, of course, is based on the definition of the zero bond).

9.1 Interest Rate Products Part 1: Products without Optionality

9.1.1 Fix, Floating, and Swap

We define a trivial generalization of the zero bond:

Definition 108 (Coupon Bond):
A coupon bond with coupons C_i, $i = 1, \ldots, n-1$ and *tenor structure* T_i, $i = 1, \ldots, n$ and maturity T_n pays

$$C_i \, (T_{i+1} - T_i) \; + \; \begin{cases} 0 & \text{if } i+1 < n \\ 1 & \text{if } i+1 = n \end{cases} \quad \text{in } T_{i+1}$$

($n-1$ payments).

Theorem 109 (Value of a Coupon Bond): The coupon bond consists of $n-1$ guaranteed payments with different payment dates. Clearly, the value of the coupon bond as seen in $t < T_2$ is given by

$$\sum_{i=1}^{n-1} C_i \, (T_{i+1} - T_i) \, P(T_{i+1}; t) \; + \; P(T_n; t). \tag{9.1}$$

Remark 110 (Dirty Price, Clean Price, Accrued Interest): The value of a coupon bond as given by Equation (9.1) is called *dirty price*. The dirty price is sometimes split into two parts, called the *clean price* and *accrued interest*.

If $T_1 < t < T_2$, i.e., the bond is evaluated within the first interest rate period, then the accrued interest is defined by

$$A(T_1, T_2; t) := \frac{t - T_1}{T_2 - T_1} C_1 (T_2 - T_1).$$

Remember that in Definition 109 it is assumed that $t < T_2$. $A(T_1, T_2; t)$ is called *accrued interest*. Dirty price and clean price are now related through

$$P_{\text{Dirty}}(t) = \sum_{i=1}^{n-1} C_i (T_{i+1} - T_i) P(T_{i+1}; t) + P(T_n; t),$$

$$P_{\text{Clean}}(t) = P_{\text{Dirty}}(t) - A(T_1, T_2; t).$$

The *accrued interest* represents the fraction of the future coupon payment that relates to the past fraction of the period. This stems from the interpretation of the coupon payment as equally distributed. The price of a bond is often quoted only by its *clean price*.

The decomposition in *clean price* and *accrued interest* may appear useless, since upon trading their sum, i.e., the dirty price, has to be paid. However, quoting the clean price has an advantage: The clean price evolves continuously in t across period end dates, while the dirty price exhibits a jump at the end of a period, due to the paid coupon.

The zero bond $P(T_1; t)$ is the time t value of a guaranteed payment of 1 in T_1. It thus represents a fixed interest rate payment. A product with variable interest rate payments is the *floater*.

Definition 111 (Floater):
Let T_i, $i = 1, \ldots, n$ denote a given time discretization (a tenor structure). A floater with notional N pays

$$N \, L(T_i, T_{i+1}; T_i) \, (T_{i+1} - T_i) \quad \text{in } T_{i+1}$$

for $i = 1, \ldots, n - 1$ ($n - 1$ payments).

Theorem 112 (Value of a Floater): At time $t \leq T_1$ the value of a floater (as in Definition 111) is given by

$$V_{\text{Floater}}(t) = N \sum_{i=1}^{n-1} L(T_i, T_{i+1}; t) \, (T_{i+1} - T_i) \, P(T_{i+1}; t)$$

$$= N \, (P(T_1) - P(T_n)).$$

Proof: *Variant 1 of the proof:* From the definition of the forward rate we have for a single payment of the floater in time T_{i+1}

$$V^i_{\text{Floater}}(T_{i+1}) := N\, L(T_i, T_{i+1}; T_i)\, (T_{i+1} - T_i) \overset{\text{Def. } L}{=} N\, \frac{P(T_i; T_i) - P(T_{i+1}; T_i)}{P(T_{i+1}; T_i)}.$$

This payment is an $\mathcal{F}_{T_i}$-measurable random variable. The interest rate of V^i_{Floater} was fixed in T_i and is no longer stochastic when observed on $[T_i, T_{i+1}]$. As seen in T_i the value of this payment is thus a multiple of $P(T_{i+1}; T_i)$, namely

$$V^i_{\text{Floater}}(T_i) = V^i_{\text{Floater}}(T_{i+1})\, P(T_{i+1}; T_i) = N\, (P(T_i; T_i) - P(T_{i+1}; T_i)).$$

Thus we see that the time T_i value of the floater is given by a portfolio of bonds. The time t value of this portfolio is known. Thus we have

$$V^i_{\text{Floater}}(t) = N\, (P(T_i; t) - P(T_{i+1}; t)) = N\, L(T_i, T_{i+1}; t)\, (T_{i+1} - T_i)\, P(T_{i+1}; t).$$

This is the value of a single floater payment. The claim follows by summation over i.

Variant 2 of the proof: We have to value $V^i_{\text{Floater}}(T_{i+1})$. Choose $N(t) = P(T_{i+1}; t)$ as numéraire and let $\mathbb{Q}^N$ denote a corresponding martingale measure. Then $N(T_{i+1}) = 1$ and

$$
\begin{aligned}
V^i_{\text{Floater}}(t) &= N(t)\, \mathrm{E}^{\mathbb{Q}^N}\!\left(\frac{V^i_{\text{Floater}}(T_{i+1})}{N(T_{i+1})} \,\Big|\, \mathcal{F}_t \right) = N(t)\, \mathrm{E}^{\mathbb{Q}^N}(V^i_{\text{Floater}}(T_{i+1}) \,|\, \mathcal{F}_t) \\
&= N(t)\, \mathrm{E}^{\mathbb{Q}^N}\!\left(N\, \frac{P(T_i; T_i) - P(T_{i+1}; T_i)}{P(T_{i+1}; T_i)} \,\Big|\, \mathcal{F}_t \right) \\
&= N(t)\, \mathrm{E}^{\mathbb{Q}^N}\!\left(N\, \frac{P(T_i; T_i) - P(T_{i+1}; T_i)}{N(T_i)} \,\Big|\, \mathcal{F}_t \right) \\
&= N(t)\, N\, \frac{P(T_i; t) - P(T_{i+1}; t)}{N(t)} \\
&= N\, (P(T_i; t) - P(T_{i+1}; t)) = N\, L(T_i, T_{i+1}; t)\, (T_{i+1} - T_i)\, P(T_{i+1}; t).
\end{aligned}
$$

This is the value of a single floater payment. The claim follows by summation over i.
□|

Definition 113 (Floating Rate Bond):
Let T_i, $i = 1, \dots, n$ denote a given time discretization (a tenor structure). A floating rate bond with notional N pays

$$N\, L(T_i, T_{i+1}; T_i)\, (T_{i+1} - T_i) \;+\; N \begin{cases} 0 & \text{if } i+1 < n \\ 1 & \text{if } i+1 = n \end{cases} \quad \text{in } T_{i+1}$$

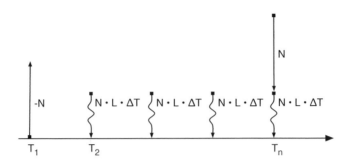

Figure 9.1. *Cash flow of a floater with exchange of notional N.*

for $i = 1, \ldots, n - 1$ ($n - 1$ payments).

The value of a floating rate bond is $N\,P(T_1)$, because it is just the sum of a floater (value $N\,P(T_1)) - N\,P(T_2)$) and a zero-coupon bond with maturity T_2 (value $N\,P(T_2)$).

 Interpretation: Definition 111 considers only the coupon payments of a floater. Normally an exchange of notional takes place at the beginning and end of the product: A payment of $-N$ is made in T_1 (receive notional) and a payment of N is made in T_2 (pay notional). Since from Theorem 112 the value of the pure coupon payments is $N\,P(T_1) - N\,P(T_n)$, the value of a floater with exchange of notional is 0.

Figure 9.1 shows a cash flow diagram for a floater with exchange of notional. At time T_1 the notional N is invested over the period $[T_1, T_2]$ with an interest rate $L(T_1, T_2; T_1)$, fixed at the beginning of the period. In T_2 the interest is paid and the notional N is reinvested over the following period (with a newly fixed rate). At the end T_n the interest for the last period is paid together with the notional. ◁

As shown, there are two different ways to derive the value of the floater. The first method uses the fact that the payment $N\,L(T_i, T_{i+1}; T_i)\,(T_{i+1} - T_i)$ is an $\mathcal{F}_{T_i}$-measurable random variable paid in T_{i+1}. Thus, its value as of time T_i is given by multiplication with $P(T_{i+1}; T_i)$. Since this value could be expressed as a portfolio of bonds, we know its time t value. Essentially, we derive a replication portfolio for each cash flow. The second method considers relative prices and applies Theorem 79.

In this context, the time T_i is called *fixing date* and the time T_{i+1} is called payment date.

Definition 114 (Fixing Date, Payment Date): ⌐

Let $T_2 \geq T_1$ and $V_{T_1}(T_2)$ be an $\mathcal{F}_{T_1}$-measurable random variable defining a payment made in T_2. Then T_1 is called *fixing date* and T_2 is called *payment date* of $V_{T_1}(T_2)$. See also Figure 9.2.

Lemma 115 (Moving the Payment Date): Let $t \geq T_1$. The value of a payment $V_{T_1}(T_2)$ with *fixing date* T_1 and *payment date* T_2 corresponds to the value of a payment of $V_{T_1}(T_2) \, P(T_2; t)$ in t for $t < T_2$ and the value of a payment of $V_{T_1}(T_2) \, \frac{1}{P(t;T_2)}$ in t for $t > T_2$.

Proof: The first part follows as in the proof of Theorem 112; the second part follows from exchanging t and T_2. □|

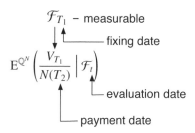

Figure 9.2. *Fixing date, payment date, and evaluation date.*

Remark 116 (On the Additivity of *Cash Flows* with Different *Payment Dates*): The value of a financial product or a single cash flow depends on its evaluation time, the time we select to observe it. Two payments with the same payment date may be added to create a single one. For two payments with different payment dates a summation is not meaningful. To calculate the total value of several products or cash flows at time t we have to move all cash flows as in Lemma 115. However, Lemma 115 applies only to times t larger than the fixing dates. The lemma is not applicable for times before the fixing date. Here, a risk-neutral evaluation has to be performed.

Relative prices behave differently: *Relative prices are additive (independent of the fixing date).* Let V_{T_1} denote an $\mathcal{F}_{T_1}$-measurable random variable defining the value of a financial product in time T_2. Then we have

- For $t < T_1$: $E^{Q^N}\!\left(\frac{V_{T_1}}{N(T_2)} \mid \mathcal{F}_t\right)$ is the $N(t)$-relative value of V_{T_1} at time t. This follows from Theorem 79.

- For $t > T_1$: $E^{Q^N}\!\left(\frac{V_{T_1}}{N(T_2)} \mid \mathcal{F}_t\right) = V_{T_1} \, E^{Q^N}\!\left(\frac{1}{N(T_2)} \mid \mathcal{F}_t\right) = \frac{V_{T_1} \, P(T_2;t)}{N(t)}$ is the $N(t)$-relative value of V_{T_1} at time t. This follows from Lemma 115.

The additivity of relative prices follows from the linearity of the expectation operator.

Definition 117 (Swap (Payer Swap)): ⌐

A swap is an exchange payment of fixed rate for a floating rate. Let $0 = T_0 < T_1 < T_2 < \cdots < T_n$ denote a given tenor structure. A swap pays

$$N \; (L(T_i, T_{i+1}; T_i) - S_i) \; (T_{i+1} - T_i) \quad \text{in } T_{i+1}$$

for $i = 1, \ldots, n - 1$ ($n - 1$ payments), where $S_i \in \mathbb{R}$ denotes the fixed *swap rate* and $L(T_i, T_{i+1}; T_i)$ denotes the forward rate from Definition 99, and N denotes the *notional*. The swap defined here is a *payer swap*; see Definition 118. ⌙

Definition 118 (Payer Swap, Receiver Swap): ⌐

The swap defined in Definition 117 with payments

$$N \; (L(T_i, T_{i+1}; T_i) - S_i) \; (T_{i+1} - T_i) \quad \text{in } T_{i+1}$$

is called *payer swap*. In contrast, the swap with reversed payments

$$N \; (S_i - L(T_i, T_{i+1}; T_i)) \; (T_{i+1} - T_i) \quad \text{in } T_{i+1}$$

is called *receiver swap*. The term payer/receiver indicates whether the holder of the swap has to pay the fixed coupon (it enters negative) or receives the fixed coupon (it enters positive). ⌙

Definition 119 (Floating Leg, Fixed Leg): ⌐

The payments of a swap may be decomposed into

$$N \, L(T_i, T_{i+1}; T_i) \, (T_{i+1} - T_i) \quad \text{in } T_{i+1} \tag{9.2}$$

and

$$N \, S_i \, (T_{i+1} - T_i) \quad \text{in } T_{i+1}. \tag{9.3}$$

The payments (9.2) of the variable rates are called *floating leg*; the payments of the constant rates (9.3) are called *fixed leg*. ⌙

Theorem 120 (Value of a Swap): At time $t \le T_1$ the value of a swap is given by

$$V_{\text{Swap}}(t) \;=\; N \sum_{i=1}^{n-1} (L(T_i, T_{i+1}; t) - S_i) \; (T_{i+1} - T_i) \, P(T_{i+1}; t).$$

Proof: The swap consists of a floater (floating leg, (9.2)) and fixed payments $-N \, S_i \, (T_{i+1} - T_i)$ in T_{i+1} for which their time t value is the corresponding multiple of $P(T_{i+1}; t)$. The claim follows by applying Theorem 112 to the floating leg. □

Remark 121 (Swap Rate): Let $T_1, \ldots, T_n$ be a given tenor structure. Consider a swap as in Definition 117. The *par swap rate* S (in t) is the unique rate for which a swap with $S_i := S$ has the time t value 0, i.e., the total time t value of the payments

$$N \ (L(T_i, T_{i+1}; T_i) - S) \ (T_{i+1} - T_i) \quad \text{in } T_{i+1}$$

is 0. Since the time t value of such a swap is given by

$$N \sum_{i=1}^{n-1} (L(T_i, T_{i+1}; t) - S) \ (T_{i+1} - T_i) \ P(T_{i+1}; t),$$

then

$$S = \frac{\sum_{i=1}^{n-1} L(T_i, T_{i+1}; t)(T_{i+1} - T_i) \ P(T_{i+1}; t)}{\sum_{i=1}^{n-1}(T_{i+1} - T_i) \ P(T_{i+1}; t)} = \frac{P(T_1; t) - P(T_n; t)}{\sum_{i=1}^{n-1}(T_{i+1} - T_i) \ P(T_{i+1}; t)}.$$

Definition 122 (Par Swap Rate):
Let $T_1 < T_2 < \cdots < T_n$. The *par swap rate* (often just called *swap rate*) $S(T_1, \ldots, T_n)$ is defined by

$$S(T_1, \ldots, T_n; t) := \frac{P(T_1; t) - P(T_n; t)}{\sum_{i=1}^{n-1}(T_{i+1} - T_i) \ P(T_{i+1}; t)}.$$

 Interpretation: Since the par swap rate is the rate for which a corresponding swap has value 0, we may see the par swap rate $S_{i,j} := S(T_i, \ldots, T_j)$ as some mean of the forward rates L_k, $k = i, \ldots, j - 1$. Indeed, the par swap rate $S_{i,j}$ is a convex combination (and thus a weighted average) of the forward rated L_k, $k = i, \ldots, j - 1$ as shown in the following lemma. ◁|

Lemma 123 (Swap Rate as Convex Combination of the Forward Rates): Let $T_i < T_{i+1} < \cdots < T_j$ denote a given tenor structure. Then we have

$$S_{i,j} = \sum_{k=i}^{j-1} \alpha_k L_k, \quad \text{with} \quad \alpha_k \geq 0, \quad \sum_{k=i}^{j-1} \alpha_k = 1.$$

The weights α_k are given by

$$\alpha_k := \frac{P(T_{k+1}) \ (T_{k+1} - T_k)}{\sum_{l=i}^{j-1} P(T_{l+1}) \ (T_{l+1} - T_l)}.$$

The weights are stochastic.

139

Proof: With $P(T_{k+1})(T_{k+1} - T_k) L_k = P(T_k) - P(T_{k+1})$ we have

$$\sum_{k=i}^{j-1} P(T_{k+1})(T_{k+1} - T_k) L_k = P(T_i) - P(T_j),$$

and thus

$$\sum_{k=i}^{j-1} \frac{P(T_{k+1})(T_{k+1} - T_k)}{\sum_{l=i}^{j-1} P(T_{l+1}) (T_{l+1} - T_l)} L_k = \frac{P(T_i) - P(T_j)}{\sum_{l=i}^{j-1} P(T_{l+1}) (T_{l+1} - T_l)} = S_{i,j}.$$

 Interpretation (Usage of the Terms *Bond* and *Swap*): The terms *bond* and *swap* are also used in a much broader sense than given. A financial product with coupon payments and final notional payments at maturity is called a *bond*. A financial product where coupon payments are exchanged (and no notional is paid) is called a *swap*. The terms are used independently of the specific structure of the coupons. A coupon may be a constant (*fix*), a variable rate (*float*), or a complex function of one or more interest rates (*structured*). In the latter case, the coupon is called a *structured coupon*, and the corresponding bond and swap are called *structured bond* and *structured swap*.

A swap may be interpreted as a portfolio of a bond long (i.e., with positive cash flow) and a bond short (i.e., with negative cash flow), where the two notional payments cancel. In Section 12.2.1 we will consider the relationship between bonds and swaps.
◁

9.1.2 Money Market Account

If we invest at time $T_0 = 0$ a unit currency over the period $[T_0, T_1]$, then we receive at T_1 the amount $1 + L(T_0, T_1; T_0) (T_1 - T_0)$. If this amount is reinvested for another period and if this process is continued for periods $[T_j, T_{j+1}]$ with $j = 1, 2, \ldots$, then we have at time T_i a value of

$$\prod_{j=0}^{i-1}(1 + L(T_j, T_{j+1}; T_j) (T_{j+1} - T_j)). \tag{9.4}$$

Equivalently, we may write this with the instantaneous forward rate as

$$\prod_{j=0}^{i-1} \exp\left(\int_{T_j}^{T_{j+1}} f(T_j, \tau) \, d\tau \right) = \exp\left(\sum_{j=0}^{i-1} \int_{T_j}^{T_{j+1}} f(T_j, \tau) \, d\tau \right). \tag{9.5}$$

If we consider a continuum of infinitesimal periods, i.e., we consider continuously compounding with the short rate, then the corresponding value will evolve as

$$B(t) = \exp\left(\int_{T_0}^{t} r(\tau)\, d\tau\right).\tag{9.6}$$

Definition 124 (Rolling Bond):
The financial product

$$R(t) := P(T_{m(t)+1}; t) \prod_{j=0}^{m(t)} (1 + L(T_j, T_{j+1}; T_j)\,(T_{j+1} - T_j)),$$

where $m(t) := \max\{i \; : \; T_i \le t\}$, is called *(single period) rolling bond*.

Definition 125 (Savings Account, Money Market Account):
The financial product B in (9.6) is called *savings account* or *money market account*.

Interpretation: The financial product B has to be interpreted as an idealization (like the short rate itself), since infinitesimal periods are an idealization.

Note that (9.4) and (9.5) are equivalent, whereas $B(T_i)$ does not co-incide with the value of (9.4) generally. The expressions (9.4) and (9.5) depend on the choice of the periods and if evaluated in T_i, they are $\mathcal{F}_{T_{i-1}}$-measurable random variables, whereas $B(T_i)$ is $\mathcal{F}_{T_i}$-measurable only.

9.2 Interest Rate Products Part 2: Simple Options

9.2.1 Cap, Floor, and Swaption

Definition 126 (Caplet): ⌐

A *caplet* is an option on the *forward rate* (LIBOR) and pays

$$V_{\text{caplet}}(T_2) = N \ \max \left(L(T_1, T_2; T_1) - K \ , \ 0 \right) \ (T_2 - T_1) \quad \text{in } T_2 \qquad (9.7)$$

where K is the *strike* rate, $L(T_1, T_2; t)$ the LIBOR, and N the *notional*. Payment date and fixing date $0 < T_1 < T_2$ coincide with the LIBOR period $[T_1, T_2]$. ⌐

Definition 127 (Cap): ⌐

A *cap* is a portfolio of caplets. Let $0 = T_0 < T_1 < T_2 < \cdots < T_n$ denote a given tenor structure. A cap pays

$$N \ \max \left(L(T_i, T_{i+1}; T_i) - K_i \ , \ 0 \right) \ (T_{i+1} - T_i) \quad \text{in } T_{i+1} \qquad (9.8)$$

for $i = 1, \ldots, n - 1$ $(n - 1$ payments), where K_i are the *strike* rates, $L(T_i, T_{i+1}; T_i)$ are the LIBOR rates, and N denotes the *notional*. ⌐

Remark 128 (Floorlet, Floor): If in (9.8) or correspondingly in (9.7) the payoff is

$$N \ \max \left(K_i - L(T_i, T_{i+1}; T_i) \ , \ 0 \right) \ (T_{i+1} - T_i) \quad \text{in } T_{i+1},$$

then the product is called *floor* or *floorlet*, respectively.

Remark 129 (Caplet, Cap): The name caplet (and thus cap) seems counterintuitive. A cap is usually an upper bound, a floor a lower bound. Indeed, the payoff

$$[L]^K := \min(L, K)$$

is called *capped* and the payoff

$$[L]_K := \max(L, K)$$

is called *floored*. The counter-intuitive name caplet for (9.7) stems from its application as a swap that exchanges a floating rate L against a capped coupon $[L]^K$:

$$L - [L]^K = \max(L - K, 0),$$

i.e.,

$$-[L]^K = -L + \max(L - K, 0).$$

If we have the obligation to pay a variable interest rate $(-L)$, buying a cap $(+ \max(L - K, 0))$ will cover the risk of an increasing interest rate, i.e., the payment is capped $(-[L]^K)$. The cap is the product one has to buy to have floating payments capped.[1]

Definition 130 (Swaption):   ⌐

A *swaption* is an option on a swap. Let $V_{\text{swap}}(t)$ denote the time t value of a swap as defined by Definition 117. Then the value of a *swaption* (with underlying V_{swap}) is given by the payout

$$V_{\text{swaption}}(T_1) := \max\left(V_{\text{swap}}(T_1), 0\right) \quad \text{in } T_1.$$

⌙

Definition 131 (Digital Caplet):   ⌐

A *digital caplet* pays

$$V_{\text{digital}}(T_2) = N\,\mathbf{1}(L(T_1, T_2; T_1) - K)\,(T_2 - T_1) \quad \text{in } T_2,$$

where K is the *strike* rate, $L(T_1, T_2; t)$ is the LIBOR, N denotes the *notional* and $\mathbf{1}$ denotes the indicator function (with $\mathbf{1}(x) := 1$ for $x > 0$ and $\mathbf{1}(x) := 0$ else).   ⌙

Lemma 132 (Digital Caplet Valuation, Call-Spread): For the value $V_{\text{digital}}(K, 0)$ of a digital caplet with strike K we have

$$V_{\text{digital}}(K; 0) = -\frac{\partial}{\partial K} V_{\text{caplet}}(K; 0).$$

The approximation (see Figure 9.3) of the differential using finite differences

$$\tilde{V}_{\text{digital}}(K; 0) = -\frac{V_{\text{caplet}}(K + \epsilon; 0) - V_{\text{caplet}}(K - \epsilon; 0)}{2\epsilon}$$

is called *call spread*.

Proof: The proof follows the lines of the proof in Lemma 81.   □|

---

[1] See [7], p. 12.

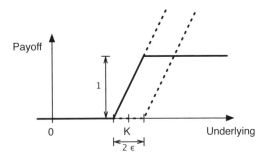

Figure 9.3. *Call spread approximation of a digital option by two call options.*

9.2.1.1 Example: Option on a Coupon Bond

Consider the option to receive at T_1 a coupon bond in exchange for a notional payment. A coupon bond with tenor structure T_i, $i = 1, \ldots, n$, coupons C_i and maturity T_n pays

$$C_i \, (T_{i+1} - T_i) \, + \, \begin{cases} 0 & \text{if } i + 1 < n \\ 1 & \text{if } i + 1 = n \end{cases} \quad \text{in } T_{i+1}.$$

The time t value of a forward starting coupon bond with an initial notional payment 1 in T_1 is

$$V_{\text{fwdCpnBnd}}(t) \; = \; \sum_{i=1}^{n-1} C_i \, (T_{i+1} - T_i) \, P(T_{i+1}; t) \; + \; P(T_n; t) - P(T_1; t),$$

for $t \leq T_1$; see (9.1). Since $P(T_1; t) - P(T_n; t)$ is the value of a floating rate bond, see Definition 113; this is just a swap

$$= \; \sum_{i=1}^{n-1} (C_i - L(T_i, T_{i+1}; t)) \, (T_{i+1} - T_i) \, P(T_{i+1}; t).$$

Consequently, an option on a forward starting coupon bond is just a swaption.

9.2.2 Foreign Caplet, Quanto

Definition 133 (Foreign Caplet): ⌐

A *foreign caplet* is a caplet in a foreign currency. From the domestic investor's point

of view it pays

$$\tilde{N} \, \max\left(\tilde{L}(T_1, T_2; T_1) - K \, , \, 0\right) (T_2 - T_1) \, FX(T_2) \quad \text{in } T_2,$$

where $K \in \mathbb{R}$ is the strike rate, $\tilde{L}(T_1, T_2; t)$ is the *foreign* LIBOR, $FX(T_2)$ is the exchange rate, and $\tilde{N}$ denotes the notional in foreign currency.[2]

Remark 134 (Units): It is useful to consider units, just as one would do in physics. A domestic bond P has the unit of one domestic currency, $[P]$ = dom. Interest *rates* have the dimension $\frac{1}{\text{time}}$, e.g., for the forward rate (LIBOR) we have $[L(T_i, T_{i+1}) \, (T_{i+1} - T_i)] = 1$, since it is the quotient of two bonds. The unit of the stochastic process FX is $[FX] = \frac{\text{dom}}{\text{for}}$, i.e., $FX(t)$ is the time t value of a foreign currency unit in domestic currency. In Definition 133 we have $[\tilde{N}]$ = for. For the following product it is crucial to consider units.

Definition 135 (Quanto, Quanto Caplet):

A *quanto* is a financial product for which a payout will be converted from a foreign currency to a domestic currency without use of the exchange rate. Instead of $FX(t)$ it uses $1\frac{\text{dom}}{\text{for}}$ or another conversion factor (the *quanto rate*) fixed a priori.

Let $0 < T_1 < T_2$ denote *fixing* and *payment date*, respectively. A *quanto caplet* pays

$$\tilde{N} \, \max\left(\tilde{L}(T_1, T_2; T_1) - K \, , \, 0\right) (T_2 - T_1) \, 1\frac{\text{dom}}{\text{for}} \quad \text{in } T_2 \, ,$$

where K is the strike rate (dimension $\frac{1}{\text{time}}$), $\tilde{L}(T_1, T_2; t)$ is the *foreign* LIBOR (dimension $\frac{1}{\text{time}}$), and $\tilde{N}$ denotes the notional in *foreign currency*.

 Further Reading: An introduction to the basics of interest rates products may be found in [4] (in German).

---

[2] FX = Foreign eXchange.

145

CHAPTER 10

The Black Model for a Caplet

We consider a caplet as defined by Definition 126 as an option on the forward rate $L_1 := L(T_1, T_2)$ for given times $0 < T_1 < T_2$.

The Black model for the valuation of a caplet postulates a lognormal dynamic of the underlying LIBOR[1]

$$dL_1(t) = \mu^{\mathbb{P}}(t)L_1(t)\,dt + \sigma(t)L_1(t)\,dW^{\mathbb{P}}(t), \quad \sigma(t) \geq 0, \text{ under } \mathbb{P}. \quad (10.1)$$

We seek the price $V(0)$ of the payoff profile

$$V(T_2) := \max\left((L_1(T_1) - K)(T_2 - T_1), 0\right),$$

where $L_1(T_2 - T_1) := P(T_1)/P(T_2) - 1$, i.e., $L_1 = L_1(T_1, T_2)$ denotes the forward rate (the (forward) LIBOR) of the period $[T_1, T_2]$. Without loss of generality we assume that the notional is 1. We choose the T_2-bond as numéraire:

$$N(t) := P(T_2; t).$$

This choice of the numéraire is the crucial trick in the derivation of a risk-neutral pricing formula. Since

$$L_1 = \frac{1}{(T_2 - T_1)}\left(\frac{P(T_1)}{P(T_2)} - 1\right) = \frac{1}{(T_2 - T_1)}\frac{P(T_1) - P(T_2)}{P(T_2)}$$

$$= \frac{\frac{1}{(T_2 - T_1)}(P(T_1) - P(T_2))}{N},$$

L_1 is the N-relative price of a traded asset.[2] From Theorem 74 we have the existence of an *equivalent martingale measure* $\mathbb{Q}^N$ such that all N-relative prices of traded

---

[1] The lognormal process is often written in the form $\frac{dL_1(t)}{L_1(t)} = \mu^{\mathbb{P}}(t)\,dt + \sigma(t)\,dW^{\mathbb{P}}(t)$.

[2] The traded asset is the portfolio $\frac{1}{(T_2 - T_1)}(P(T_1) - P(T_2))$ consisting of $\frac{1}{(T_2 - T_1)}$ fractions of a T_1-bond (long) and T_2-bond (short).

assets are martingales. Thus L_1 is drift-free (see Lemma 53), i.e.,

$$dL_1(t) = \sigma(t)L_1(t) \, dW^{\mathbb{Q}^N}(t), \qquad \text{under } \mathbb{Q}^N.$$

For the process $Y := \log(L_1)$ we have from Lemma 50 that

$$d(\log(L_1(t))) = -\frac{1}{2}\sigma(t)^2 \, dt + \sigma(t) \, dW^{\mathbb{Q}^N}(t),$$

i.e., $\log(L_1(T))$ is normally distributed with mean $\log(L_1(0)) - \frac{1}{2}\bar{\sigma}^2 T$ and standard deviation $\bar{\sigma}\sqrt{T}$, with $\bar{\sigma} := (\frac{1}{T}\int_0^T \sigma^2(t)\,dt)^{1/2}$; see Section 4.

For the option value we now have

$$V(T_2) = \max\left((L_1(T_1) - K)(T_2 - T_1), 0\right) \qquad \text{in } T_2$$

and from $N(T_2) = 1$ we have[3]

$$\frac{V(0)}{N(0)} = \mathrm{E}^{\mathbb{Q}^N}\left(\frac{V(T_2)}{N(T_2)}\mid \mathcal{F}_0\right) = \mathrm{E}^{\mathbb{Q}^N}\left(V(T_2)\mid \mathcal{F}_0\right)$$

$$= \mathrm{E}^{\mathbb{Q}^N}\left(\max((L_1(T_1) - K)(T_2 - T_1), 0)\mid \mathcal{F}_0\right)$$

i.e.,

$$V(0) = P(T_2; 0)\,\mathrm{E}^{\mathbb{Q}^N}\left(\max((L_1(T_1) - K), 0)\right)(T_2 - T_1).$$

Knowing the distribution of L_1 under $\mathbb{Q}^N$ this expectation may be represented as

$$V(0) = P(T_2; 0)\,[L_1(0)\Phi(d_+) - K\Phi(d_-)]\,(T_2 - T_1), \tag{10.2}$$

where

$$\Phi(x) := \frac{1}{\sqrt{2\pi}}\int_{-\infty}^{x}\exp(-\frac{y^2}{2})\,dy$$

and

$$d_\pm = \frac{\log(\frac{L_1(0)}{K}) \pm \frac{1}{2}\bar{\sigma}^2 T_1}{\bar{\sigma}\sqrt{T_1}},$$

see Chapter 4. Equation (10.2) is termed Black formula (for caplets).

Remark 136 (Implied Black Volatility): Similar to Remark 80 in Chapter 4 we have: Equation (4.3) gives us the price of the option under the model (10.1) as a function of the model parameter $\bar{\sigma}$. In this context $\bar{\sigma}$ is called the Black volatility.

---

[3] This is the point where the specific choice of numéraire comes in handy for the second time.

Taking the other model and product parameters (r, K, T_1, T_2) as constants, the pricing formula (10.2) represents a bijection:

$$\bar{\sigma} \mapsto V(0)$$
$$[0, \infty) \to [P(T_2; 0) \ \max(L(0) - K, 0) \, , \ P(T_2; 0) \ L(0)].$$

The $\bar{\sigma}$ calculated for a given price $V(0)$ through inversion of the pricing formula is called the *implied Black volatility*.

Lemma 137 (Price of a Digital Caplet under the Black Model Dynamics): The price of a digital caplet under the Black model is

$$V_{\text{digital}}(K; 0) = N \, P(T_2; 0) \, \Phi(d_-) \, (T_2 - T_1),$$

where

$$\Phi(x) := \frac{1}{\sqrt{2\pi}} \int_{-\infty}^{x} \exp\left(-\frac{y^2}{2}\right) dy$$

and

$$d_- = \frac{\log\left(\frac{L_1(0)}{K}\right) - \frac{1}{2}\bar{\sigma}^2 T_1}{\bar{\sigma}\sqrt{T_1}}.$$

149

CHAPTER 11

Pricing of a Quanto Caplet (Modeling the FFX)

In this chapter all quantities related to a foreign currency are marked with a tilde (~).
Let $0 < T_1 < T_2$ denote *fixing* and *payment date*, respectively. The payoff profile of a
quanto caplet is given by

$$V(T_2) = \max\left(\tilde{L}(T_1, T_2; T_1) - K,\ 0\right)(T_2 - T_1)\,1\frac{\text{dom}}{\text{for}} \quad \text{in } T_2,$$

where K is the given strike rate and $\tilde{L}(T_1, T_2; t)$ is the foreign forward rate. The
notional and quanto rate are assumed to be 1.

We assume a lognormal dynamic for the foreign LIBOR, i.e., we model it as[1]

$$d\tilde{L}(t) = \mu^{\mathbb{P}}(t)\tilde{L}(t)\,dt + \sigma_{\tilde{L}}(t)\tilde{L}(t)\,dW_3^{\mathbb{P}}(t).$$

11.1 Choice of Numéraire

If we choose the foreign T_2-bond converted to domestic currency, i.e., $\tilde{P}(T_2; t)FX(t)$,
as numéraire, then from

$$\tilde{L}(T_1, T_2; t) = \frac{1}{T_2 - T_1}\frac{\tilde{P}(T_1) - \tilde{P}(T_2)}{\tilde{P}(T_2)} = \frac{1}{T_2 - T_1}\frac{\tilde{P}(T_1)FX(t) - \tilde{P}(T_2)FX(t)}{\tilde{P}(T_2)FX(t)}$$

(see Chapter 10) we find

$$d\tilde{L}(t) = \sigma_{\tilde{L}}(t)\tilde{L}(t)\,dW_3(t) \quad \text{under } \mathbb{Q}^{\tilde{P}(T_2)\,FX}.$$

---

[1] We write $\tilde{L}$ for $\tilde{L}(T_1, T_2)$.

Remark 138 (Foreign Market, Cross Currency Change of Numéraire): Note that the foreign LIBOR is *not* a martingale with respect to $\mathbb{Q}^{\tilde{P}(T_2)}$, since we are based in the domestic market. For the domestic investor the foreign bond $\tilde{P}(T_i)$ is not a traded asset, but the foreign bond converted to domestic currency $\tilde{P}(T_i) \, FX$ is a traded asset. Although we have

$$d\tilde{L}(t) = \sigma_{\tilde{L}}(t)\tilde{L}(t) \, dW_3(t) \quad \text{under } \mathbb{Q}^{\tilde{P}(T_2)},$$

we cannot use this change of measure, since $\tilde{P}(T_2)$ is not a traded asset and thus not a numéraire in the domestic market. Choosing the domestic bond $P(T_2)$ (a traded asset) as numéraire, we generally have

$$d\tilde{L}(t) \neq \sigma_{\tilde{L}}(t)\tilde{L}(t)(t) \, dW_3(t) \quad \text{under } \mathbb{Q}^{P(T_2)}.$$

Since the payoff profile of the quanto caplet is

$$V(T_2) = \max\left(\tilde{L}(T_1, T_2; T_1) - K, \, 0\right)(T_2 - T_1)\, 1\frac{\text{dom}}{\text{for}} \quad \text{in } T_2,$$

it is advantageous to know the dynamics of $\tilde{L}(T_1, T_2)$ under the measure $\mathbb{Q}^{P(T_2)}$, the *domestic T_2 terminal measure*. Choosing $P(T_2)$ as numéraire, the numéraire is 1 at payment date and will not show up in the expectation operator above. This trick has already been used in Chapter 10, where we were lucky that additionally the underlying was a martingale under this measure.

By the change of measure from $\mathbb{Q}^{\tilde{P}(T;t)\, FX(t)}$ to $\mathbb{Q}^{P(T_2)}$ we have a change of the drift; see Theorem 59 (Girsanov, Cameron, Martin), i.e.,

$$d\tilde{L}(t) = \mu^{P(T_2)}(t)\tilde{L}(t) \, dt + \sigma_{\tilde{L}}(t)\tilde{L}(t) \, dW_3^{P(T_2)}(t) \quad \text{under } \mathbb{Q}^{P(T_2)}.$$

In other words, the dynamics of the underlying is known under the measure $\mathbb{Q}^{\tilde{P}(T;t)\, FX(t)}$. From the shape of the payoff function a change of numéraire from $\tilde{P}(T;t)\, FX(t)$ to $P(T;t)$, thus a change of measure from $\mathbb{Q}^{\tilde{P}(T;t)\, FX(t)}$ to $\mathbb{Q}^{P(T;t)}$ is desirable. We thus define:

Definition 139 (Forward FX Rate): ⌐

Let $0 < t < T$. The *forward FX rate* $FFX(T)$ is defined as

$$FFX(T;t) := \frac{\tilde{P}(T;t)}{P(T;t)} FX(t).$$

⌐

Remark 140 (Forward FX Rate): The forward FX rate (also known as FX forward) is a relative price of two domestic traded assets. It is dimensionless, since $[\tilde{P}(T;t)] = 1$ for, $[P(T;t)] = 1$ dom and $[FX(t)] = 1 \frac{\text{dom}}{\text{for}}$. It is a $\mathbb{Q}^{P(T)}$ martingale.

We assume lognormal dynamics for $FFX(T_2)$, i.e.

$$dFFX(T_2; t) = \sigma_{FFX}(t) FFX(T_2; t)\, dW_2^{P(T_2)}(t) \quad \text{under } \mathbb{Q}^{P(T_2)}.$$

Since

$$
\begin{aligned}
\tilde{L}(T_1, T_2)\, FFX(T_2) &= \frac{1}{T_2 - T_1} \frac{\tilde{P}(T_1) - \tilde{P}(T_2)}{\tilde{P}(T_2)} \frac{\tilde{P}(T_2)}{P(T_2)} FX(t) \\
&= \frac{1}{T_2 - T_1} \frac{(\tilde{P}(T_1) - \tilde{P}(T_2)) FX(t)}{P(T_2)}
\end{aligned}
$$

is a $P(T_2)$-relative price of domestic traded assets (namely a portfolio of foreign bonds), we have that $\tilde{L}(T_1, T_2)\, FFX(T_2)$ is a martingale under $\mathbb{Q}^{P(T_2)}$, i.e.,

$$\text{Drift}^{\mathbb{Q}^{P(T_2)}}(\tilde{L}(T_1, T_2)\, FFX(T_2)) = 0. \tag{11.1}$$

On the other hand

$$
\begin{aligned}
d(\tilde{L}\, FFX) =\ & d\tilde{L}\, FFX + \tilde{L}\, dFFX + d\tilde{L}\, dFFX \\
=\ & FFX\, \tilde{L}\, \mu^{P(T_2)}\, dt + FFX\, \tilde{L}\, \sigma_{\tilde{L}}\, dW_3^{P(T_2)} \\
& + FFX\, \tilde{L}\, \sigma_{FFX}\, dW_2^{P(T_2)} + FFX\sigma_{FFX}\, dW_2^{P(T_2)}\, \sigma_{\tilde{L}}\tilde{L}\, dW_3^{P(T_2)},
\end{aligned}
$$

and assuming an instantaneous correlation $\rho(t)$ for $dW_2^{P(T_2)}$ and $dW_3^{P(T_2)}$

$$
= FFX\, \tilde{L}\left((\mu^{P(T_2)} + \rho\sigma_{FFX}\sigma_{\tilde{L}})\, dt + \sigma_{\tilde{L}}\, dW_3^{P(T_2)} + \sigma_{FFX}\, dW_2^{P(T_2)}\right).
$$

From (11.1) we thus find

$$\mu^{P(T_2)}(t) = -\rho(t)\sigma_{FFX}(t)\sigma_{\tilde{L}}(t).$$

We now know the dynamics of $\tilde{L}$ under $\mathbb{Q}^{P(T_2)}$

$$d\tilde{L} = -\rho(t)\sigma_{FFX}(t)\sigma_{\tilde{L}}(t)\tilde{L}(t)\, dt + \sigma_{\tilde{L}}(t)\tilde{L}(t)\, dW_3^{P(T_2)}(t)$$

and (as in Chapter 10) we know the distribution of $\tilde{L}(T_1, T_2)$ (under $\mathbb{Q}^{P(T_2)}$) is lognormal with

$$
\begin{aligned}
\log(\tilde{L}(T)) &\sim \mathcal{N}\left(\log(\tilde{L}(0)) - \int_0^T \rho(t)\sigma_{FFX}(t)\sigma_{\tilde{L}}(t)\, dt - \frac{1}{2}\bar{\sigma}_{\tilde{L}}^2\, T,\ \bar{\sigma}_{\tilde{L}}^2\, T\right) \\
&= \mathcal{N}\left(\log\left(\tilde{L}(0)\, e^{-\int_0^T \rho(t)\sigma_{FFX}(t)\sigma_{\tilde{L}}(t)\, dt}\right) - \frac{1}{2}\bar{\sigma}_{\tilde{L}}^2\, T,\ \bar{\sigma}_{\tilde{L}}^2\, T\right),
\end{aligned}
$$

where

$$\bar{\sigma}_{\tilde{L}}^2 = \frac{1}{T} \int_0^T \sigma_{\tilde{L}}^2(t) \, dt \qquad \text{(mean variance)}.$$

Altogether it follows an (adjusted) Black formula, where in contrast to the Black formula from Chapter 10 $\tilde{L}(0)$ is replaced by $\tilde{L}(0) \, e^{-\int_0^T \rho(t)\sigma_{FFX}(t)\sigma_{\tilde{L}}(t) \, dt}$. The factor

$$e^{-\int_0^T \rho(t)\sigma_{FFX}(t)\sigma_{\tilde{L}}(t) \, dt}$$

is called the *quanto adjustment*.

CHAPTER 12

Exotic Derivatives

We have already introduced some simple interest rate derivatives. In this section we give a selection of so-called *exotic derivatives*. The name "exotic" does not mean that these derivative products are of less importance. With respect to evaluation models the converse is true: The value of exotic derivatives usually depends on a multitude of model properties, which may not even play a role in the pricing of simple derivatives. An example is the time structure $t \mapsto \sigma(t)$ of the volatility in the Black-(Scholes) model (4.2), (10.1): Its distribution over time does not play a role in the pricing of a European option, only the integrated variance enters into the pricing formula. It will, however, play a role for the pricing of a Bermudan option.

Thus, in this sense we may view certain prototypical properties of exotic derivatives (e.g., having more than one exercise date) as test functions for prototypical properties of (complex) models (e.g., the term structure of volatility).[1]

12.1 Prototypical Product Properties

The list of exotic interest rate derivatives we give in Section 12.2 does not claim to be complete or representative. It is exemplary for *prototypical product properties* and for applications of the pricing models and methodologies which we will discuss later. We focus a bit on more recent products, where we will discuss the relationship of prototypical product properties to models and their implementation.

Some product properties, like *path dependency* or *early exercise* characterize a whole class of products. To evaluate a product of the respective class, the object *path*, i.e., the history, for path dependency and *conditional expectation* for early

---

[1] To clarify the meaning of this sentence we remind the reader that a digital option may be used to extract the model-implied terminal distribution function of its underlying; see Chapter 5. Thus digital options may be viewed as test functions of a model's terminal distributions.

exercise are central. The path dependency is best represented in a *path simulation* (*forward algorithm*). The conditional expectation is best represented in a *state lattice* (*backward algorithm*). See Table 12.1.

Furthermore, some models have a preferred mode of implementation, i.e., as path simulation or as state lattice. Whether a model can be implemented on a state lattice is often decided by its Markov dimension; see Table 12.2.

Thus, prototypical product properties impose requirements on model and implementation.

Prototypical Product Property	Model Requirement / Implementation
Early exercise/bermudan, low Markov dimension	Backward algorithm, coarse time discretization → state lattice/tree, Section 13.3
Early exercise/american, low Markov dimension	Backward algorithm, fine time discretization → PDE, Chapter 14
Path Dependency, high Markov dimension	Forward algorithm → path simulation, Section 13.1
Path dependency, model: low Markov dimension; product: high Markov dimension	Path simulation through a lattice → Section 13.4
Early exercise, high Markov dimension	Forward algorithm with estimator of conditional expectation → Chapter 15
Path dependency, low Markov dimension	Backward algorithm with extension of state space → Chapter 16

Table 12.1. *Prototypical product properties and corresponding model requirements and implementation techniques.*

Model		Property
Short rate models	(→ Chapter 23)	Low Markov dimension
Market models	(→ Chapter 19)	High Markov dimension
Markov functional models	(→ Chapter 27)	Low Markov dimension

Table 12.2. *Markov dimension of some models.*

12.2 Interest Rate Products Part 3: Exotic Interest Rate Derivatives

 Motivation ("Why Exotic Derivatives?"): A simple European option with payoff $\max(L(T) - K, 0)$ may be interpreted as an insurance against an increase of the interest rate $L(T)$. In case of an increasing interest rate it pays the corresponding compensation.

To interpret an exotic derivative, e.g., one with payoff

C_i	if $L(T_j) < K \ \forall \ j \le i$	and $i + 1 < n$
$C_i + 1$	if $L(T_{i-1}) < K \le L(T_i)$	or $i + 1 = n$
0	else.	

as an insurance is not intuitive. The payoff above constitutes a coupon bond which matures if $L(T_i)$ exceeds the rate K. Such a structure is usually offered with an above-average coupon C_1 and below-average coupons C_i, $i > 1$.[2] Thus, this product is appealing if the investor would like to receive a high initial coupon (this could be done with a standard coupon bond) and at the same time expects that the interest rate will rise faster than the market predicts. Since the coupon bond will mature early if interest rates rise, the lower than average coupons C_i, $i > 1$ do not take effect. Thus, in this case, the investor would have a coupon bond with an above-average coupon. Since the investor takes the risk that he will receive lower than average coupons if interest rates do not rise, the product will be much cheaper than a standard coupon bond paying a coupon C_1 and maturing early. The investor is financing the initial above-average coupon by taking the risk of losing his bet on rising interest rates. He is a risk taker. The product is appealing to the investor since he has a different view on the future than the market (i.e., the average).

Exotic derivatives interpreted as an investment usually link a guaranteed high initial payment with a risky structure, which extracts the favorable case of the investor's market view.

An exotic derivative may both reward for taking risk as well as cover risk (in the sense of an insurance). Both interpretations jointly exist. For example, the structure above has an insurance against total loss of investment. The worst-case scenario is a coupon bond with below-average coupons. ◁|

---

[2] A similar structure is given by the target redemption note.

12.2.1 Structured Bond, Structured Swap, and Zero Structure

Exotic interest rate options mainly come in two different forms: as a (structured) bond or a (structured) swap. The products bond and swap are closely related. For a given (coupon) bond we may define a swap, such that the swap together with a floating rate bond replicates the coupon bond. We consider this relationship first for the trivial case of a simple coupon bond, then for the more tricky case of a zero-coupon bond. All coupons may be structured coupons.

A structured bond is a bond for which the coupons C_i are arbitrary complex functions of interest rates or other market observables.[3] In this case the coupons are called *structured coupons*. A corresponding swap exchanges the structured coupon payments of the bond by coupon payments of a corresponding floating rate bond; see Figures 12.1 and 12.2. Taking the swap and the floating rate bond (which just pays the current market rate on the notional) together, we may hedge the structured bond. For the values of the products in T_0 we thus have $N_1 + V_{\text{swap}} = V_{\text{bond}}$.

The structured bond and its (hedge-)swap are separate products, since they are often offered by separate institutions.[4]

Zero Structures

Besides (structured) coupon paying bonds, another common type of bond is that for which the coupon is accrued instead of paid. Then, as for the zero-coupon bond, there will be a single payment at maturity; see Figure 12.3, right. An accruing product is called a *zero structure*. A bond with accruing coupons is sometimes called a zero-coupon bond.

For an accruing zero-coupon bond we may define a corresponding swap too. To do so we consider the following (equivalent) representation of the bond: At the end of each coupon period the notional and the period's coupon is paid. The amount defines the new notional for the following period and is reinvested (this corresponds to accruing the coupon); see Figure 12.3, left. The swap is then defined such that it exchanges the structured coupon (on the various notionals) by a corresponding coupon with a given market rate.

The swap will then allow to build up the bond's payment at maturity using the starting notional invested at market rates; see Figure 12.4.

For the valuation in T_0 we again have

$$N_1 + V_{\text{swap}} = V_{\text{bond}}, \tag{12.1}$$

---

[3] Common coupons are options on interest rates, e.g., a guaranteed minimum rate in the form of $C_i = \max(L_i(T_i), K)$, or even coupons which depend on the performance of one or more stocks, in which case the bond would be a hybrid interest rate product.

[4] E.g., a mortgage bank and an investment bank.

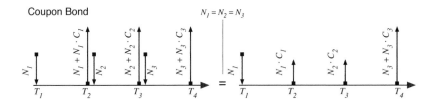

Figure 12.1. *Cash flows for a coupon bond. Left: with imaginary exchange of notionals at the end of each period; right: with effective cash flow only.*

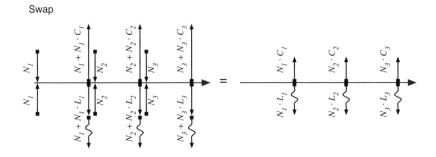

Figure 12.2. *Cash flows for a swap whose fixed leg corresponds to the coupon bond in Figure 12.1. Left: with imaginary exchange of notionals at the end of each period; right: with effective cash flow only.*

where the swap is interpreted as payer swap, i.e., it pays $C_i - L_i$. That Equation (12.1) holds, follows iteratively by considering a single period: The notional N_1 is invested at the market rate like a floating rate bond. The floating rate bond pays $N_1 + N_1 L_1$ in T_2. The swap exchanges the coupon $N_1 L_1$ for the structured coupon. Thus we have $N_1 + N_1 C_1 =: N_2$ which is the notional N_2 which is used for the same construction for the following period.

Such a construction is especially meaningful if the zero-coupon bond may be canceled at the end of a coupon period. In this case it would pay the accrued notional up to this period. For a cancelable bond the swap has the same cancellation right and

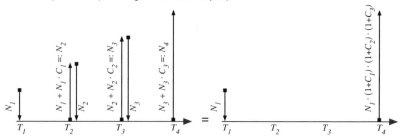

Figure 12.3. *Cash flows for a zero coupon bond. Left: with imaginary exchange of notionals at the end of each period; right: with effective cash flow only.*

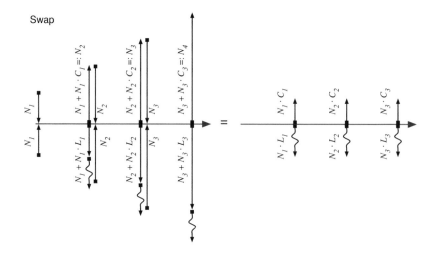

Figure 12.4. *Cash flows for a swap whose fixed leg corresponds to the zero coupon bond in Figure 12.3. Left: with imaginary exchange of notionals at the end of each period; right: with effective cash flow only.*

is canceled simultaneously.[5]

We repeat the notions *structured bond* and *structured swap* in a definition, although these definitions differ only in one minor aspect from the ones for coupon bond or swap, respectively: the coupon C_i may be an arbitrary $\mathcal{F}_{T_{i+1}}$-measurable random variable.

Definition 141 (Structured Bond):
Let $0 = T_0 < T_1 < T_2 < \cdots < T_n$ denote a given tenor structure.

For $i = 1, \ldots, n - 1$ let C_i denote a (generalized) interest rate for the periods $[T_i, T_{i+1}]$, respectively. Let C_i be an $\mathcal{F}_{T_{i+1}}$-measurable random variable.

Furthermore let N_i denote a constant value (*notional*). The *structured bond* pays

$$N_i\, C_i\, (T_{i+1} - T_i) + N_i \begin{cases} 0 & (i \neq n) \\ 1 & (i = n - 1) \end{cases}$$

in T_{i+1}. The value of the *structured bond* seen in $t < T_2$ is

$$V_{\text{Bond}}(T_1, \ldots, T_n; t) = N(t)\, \mathrm{E}^{\mathbb{Q}^N}\!\left(\sum_{k=1}^{n-1} \frac{N_k\, C_k\, (T_{k+1} - T_k)}{N(T_{k+1})} \,\Big|\, \mathcal{F}_t \right) + N_{n-1}\, P(T_n; t).$$

Definition 142 (Structured Swap (Structured Receiver Swap)):
Let $0 = T_0 < T_1 < T_2 < \cdots < T_n$ denote a given tenor structure.

For $i = 1, \ldots, n - 1$ let C_i denote a (generalized) interest rate for the periods $[T_i, T_{i+1}]$, respectively. Let C_i be an $\mathcal{F}_{T_{i+1}}$-measurable random variable.

Furthermore let s_i denote a constant interest rate (*spread*) and N_i a constant value (*notional*). The *structured swap* pays

$$X_i := N_i\, (C_i - (L(T_i, T_{i+1}; T_i) + s_i))\, (T_{i+1} - T_i).$$

in T_{i+1}. The value of the *structured swap* seen in $t < T_2$ is

$$V_{\text{Swap}}(T_1, \ldots, T_n; t) = N(t)\, \mathrm{E}^{\mathbb{Q}^N}\!\left(\sum_{k=1}^{n-1} \frac{X_k}{N(T_{k+1})} \,\Big|\, \mathcal{F}_t \right).$$

Remark 143 (Structured Coupon): By C_i we denote an arbitrary, generalized interest rate. In general it will be a function of $L(T_k, T_{k+1})$ with fixing date T_i, and thus even $\mathcal{F}_{T_i}$-measurable. We allow that the generalized interest rate C_i depends on

---

[5] Other arguments for this construction are reduction of default and market risk.

events within the period $[T_i, T_{i+1}]$ and requires $\mathcal{F}_{T_{i+1}}$-measurability only. An example of C_i is a constant rate $C_i = \text{const.}$, the forward rate $C_i = L(T_i, T_{i+1}; T_i)$, or a swap rate $C_i = S(T_i, \ldots, T_k; T_i)$.

Remark 144 (Zero Structure): The swap in Definition 142 is called *zero structure*, if the notional N_i if given by

$$N_{i+1} := N_i \, C_i \, (T_{i+1} - T_i) \qquad i = 1, \ldots, n-1.$$

Remark 145 (Structured Payer Swap/Structured Receiver Swap): The swap defined in Definition 142 is a *receiver swap*. A swap with reversed payments

$$X_i := N_i \left((L(T_i, T_{i+1}; T_i) + s_i) - C_i \right) (T_{i+1} - T_i)$$

is called *structured payer swap*. See Definition 118.

12.2.2 Bermudan Option

Definition 146 (Bermudan): ⌐

A financial product is called *Bermudan* if it has multiple exercise dates (options), i.e., there are times T_i at which the holder of a Bermudan may choose between different payments or financial products (underlyings). ⌐

A more formal definition of the Bermudan option, anticipating the result that the optimal exercise is to choose the maximum of the exercise and nonexercise value, is given in the following definition:

Definition 147 (Bermudan Option): ⌐

Let $\{T_i\}_{i=1,\ldots,n}$ denote a set of exercise dates and $\{V_{\text{underl},i}\}_{i=1,\ldots,n}$ a corresponding set of underlyings. The Bermudan option is the right to receive at one and only one time T_i the corresponding underlying $V_{\text{underl},i}$ (with $i = 1, \ldots, n$) or receive nothing.

At each exercise date T_i, the optimal strategy compares the value of the product upon exercise with the value of the product upon nonexercise and chooses the larger one. Thus the value of the Bermudan is given recursively

$$\underbrace{V_{\text{berm}}(T_i, \ldots, T_n; T_i)}_{\substack{\text{Bermudan with} \\ \text{exercise dates} \\ T_i, \ldots, T_n}} := \max \big(\underbrace{V_{\text{berm}}(T_{i+1}, \ldots, T_n; T_i)}_{\substack{\text{Bermudan with} \\ \text{exercise dates} \\ T_{i+1}, \ldots, T_n}}, \underbrace{V_{\text{underl},i}(T_i)}_{\substack{\text{Product} \\ \text{received upon} \\ \text{exercise in } T_i}} \big),$$

where $V_{\text{berm}}(T_n; T_n) := 0$ and $V_{\text{underl},i}(T_i)$ denotes the value of the underlying $V_{\text{underl},i}$ at exercise date T_i. ⌐

An example is given by the Bermudan swaption. Here the option holder has the right to enter a swap at several different times. The optimal exercise strategy chooses the maximal value from either the swap or the Bermudan with the remaining exercise dates.

Definition 148 (Bermudan Swaption):

Let $0 = T_0 < T_1 < T_2 < \cdots < T_n$ denote a given tenor structure. The value $V_{\text{BermSwpt}}(T_1, \ldots, T_n; T_0)$ of a Bermudan swaption seen at time T_0 is defined recursively by

$$V_{\text{BermSwpt}}(T_i, \ldots, T_n; T_i)$$
$$:= \max\left(V_{\text{BermSwpt}}(T_{i+1}, \ldots, T_n; T_i), \ V_{\text{Swap}}(T_i, \ldots, T_n; T_i)\right), \tag{12.2}$$

where $V_{\text{BermSwpt}}(T_n; T_n) := 0$ and $V_{\text{Swap}}(T_i, \ldots, T_n; T_i)$ denotes the value of a swap with fixing dates $T_i, \ldots, T_{n-1}$ and payment dates $T_{i+1}, \ldots, T_n$, seen in T_i. Furthermore, with a given numéraire N and a corresponding equivalent martingale measure $\mathbb{Q}^N$

$$V_{\text{BermSwpt}}(T_{i+1}, \ldots, T_n; T_i) = N(T_i) \, \mathrm{E}^{\mathbb{Q}^N}\left(\frac{V_{\text{BermSwpt}}(T_{i+1}, \ldots, T_n; T_{i+1})}{N(T_{i+1})} \,\bigg|\, \mathcal{F}_{T_i}\right). \tag{12.3}$$

Interpretation: The *Bermudan swaption* $V_{\text{BermSwpt}}(T_{n-1}, T_n)$ is simply a *swaption* (option on a swap) and since the *swap* has only a single period $[T_{n-1}, T_n]$ it is actually a *caplet*. The Bermudan swaption $V_{\text{BermSwpt}}(T_{n-2}, T_{n-1}, T_n)$ is an option which allows a choice in time T_{n-2} between a swaption (with later exercise date) or a (longer) swap. Thus it is an option on an option. Iteratively the Bermudan swaption is an option on an option (on an option, etc.).

Taking the underlying swap as the defining object, we see that the Bermudan swaption $V_{\text{BermSwpt}}(T_1, \ldots, T_n)$ can also be interpreted as an option either to enter at times $T_1, \ldots, T_{n-1}$ a swap with remaining periods up to T_n, or to wait. Options with multiple exercise times are called *Bermudan*. Options with a single exercise time are called *European*. ◁|

Remark 149 (Bermudan Swaption): It is key to the evaluation of the Bermudan swaption that by Equation (12.3) we have at each exercise date T_i an evaluation of a derivative product. For this the *conditional expectation* has to be calculated. Depending on the model and the implementation, the calculation of conditional expectations may be nontrivial. In Chapter 15 we give an in-depth discussion on how to calculate a conditional expectation in a path simulation (Monte Carlo simulation).

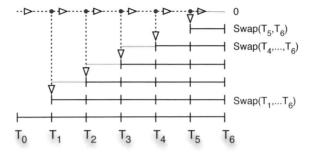

Figure 12.5. *Bermudan swaption.*

12.2.3 Bermudan Callable and Bermudan Cancelable

We will now define a product class that generalizes the structure of a Bermudan swaption.

Definition 150 (Bermudan Callable Structured Swap[6]):

Let $0 = T_0 < T_1 < T_2 < \cdots < T_n$ denote a given tenor structure.

For $i = 1, \ldots, n - 1$ let C_i denote a (generalized) interest rate for the periods $[T_i, T_{i+1}]$, respectively. Let C_i be an $\mathcal{F}_{T_{i+1}}$-measurable random variable.

Furthermore let s_i denote a constant interest rate (*spread*), N_i a constant value (*notional*) and

$$X_i := N_i \left(C_i - (L(T_i, T_{i+1}) + s_i) \right) (T_{i+1} - T_i).$$

Let $V_{\text{underl}}(T_i, \ldots, T_n; T_i)$ denote the value of the product paying X_k in T_{k+1} for $k = i, \ldots, n - 1$, seen in T_i. If N denotes a numéraire and $\mathbb{Q}^N$ a corresponding equivalent martingale measure, then

$$V_{\text{underl}}(T_i, \ldots, T_n; T_i) = N(T_i) \, \mathrm{E}^{\mathbb{Q}^N} \left(\sum_{k=i}^{n-1} \frac{X_k}{N(T_{k+1})} \, \Big| \mathcal{F}_{T_i} \right). \tag{12.4}$$

The value of a *Bermudan callable swap with structured leg* C_i[7] is recursively defined by

$$V_{\text{BermCall}}(T_i, \ldots, T_n; T_i)$$
$$:= \max \left(V_{\text{BermCall}}(T_{i+1}, \ldots, T_n; T_i), \, V_{\text{underl}}(T_i, \ldots, T_n; T_i) \right), \tag{12.5}$$

---

[6] Compare [89].

[7] On the naming see Remark 154.

164

where $V_{\text{BermCall}}(T_n; T_n) := 0$ and

$$V_{\text{BermCall}}(T_{i+1}, \ldots, T_n; T_i) = N(T_i) \, \text{E}^{Q^N} \left(\frac{V_{\text{BermCall}}(T_{i+1}, \ldots, T_n; T_{i+1})}{N(T_{i+1})} \Big| \mathcal{F}_{T_i} \right). \quad (12.6)$$

Remark 151 (Bermudan Callable, Structured Leg): For $C_i = S_i = \text{const.}$ the product defined in Definition 150 is a Bermudan swaption.

The payments (cash flows) X_i consist of some part $N_i \, C_i \, (T_{i+1} - T_i)$ which is called the *structured leg* and another part $N_i \, (L(T_i, T_{i+1}) + s_i) \, (T_{i+1} - T_i)$ which is called the *floating leg*.

The *underlying* $V_{\text{underl}}(T_i, \ldots, T_n)$ is a swap swapping the rate C_i against $L(T_i, T_{i+1}) + s_i$, with *fixing dates* $T_i, \ldots, T_{n-1}$ and *payment dates* $T_{i+1}, \ldots, T_n$.

To evaluate a Bermudan option we need to calculate at most two conditional expectations (12.4), and (12.6) at each exercise time T_i. Under the assumption of optimal exercise these two values are linked by (12.5), going backward from exercise date to exercise date. In Chapter 15 we give an in-depth discussion of the calculation of conditional expectation in a path simulation (Monte Carlo simulation).

The Bermudan swaption (or a Bermudan callable) allows an (possibly structured) swap to be entered into at one single time of predefined times T_i. In contrast to this, the *Bermudan cancelable* swap allows the cancelation of the underlying swap at one single time of the predefined times T_i.

Definition 152 (Bermudan Cancelable Swap):
With the notation from Definition 150 let

$$V_{\text{coupons}}(T_i; T_i) = N(T_i) \, \text{E}^{Q^N} \left(\frac{X_i}{N(T_{k+1})} \Big| \mathcal{F}_{T_i} \right).$$

the value of the coupon payments (cash flows) for the period $[T_i, T_{i+1}]$, seen in T_i.

The value of a *Bermudan cancelable with structured leg C_i* is recursively defined by

$$V_{\text{BermCancel}}(T_i, \ldots, T_n; T_i)$$
$$:= \max \left(V_{\text{coupons}}(T_i; T_i) + V_{\text{BermCancel}}(T_{i+1}, \ldots, T_n; T_i) \,, \, 0 \right),$$

where $V_{\text{BermCancel}}(T_n; T_n) := 0$ and

$$V_{\text{BermCancel}}(T_{i+1}, \ldots, T_n; T_i) = N(T_i) \, \text{E}^{Q^N} \left(\frac{V_{\text{BermCancel}}(T_{i+1}, \ldots, T_n; T_{i+1})}{N(T_{i+1})} \Big| \mathcal{F}_{T_i} \right).$$

165

Remark 153 (Bermudan Cancelable):   With the notation from Definitions 150 and 152 we have

$$V_{\text{BermCancel}}(T_i,\ldots,T_n;T_i) = V_{\text{underl}}(T_i,\ldots,T_n;T_i) + V_{\text{BermCallPayer}}(T_i,\ldots,T_n;T_i),$$
(12.7)

where $V_{\text{BermCallPayer}}$ denotes a Bermudan callable with reversed sign in the underlying. The right to cancel the structured swap V_{underl} corresponds to the right to enter such a structured swap with reversed cash flow.

Likewise we have

$$V_{\text{BermCall}}(T_i,\ldots,T_n;T_i) = V_{\text{underl}}(T_i,\ldots,T_n;T_i) + V_{\text{BermCancelPayer}}(T_i,\ldots,T_n;T_i),$$
(12.8)

where $V_{\text{BermCancelPayer}}$ denotes a Bermudan cancelable with reversed sign in the underlying.

From this we conclude that the problem of evaluating a Bermudan cancelable corresponds to the problem of evaluating a Bermudan callable—and vice versa.

Remark 154 (Bermudan Callable):   Our definition of a Bermudan option is a general one: For each exercise date the corresponding underlying can be specified arbitrarily. The Bermudan callable is a special variant of a Bermudan option, where the underlyings share the same cash flow after exercise.[8] The Bermudan callable is the right to enter a financial product at some later time. The Bermudan cancelable is the right to terminate a financial product at some later time. Bermudan callable and Bermudan cancelable are counterparts in the sense of Equation (12.7).

For (structured) bonds it is usually the case that the issuer (i.e. the party that pays the coupons) has the right to cancel the bond.[9] Due to relationship (12.7) it is the case that the issuer of a bond having the right to cancel the issued bond essentially has a callable bond. Therefore the use of the words *callable* and *the call right* are often used where our definition would seem to point to a cancelable contract.

12.2.4 Compound Options

A *compound option* is an option on an option. Popular are a European call option on a European call option, a call on a put, a put on a call, and a put on a put. The compound option is closely related to the Bermudan option with two exercise dates. The compound option may be viewed as a special variant of a Bermudan option.

As for the Bermudan option, the evaluation of a compound option requires the evaluation of an option at a future time (the exercise date of the first option). The

---

[8] Our definition of a Bermudan callable is the same as, for example, in Piterbarg [89].

[9] Upon cancelation the notional is repaid.

methods used for the pricing of Bermudan options can thus be applied to the evaluation of compound options as well.

12.2.5 Trigger Products

For a Bermudan option and a Bermudan cancelable the exercise criterion is given by optimal exercise: The option holder chooses the maximum value. Thus the recursive definition of the product value uses the maximum function on the values of the two alternatives nonexercise and exercise. For a trigger product the exercise is given by some criterion, the *trigger*, which does not necessarily represent an optimal exercise. An example of such a product is the autocap, which we will define in Section 12.2.6.4, or the following target redemption note.

12.2.5.1 Target Redemption Note

Definition 155 (Target Redemption Note):
Let $0 = T_0 < T_1 < T_2 < \cdots < T_n$ denote a given tenor structure.

For $i = 1, \ldots, n - 1$ let C_i denote a (generalized) interest rate for the periods $[T_i, T_{i+1}]$, respectively. Let C_i be an $\mathcal{F}_{T_{i+1}}$-measurable random variable.

Furthermore let N_i denote a constant value (*notional*). A *target redemption note* pays

$$N_i\, X_i \qquad \text{in} \qquad T_{i+1}$$

with

$$X_i := \begin{cases} C_1 & \text{for } i = 1, \\ \min(C_i\,, K - \sum_{k=1}^{i-1} C_k) & \text{for } i > 1 \end{cases} \quad \text{(structured coupon)}$$

$$+ \begin{cases} 1 & \text{for } \sum_{k=1}^{i-1} C_k < K <= \sum_{k=1}^{i} C_k \text{ or } i = n, \\ 0 & \text{else.} \end{cases} \quad \text{(redemption)}$$

$$+ \begin{cases} \max(0\,, K - \sum_{k=1}^{i} C_k) & \text{for } i = n \\ 0 & \text{else.} \end{cases} \quad \text{(target coupon guarantee).}$$

Remark 156 (Target Redemption Note Coupon): The rule that defines the coupon C_i in Definition 155 may vary, and Definition 155 is only the framework of the product. With a coupon, e.g., like

$$C_i = K - a\, L_i\, (T_{i+1} - T_i) \qquad \text{for } i > 1. \tag{12.9}$$

the product is also called *variable maturity inverse floater (VMIF)*. We will consider structured coupons in Section 12.2.6.

Interpretation: The holder of a target redemption note receives the coupon C_i until the sum of the coupons has reached K, the target coupon. If the accumulated coupon exceeds the target coupon ($\sum_{k=1}^{i} C_k >= K$), then the difference between the target coupon and the notional is paid. After this no coupon payments will be made. The structure is canceled if the target coupon has been reached. If the target coupon has not been reached over the full life time of the product, then the difference between the target coupon and the notional is paid at maturity.

Thus, the target redemption note guarantees the payment of a coupon K and the redemption of the notional. What is uncertain is the time of payment and the maturity, and thus the yield of the product. The yield of the product depends on when the condition

$$\sum_{k=1}^{i} C_k = K \quad \text{and} \quad \sum_{k=1}^{i-1} C_k < K \qquad (12.10)$$

is fulfilled.

It is normal to have an above-average initial coupon C_1. The life time of the product varies between T_2 and T_n, giving yields between K/T_2 and K/T_n. An investor would buy this product if he assumed that the condition (12.10) would be fulfilled early, such that the yield of the product (is expected to be) above average.

An example of a target redemption note, actually offered in 2004, is

$$C_i = 7.5\% - 2\,L_i\,(T_{i+1} - T_i) \qquad \text{for } i > 1.$$

Here the option holder profits from the product if the interest rates L_i decline faster than expected (and thus the product will be redeemed early). ◁|

12.2.6 Structured Coupons

In the previous definitions we have defined the structured bond (Definition 141), the structured swap (Definition 142), the Bermudan option (Definition 150), the Bermudan cancelable (Definition 152), and the target redemption note (Definition 155) without specifying the coupons C_i. We now refine our definitions by defining some of the most common structured coupons C_i. In the respective definitions we only give the characteristic that describes the coupon. The characteristics defined in the following exist for bonds and swaps and for Bermudan callable and Bermudan cancelables alike. For example, we define a coupon of a *CMS spread* product and the name of the corresponding Bermudan callable swap as a *Bermudan callable CMS spread swap*.

12.2.6.1 Capped, Floored, Inverse, Spread, CMS

Definition 157 (Capped Floater): ⌐
A product as in Definition 141, 142, 150, or 152 with

$$C_i := \min(L(T_i, T_{i+1}) + s_i, c_i)$$

is called a *capped floater*. Here c_i denotes a constant (*cap*). ⌐

Definition 158 (Floored Floater): ⌐
A product as in Definition 141, 142, 150, or 152 with

$$C_i := \max(L(T_i, T_{i+1}) + s_i, f_i)$$

is called a *floored floater*. Here f_i denotes a constant (*floor*). ⌐

Definition 159 (Inverse Floater): ⌐
A product as in Definition 141, 142, 150, or 152 with

$$C_i = \min\left(\max\left(k_i - L(T_i, T_{i+1}), f_i\right), c_i\right)$$

is called an *inverse floater*. Here $f_i < c_i$ (*floor, cap*) and k_i are constants. ⌐

Definition 160 (Capped CMS Floater): ⌐
A product as in Definition 141, 142, 150, or 152 with

$$C_i := \min(S_{i,i+m} + s_i, c_i),$$

where s_i, c_i denote constants (*spread, cap*) and $S_{i,i+m} = S(T_i, \ldots, T_{i+m})$ denotes a swap rate as in Definition 122, is called a *capped CMS[10] floater*. The rate $S_{i,i+m}$ is called *constant maturity swap (CMS) rate* since the maturity relative to the period start T_i, is constant ($T_{i+m} - T_i$ is constant, i.e., independent of i), assuming an equidistant tenor structure. ⌐

Definition 161 (Inverse CMS Floater): ⌐
A product as in Definition 141, 142, 150, or 152 with

$$C_i = \min\left(\max\left(k_i - S_{i,i+m}, f_i\right), c_i\right),$$

where k_i, f_i, c_i denote constants (*strike, floor, cap*) and $S_{i,i+m} = S(T_i, \ldots, T_{i+m})$ denotes a swap rate as in Definition 122, is called an *inverse CMS floater*. ⌐

---

[10] CMS stands for *constant maturity swap*, i.e., the maturity of the underlying swap relative to the swap start ($T_{i+m} - T_i$) is constant. For simplicity we assume here that the tenor structure is equidistant.

Definition 162 (CMS Spread): ⌐

A product as in Definition 141, 142, 150, or 152 with

$$C_i = \min\left(\max\left(S^1_{i,i+m_1} - S^2_{i,i+m_2}, f_i\right), c_i\right),$$

where f_i, c_i denote constants (*floor, cap*) and $S_{i,i+m_1} = S(T_i, \ldots, T_{i+m_1})$, $S_{i,i+m_2} = S(T_i, \ldots, T_{i+m_2})$ denotes a swap rate as in Definition 122, is called a *CMS spread*. ⌐

12.2.6.2 Range Accruals

Definition 163 (Range Accrual): ⌐

Let $t_{i,k} \in [T_i, T_{i+1})$ denote given observation points for the period $[T_i, T_{i+1})$. Furthermore let K_i and $b^l_i < b^h_i$ denote given interest rates (constants). A product as in Definition 141, 142, 150, or 152 with

$$C_i := K_i \frac{1}{n_i} \sum_{k=1}^{n_i} \mathbf{1}_{[b^l_i, b^h_i]}(L(t_{i,k}, t_{i,k} + \Delta T_i; t_{i,k})) \quad \text{in } T_{i+1}.$$

is called a *range accrual*. Here $\Delta T_i := T_{i+1} - T_i$. ⌐

 Interpretation: The interval $[b^l_i, b^h_i]$ describes an interest rate corridor. It is calculated how often the reference rate $L(t, t + \Delta T_i; t)$ stays within this corridor at the times $t_{i,k}$. The product pays the corresponding fractional amount of the rate K_i at the end of the period.

Since the interest rate corridor may be chosen as a function of the period i, the product makes it possible to profit from a specific evolution of the rate. ◁|

12.2.6.3 Path-Dependent Coupons

Definition 164 (Snowball/Memory): ⌐

A product as in Definition 141, 142, 150, or 152 with

$$C_i = \min\left(\max\left(C_{i-1} + X_i, f_i\right), c_i\right),$$

is called *snowball*, where X_i is some coupon as in Definitions 157 to 163 and f_i, c_i denote constants (*floor, cap*) and C_{i-1} denotes the coupon of the previous periods with $C_0 := 0$. ⌐

Example: A coupon

$$C_i = \min\left(\max\left(C_{i-1} + k_i - L(T_i, T_{i+1}), f_i\right), c_i\right)$$

is called an *inverse floater memory*.

Definition 165 (Power Memory):

With the notation from Definition 164 the coupon

$$C_i = \min\left(\max\left(\alpha C_{i-1} + X_i, f_i\right), c_i\right)$$

is called a *power memory* (for $\alpha \neq 1$).

Remark 166 (Snowball, Path Dependency):   The snowball is called a path-dependent product, since its coupon depends on the previous coupon, i.e., on the history.

Definition 167 (Ratchet Cap):

The *ratchet cap* pays in T_{i+1}

$$X_i = N \, \max\left(L(T_i, T_{i+1}; T_i) - K_i, \, 0\right) (T_{i+1} - T_i)$$

for times $T_1, \ldots, T_n$, where

$$K_i := \min\left(K_{i-1} + R, \, L(T_i, T_{i+1}; T_i)\right) \quad \text{for } i > 1$$

and K_1, R denote constants (*strike* and *ratchet*), $L(T_i, T_{i+1}; T_i)$ denote the forward rate (LIBOR), and N denotes the *notional*.

Remark 168 (Ratchet Cap):   The ratchet cap has an automatic adjustment of the strike K_i. Since the adjustment depends on past realizations, the ratchet cap is a path-dependent product.

12.2.6.4 Flexi-Cap

The *flexi cap* comes in two variants: As an autocap with a simple (automatic) exercise criterion and as an chooser cap with an assumed optimal exercise (like for the Bermudan). Both caps have in common that the maximum number of exercises is limited.

Definition 169 (Autocap): ⌐

Let $n_{\text{maxEx}} \in \mathbb{N}$. An *autocap* pays in T_{i+1}

if $|\{j : j < i \text{ and } X_j > 0\}| < n_{\text{maxEx}}$:

$$X_i := N \, \max\left(L(T_i, T_{i+1}; T_i) - K_i \, , \, 0\right)(T_{i+1} - T_i)$$

or else:

$$X_i := 0$$

for times $T_1, \ldots, T_n$, where K_i denotes the *strike* rate, $L(T_i, T_{i+1}; T_i)$ denotes the forward rate (LIBOR), and N denotes the *notional*. The autocap pays in the same way as a normal cap as long as the number of past (positive) payments is below n_{maxEx}. ⌐

Definition 170 (Chooser Cap): ⌐

Let $n_{\text{maxEx}} \in \mathbb{N}$. A *chooser cap* pays in T_{i+1}

if $|\{j : j < i \text{ and } X_j > 0\}| < n_{\text{maxEx}}$ and payment is chosen:

$$X_i := N \, \max\left(L(T_i, T_{i+1}; T_i) - K_i \, , \, 0\right)(T_{i+1} - T_i)$$

otherwise:

$$X_i := 0$$

for times $T_1, \ldots, T_n$, where K_i denotes the *strike* rate, $L(T_i, T_{i+1}; T_i)$ denotes the forward rate (LIBOR), and N denotes the *notional*. ⌐

Remark 171 (Autocap): The autocap is a path-dependent product, since at a future time the number of exercises allowed depends on the history. It is also a trigger product, since the exercise is not optimal, but triggered by a simple trigger. The corresponding optimal exercise is given by the chooser cap.

Remark 172 (Chooser Cap (Backward Algorithm)): Since the option holder chooses to exercise optimally, we have that the value $V_{\text{Chooser}}^{(n_{\text{maxEx}}, T_1, T_n)}(T_0)$ of the chooser cap is given by

$$V_{\text{Chooser}}^{(n_{\text{Ex}}, T_i, T_n)}(T_i) = \max\left(X_i + V_{\text{Chooser}}^{(n_{\text{Ex}}-1, T_{i+1}, T_n)}(T_i)\,,\, V_{\text{Chooser}}^{(n_{\text{Ex}}, T_{i+1}, T_n)}(T_i)\right),$$

where

$$X_i := N \, \max\left(L(T_i, T_{i+1}; T_i) - K_i \, , \, 0\right)(T_{i+1} - T_i)$$

and

$$V_{\text{Chooser}}^{(n_{\text{Ex}},T_n,T_n)}(T_n) = 0, \qquad V_{\text{Chooser}}^{(0,T_i,T_n)}(T_i) = 0,$$

where $V_{\text{Chooser}}^{(n_{\text{Ex}},T_i,T_n)}(T_k)$ denotes the value of a chooser cap with at most n_{Ex} exercises in $T_i, \ldots, T_n$, seen in T_k.

Remark 173 (Chooser Cap as Bermudan): A chooser cap with $n_{\text{maxEx}} = 1$ exercises is a *Bermudan cap*.

12.2.7 Shout Options

Definition 174 (Shout Option):
Assume that a financial product pays an underlying $S(t^*)$ at time T, i.e., t^* is the fixing date and T the payment date. The owner of the financial product has a *shout right* on the underlying, if the holder of the right can determine once at any time t with $T_1 \leq t \leq T_2 \leq T$ the fixing date t^* as $t^* := t$. The holder determines by *shouting* that the underlying should be fixed.

Remark 175: While for an American option or a Bermudan option, e.g., on $S(t) - K$, the fixing date and the payment date are both determined upon exercise, i.e., $S(t^*) - K$ is paid in t^*, for a shout option the fixing date is determined upon exercise, while the payment date stays predefined.

Theorem 176 (Value of a Shout Right): A shout right on a convex function of a submartingale (under terminal measure) is worthless.

Proof: Let T denote the payment date of $f(S(t))$, where t is fixed as $t = t^*$ by shouting. For the chosen numéraire N we assume $N(T) = 1$ (terminal measure). Let f be convex and S a submartingale under $\mathbb{Q}^N$. Then we have:

$$
\begin{aligned}
& S(t) \leq \mathrm{E}(S(T_2) \mid \mathcal{F}_t) && \text{(submartingale property)} \\
\Rightarrow \quad & f(S(t)) \leq \mathrm{E}(f(S(T_2)) \mid \mathcal{F}_t) && \text{(Jensen's inequality)} \\
\Rightarrow \quad & \mathrm{E}(f(S(t)) \mid \mathcal{F}_{T_0}) \leq \mathrm{E}(f(S(T_2)) \mid \mathcal{F}_{T_0}) && \text{(tower law)} \\
\Rightarrow \quad & \mathrm{E}\left(\frac{f(S(t))}{N(T)} \mid \mathcal{F}_{T_0}\right) \leq \mathrm{E}\left(\frac{f(S(T_2))}{N(T)} \mid \mathcal{F}_{T_0}\right) && \text{(terminal measure).}
\end{aligned}
$$

Thus, to maximize the value, the option holder will always exercise the shout right at $t^* = T_2$. □

12.3 Product Toolbox

The terms used in the previous section, like *capped*, *inverse*, *ratchet*, etc., describe properties of the payoff function. In practice the terms are used less strictly and the name of a product corresponds to its mathematical definition only loosely. Here marketing aspects are more important. Key product features are, nevertheless, indicated by the name of a product.

Table 12.3 gives a rough idea of some of the most common terms used to denote properties of the product or payoff function.

 Experiment: At http://www.christian-fries.de/finmath/applets/LMMPricing.html several interest rate products can be priced, among them a *cancelable swap*. The model used is a LIBOR market model implemented in a Monte Carlo simulation. ◁|

 Further Reading: An overview of exotic derivatives can be obtained from the customer information service of some investment banks or the *term sheets* of the products. They contain descriptions of the product and definitions of the payoffs, similar to the definitions in this chapter, as well as a short discussion of product properties.

Zhang's book [43] presents some of the most important exotic options, in particular, exotic equity options. ◁|

Attribute	Product Property
Bond (a.k.a. Note)	Receive coupons. At maturity receive the notional. See Definition 141
Swap	Exchange coupons (usually against *float*). See Definition 142
Bermudan option	Receive an underlying at one of multiple exercise dates. See Definition 146
Bermudan cancelable	Cancel product (e.g. bond or swap) at one of multiple cancelation dates. See Definition 152
Target redemption	$\min(C_i, K - \sum_{k=1}^{i-1} C_k)$ with cancelation and notional redemption at (12.10) (*trigger*). See Definition 155
Chooser	Receive an underlying at some of multiple exercise dates. See Definition 170

Attribute	Payoff Function
LIBOR / floater	$C_i = L(T_i, T_{i+1}; T_i)$
CMS	C_i is swap rate with constant time to maturity
Capped	$\min(C_i - K_i, c_i)$
Floored	$\max(C_i - K_i, f_i)$
Inverse	$K_i - C_i$
Spread	$C_i^1 - C_i^2$
Ratchet	$K_i = \min(K_{i-1} + R, C_i)$
Snowball	$C(T_i) = C(T_{i-1}) + K_i - L_i(T_i)$
Range accrual	$K_i \frac{1}{n_i} \sum_{k=1}^{n_i} \mathbf{1}_{[b_i^l, b_i^h]}(L(t_{i,k}, t_{i,k} + \Delta T_i; t_{i,k}))$

Table 12.3. *Product Toolbox: Common attributes and their representation in product property and payoff function.*

Part IV

Discretization of Stochastic Differential Equations and Numerical Valuation Methods

Motivation and Overview - Part IV

In Chapter 4 we presented the Black-Scholes model of a stock S and a riskless account B:

$$dS(t) = \mu^{\mathbb{P}}(t)S(t)\, dt + \sigma(t)S(t)\, dW^{\mathbb{P}}(t),$$
$$dB(t) = r(t)B(t)\, dt.$$

In Chapter 10 we presented the corresponding Black model for a forward interest rate $L_1 = L(T_1, T_2)$:

$$dL_1(t) = \mu^{\mathbb{P}}(t)L_1(t)\, dt + \sigma(t)L_1(t)\, dW^{\mathbb{P}}(t), \quad \sigma(t) \geq 0.$$

Using these models we could derive analytic pricing formulas for the corresponding European options. An obvious generalization of these models consists in modeling multiple stocks, e.g.,

$$dS_i(t) = \mu_i^{\mathbb{P}}(t)S_i(t)\, dt + \sigma_i(t)S_i(t)\, dW_i^{\mathbb{P}}(t),$$
$$dB(t) = r(t)B(t)\, dt,$$

respectively, a model for multiple forward rates, e.g., $L_i = L(T_i, T_{i+1})$ for $T_1 < T_2 <$ ... with, e.g.,

$$dL_i(t) = \mu_i^{\mathbb{P}}(t)L_i(t)\, dt + \sigma_i(t)L_i(t)\, dW_i^{\mathbb{P}}(t), \quad \sigma(t) \geq 0.$$

This model is called the LIBOR market model.

Such models may then be used for the evaluation of complex derivatives, e.g. a spread option, where the payout depends on two or more forward rates. The pricing, i.e., the calculation of the expectation $E^{\mathbb{Q}^N}(\frac{V(T)}{N(T)} \mid \mathcal{F}_0)$, where N is a chosen numéraire and $\mathbb{Q}^N$ is a corresponding martingale measure, often requires a numerical method, e.g., a Monte Carlo integration

$$E^{\mathbb{Q}^N}\left(\frac{V(T)}{N(T)} \mid \mathcal{F}_0\right) \approx \frac{1}{n}\sum_{i=1}^{n}\frac{\tilde{V}(T, \omega_i)}{\tilde{N}(T, \omega_i)}$$

for a sampling $\omega_1, \ldots, \omega_n$. Here, $\tilde{V}$ and $\tilde{N}$ denote approximations of V and N, respectively, since the corresponding Monte Carlo samples are generally generated for an approximating model, namely a time-discrete model.

In Chapter 13 we start by examining the approximation of time-continuous stochastic processes through time-discrete stochastic processes. We then consider the approximation of the random variables by a Monte Carlo simulation or a discretization of the state space.

179

These discretizations allow us to calculate (approximate) expectations and thus derivative prices. It turns out that within the discrete setup some calculations are difficult. In a Monte Carlo simulation the calculation of a conditional expectation is nontrivial. A conditional expectation may be required in the pricing of Bermudan products; see Chapter 15. In a discretization of a state space the calculation of path-dependent quantities is nontrivial. Path-dependent quantities appear in path-dependent options; see Chapter 16.

The treatment of complex models, like the LIBOR market model, will be given in Part V.

CHAPTER 13

Discretization of Time and State Space

In this chapter we present methods for discretization and implementation of Itô stochastic processes. We give an integrated presentation of *path simulation* (Monte Carlo simulation) and *lattice methods* (e.g., trees). Finally, we show how both methods can be combined; see Section 13.4.

Throughout our discussion of the discretization and implementation we will repeat some of the terms from Chapter 2, e.g., path, σ-algebra, filtration, process, and $\mathcal{F}_t$-adapted. Thus, this chapter will also serve as an illustration of some of the mathematical concepts from Chapter 2.

The discretization and implementation should not be seen as a minor additional step after the mathematical analysis and it should not be underestimated. The discretization and implementation allow us a second look, possibly providing further insights into a model.[1]

In Figure 13.1 we give an overview of the steps involved in the discretization and implementation of Itô processes.

13.1 Discretization of Time: The Euler and the Milstein Schemes

As a first step we shall consider the discretization of time and present the Euler scheme and the Milstein scheme.

---

[1] Indeed, it is common in mathematics to prove analytical results as a limit of a numerical, i.e., discrete, procedure.

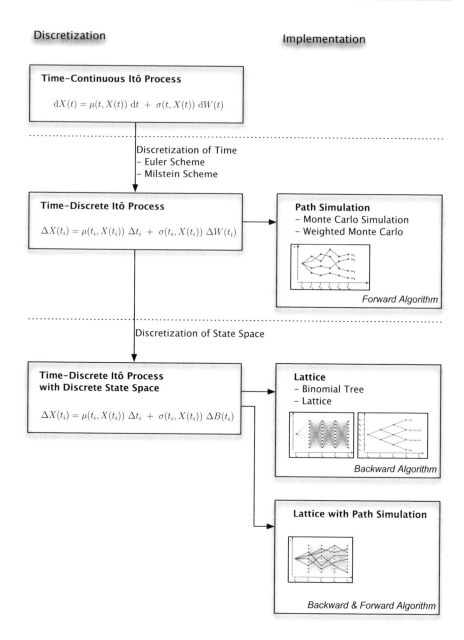

Figure 13.1. *Discretization and implementation of Itô processes*

13.1.1 Definitions

Definition 177 (Euler Scheme):
Given an Itô process

$$dX(t) = \mu(t, X(t))\, dt + \sigma(t, X(t))\, dW(t),$$

and a time discretization $\{t_i \mid i = 0, \ldots, n\}$ with $0 = t_0 < \ldots < t_n$, then the time-discrete stochastic process $\tilde{X}$ defined by

$$\tilde{X}(t_{i+1}) = \tilde{X}(t_i) + \mu(t_i, \tilde{X}(t_i))\, \Delta t_i + \sigma(t_i, \tilde{X}(t_i))\, \Delta W(t_i)$$

is called an *Euler scheme* of the process X (where $\Delta t_i := t_{i+1} - t_i$ and $\Delta W(t_i) := W(t_{i+1}) - W(t_i)$).

Interpretation (Euler Discretization): The Euler scheme derives from a simple integration rule. From the definition of the Itô process we have

$$X(t_{i+1}) = X(t_i) + \int_{t_i}^{t_{i+1}} \mu(t, X(t))\, dt + \int_{t_i}^{t_{i+1}} \sigma(t, X(t))\, dW(t).$$

Obviously, the Euler scheme is given by the approximation of the integrals

$$\int_{t_i}^{t_{i+1}} \mu(t, X(t))\, dt \approx \int_{t_i}^{t_{i+1}} \mu(t_i, X(t_i))\, dt = \mu(t_i, X(t_i))\, \Delta t_i$$

$$\int_{t_i}^{t_{i+1}} \sigma(t, X(t))\, dW(t) \approx \int_{t_i}^{t_{i+1}} \sigma(t_i, X(t_i))\, dW(t) = \sigma(t_i, X(t_i))\, \Delta W(t_i).$$

◁|

The following Milstein scheme improves the approximation of the stochastic integral $\int dW$.

Definition 178 (Milstein Scheme):
Given an Itô process

$$dX(t) = \mu(t, X(t))\, dt + \sigma(t, X(t))\, dW(t),$$

and a time discretization $\{t_i \mid i = 0, \ldots, n\}$ with $0 = t_0 < \ldots < t_n$, then the time-discrete stochastic process $\tilde{X}$ defined by

$$\tilde{X}(t_{i+1}) = \tilde{X}(t_i) + \mu(t_i, \tilde{X}(t_i))\, \Delta t_i + \sigma(t_i, \tilde{X}(t_i))\, \Delta W(t_i)$$
$$+ \frac{1}{2}\sigma(t_i, \tilde{X}(t_i))\sigma'(t_i, \tilde{X}(t_i))(\Delta W(t_i)^2 - \Delta t_i)$$

183

is called a *Milstein Scheme* of the process X (where $\Delta t_i := t_{i+1} - t_i$ and $\Delta W(t_i) := W(t_{i+1}) - W(t_i)$ and $\sigma' := \frac{\partial}{\partial X}\sigma$).

Remark 179 (Milstein Scheme):   The Milstein scheme gives an "improvement" only if σ depends on X.

Let us consider another discretization scheme:

Definition 180 (Euler Scheme with Predictor-Corrector Step):
Given an Itô process

$$dX(t) = \mu(t, X(t))\, dt + \sigma(t, X(t))\, dW(t),$$

and a time discretization $\{t_i \mid i = 0, \ldots, n\}$ with $0 = t_0 < \ldots < t_n$, then the time-discrete stochastic process $\tilde{X}$ defined by

$$\tilde{X}(t_{i+1}) = \tilde{X}(t_i) + \frac{1}{2}(\mu(t_i, \tilde{X}(t_i)) + \mu(t_{i+1}, \tilde{X}^*(t_{i+1})))\, \Delta t_i + \sigma(t_i, \tilde{X}(t_i))\, \Delta W(t_i)$$

with

$$\tilde{X}^*(t_{i+1}) = \tilde{X}(t_i) + \mu(t_i, \tilde{X}(t_i))\, \Delta t_i + \sigma(t_i, \tilde{X}(t_i))\, \Delta W(t_i)$$

is called an *Euler scheme with predictor-corrector step* of the process X (where $\Delta t_i := t_{i+1} - t_i$ and $\Delta W(t_i) := W(t_{i+1}) - W(t_i)$).

 Interpretation (Predictor-Corrector Scheme):   The predictor-corrector scheme improves the integration of the drift term $\int dt$, not of the stochastic integral $\int dW$. Instead of approximating the integral $\int_{t_i}^{t_{i+1}} \mu(t, X(t))\, dt$ by a rectangular rule $\mu(t_i, X(t_i))\, \Delta t_i$ the method aims to use a trapezoidal rule. With a trapezoidal rule the integral $\int_{t_i}^{t_{i+1}} \mu(t, X(t))\, dt$ would be approximated as $\frac{1}{2}(\mu(t_i, X(t_i)) + \mu(t_{i+1}, X(t_{i+1})))\, \Delta t_i$. Since the realization $X(t_{i+1})$ and thus $\mu(t_{i+1}, X(t_{i+1}))$ is unknown, it is approximated by an Euler step $\tilde{X}^*(t_{i+1})$ (*predictor* step) and the trapezoidal rule is applied with this approximation. This corresponds to correcting $\tilde{X}^*(t_{i+1})$ (*corrector* step). We have:

$$\tilde{X}(t_{i+1}) = \tilde{X}^*(t_{i+1}) - \mu(t_i, \tilde{X}(t_i))\, \Delta t_i + \frac{1}{2}(\mu(t_i, \tilde{X}(t_i)) + \mu(t_{i+1}, \tilde{X}^*(t_{i+1})))\, \Delta t_i$$

$$= \tilde{X}^*(t_{i+1}) + \underbrace{\frac{1}{2}(\mu(t_{i+1}, \tilde{X}^*(t_{i+1})) - \mu(t_i, \tilde{X}(t_i)))\, \Delta t_i}_{\text{correction term}}. \tag{13.1}$$

Tip (Implementation of the Predictor-Corrector Scheme):
Note that for an implementation formula (13.1) is more efficient than
the two Euler steps in the original Definition (180) of the scheme. The
second Euler step is replaced by a correction term applied to $\tilde{X}^*(t_{i+1})$
and requires only the additional calculation of $\mu(t_{i+1}, \tilde{X}^*(t_{i+1}))$. ◁|

The schemes presented give a time-discrete stochastic process $\tilde{X}$ such that $\tilde{X}(t_i)$ is
an approximation of $X(t_i)$. An in-depth discussion of numerical methods of approximating stochastic processes can be found in [21].

13.1.2 Time Discretization of a Lognormal Process

Consider the process

$$dX = \mu(t, X(t))X(t)\, dt + \sigma(t)X(t)\, dW(t), \tag{13.2}$$

where $(t, x) \mapsto \mu(t, x)$ and $t \mapsto \sigma(t)$ are given deterministic functions. With Lemma 50
we have

$$d\log(X) = (\mu(t, X(t)) - \frac{1}{2}\sigma^2(t))\, dt + \sigma(t)\, dW(t). \tag{13.3}$$

In the following we discuss several possible time discretizations of the process X. The
discussion is of special importance since the Black-Scholes model, the Black model,
and the LIBOR market model are all of the form (13.2).

13.1.2.1 Discretization via Euler Scheme

The Euler scheme for the stochastic differential equation (13.2) is given by

$$\tilde{X}(t_{i+1}) = \tilde{X}(t_i) + \mu(t_i, \tilde{X}(t_i))\tilde{X}(t_i)\,\Delta t_i + \sigma(t_i)\tilde{X}(t_i)\,\Delta W(t_i).$$

The random variables $\tilde{X}(t)$ generated by this scheme differ from the random variables $X(t)$ of the time-continuous process by a discretization error $X(t) - \tilde{X}(t)$. This
discretization error might be relatively large. Take, for example, the even simpler
case of a vanishing drift $\mu = 0$. Then $\tilde{X}(t_1)$ is normally distributed, while $X(t_1)$ is
lognormally distributed. Note that $\tilde{X}$ can attain negative values, while X cannot (this
follows from (13.3)).

13.1.2.2 Discretization via Milstein scheme

One way of reducing the discretization error is to use the Milstein scheme (Definition 178):

$$\tilde{X}(t_{i+1}) = \tilde{X}(t_i) + (\mu(t_i, \tilde{X}(t_i)) - \frac{1}{2}\sigma(t_i)^2)\tilde{X}(t_i)\,\Delta t_i$$
$$+ \sigma(t_i)\tilde{X}(t_i)\,\Delta W(t_i) + \frac{1}{2}\sigma(t_i)^2\tilde{X}(t_i)\,\Delta W(t_i)^2.$$

13.1.2.3 Discretization of the Log Process

A much better discretization than the two previous schemes is given by the Euler discretization of the Itô process of $\log(X)$. The Euler scheme of (13.3) is given by

$$\log(\tilde{X}(t_{i+1})) = \log(\tilde{X}(t_i)) + (\mu(t_i, \tilde{X}(t_i)) - \frac{1}{2}\sigma(t_i)^2)\,\Delta t_i + \sigma(t_i)\,\Delta W(t_i), \qquad (13.4)$$

and applying the exponential we have

$$\tilde{X}(t_{i+1}) = \tilde{X}(t_i)\,\exp\left((\mu(t_i, \tilde{X}(t_i)) - \frac{1}{2}\sigma(t_i)^2)\,\Delta t_i + \sigma(t_i)\,\Delta W(t_i)\right).$$

Using this scheme will give a lognormal random variable $\tilde{X}(t_1)$.

13.1.2.4 Exact Discretization

For the special case where μ does not depend on X, e.g., if X is a relative price under the corresponding martingale measure and thus even drift-free, then we can take the exact solution as a discretization scheme. We then have

$$X(t_{i+1}) = X(t_i)\,\exp\left((\mu_i - \frac{1}{2}\sigma_i^2)\,\Delta t_i + \sigma_i\,\Delta W(t_i)\right), \qquad (13.5)$$

where

$$\mu_i := \frac{1}{\Delta t_i}\int_{t_i}^{t_{i+1}}\mu(\tau)\,d\tau, \qquad \sigma_i := \sqrt{\frac{1}{\Delta t_i}\int_{t_i}^{t_{i+1}}\sigma^2(\tau)\,d\tau}.$$

13.2 Discretization of Paths (Monte Carlo Simulation)

Consider the time-discrete stochastic process

$$X(t_{i+1}) = X(t_i) + \mu(t_i, X(t_i))\,\Delta t_i + \sigma(t_i, X(t_i))\,\Delta W(t_i). \qquad (13.6)$$

This is an Euler scheme. The considerations below apply to any other discretization scheme. Furthermore, we do not apply a tilde to the process X since we are only considering the time-discrete process, and so do not have to distinguish it from the original time-continuous process.

13.2.1 Monte Carlo Simulation

The random variables $\Delta W(t_i)$ of the respective time steps are mutually independent; see Definition 29. At every time step t_i a random number is drawn according to the distribution of $\Delta W(t_i)$, (i.e., a vector of random numbers if $\Delta W(t_i)$ is vector valued), which we denote by $\Delta W(t_i, \omega_j)$. Then

$$X(t_{i+1}, \omega_j) = X(t_i, \omega_j) + \mu(t_i, X(t_i, \omega_j))\, \Delta t_i + \sigma(t_i, X(t_i, \omega_j))\, \Delta W(t_i, \omega_j)$$

determines the process X on a path, which we denote by ω_j. Here $\Delta W(t_i, \omega_j)$ and $\Delta W(t_k, \omega_j)$ $(i \neq k)$ are independent random numbers, following the definition of the Brownian motion. If we follow this rule to generate paths $\omega_1, \ldots, \omega_{n_{\text{paths}}}$, where $\Delta W(t_i, \omega_j)$ and $\Delta W(t_i, \omega_k)$ $(j \neq k)$ are independent, then we say that the set

$$\{X(t_i, \omega_k) \mid i = 0, 1, \ldots, n_{\text{times}};\ k = 0, 1, \ldots, n_{\text{paths}}\}$$

is a *Monte Carlo simulation* of the process X.

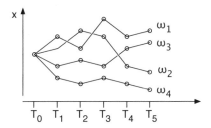

Figure 13.2. *Monte-Carlo Simulation*

An approximation of the expectation of some function f of the $X(t_i)$'s is then given by

$$\mathrm{E}^{\mathbb{P}}(f(X(t_0), \ldots, X(t_{n_{\text{times}}})) \mid \mathcal{F}_{t_0}) \approx \sum_{j=1}^{n_{\text{paths}}} f(X(t_0, \omega_j), \ldots, X(t_{n_{\text{times}}}, \omega_j)) \left(\frac{1}{n_{\text{paths}}}\right).$$

The generation of random numbers is discussed in Section B.1.

13.2.2 Weighted Monte Carlo Simulation

A generalization of the procedure is to generate the random numbers $\Delta W(t_i, \omega_j)$ not according to the distribution $\Delta W(t_i)$, which means that all paths ω_j are generated with nonuniform weights p_j ($\sum_{j=1}^{n_{\text{paths}}} p_j = 1$). In this case we call the simulation *weighted Monte Carlo simulation*. For the expectation we have

$$E^{\mathbb{P}}(f(X(t_0),\ldots,X(t_{n_{\text{times}}})) \mid \mathcal{F}_{t_0}) \approx \sum_{j=1}^{n_{\text{paths}}} p_j f(X(t_0,\omega_j),\ldots,X(t_{n_{\text{times}}},\omega_j)).$$

To summarize, the Monte Carlo simulation consists of the time-discrete process X in (13.6), represented over a discrete probability space $(\tilde{\Omega},\tilde{\mathcal{F}},\tilde{P})$, where

$$\tilde{\Omega} = \{\omega_1,\ldots,\omega_{n_{\text{paths}}}\} \subset \Omega, \quad \tilde{\mathcal{F}} = \sigma(\{\omega_j\} | j = 1,\ldots,n_{\text{paths}}), \quad \tilde{P}(\{\omega_j\}) = p_j.$$

13.2.3 Implementation

Figures 13.3 and 13.4 show an example for an object-oriented design. The figures follow the *Unified Modeling Language* (UML) 1.3; see [28].

The generation of a Monte Carlo simulation of a lognormal process is realized through the abstract base class LOGNORMALPROCESS. The class defines abstract methods for initial conditions, drift, and volatility. A specific model has to be derived from this class and implement the three methods. The abstract base class LOGNORMALPROCESS provides the implementation of the discretization scheme, using the methods for initial conditions, drift, and volatility.

The calculation of the Brownian increments, i.e., the random numbers, is given by an additional class: BROWNIANMOTION.

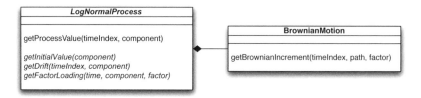

Figure 13.3. *UML Diagram: Monte Carlo simulation/lognormal process.*

13.2.3.1 Example: Valuation of a Stock Option under the Black-Scholes Model Using Monte Carlo Simulation

Consider the model from Chapter 4, the Black-Scholes model: We have to simulate the process

$$dS(t) = r(t)S(t)\,dt + \sigma(t)S(t)\,dW^{\mathbb{P}}(t) \quad \text{under the measure } \mathbb{Q}^N, \quad S(0) = S_0$$

together with the numéraire

$$dN(t) = r(t)N(t)\, dt, \qquad\qquad N(0) = 1,$$

which is not stochastic here. For this example we have to set $X = (X_1, X_2) = (S, N)$ in the previous section. We choose r and σ constant and apply the Euler scheme to $\log(S)$, following Section 13.1.2.3:

$$S(t_{i+1}) = S(t_i)\, \exp\left((r - \frac{1}{2}\sigma^2)\, \Delta t_i + \sigma\, \Delta W(t_i)\right), \qquad S(0) = S_0,$$

$$N(t_{i+1}) = N(t_i)\, \exp\left(r\, \Delta t_i\right), \qquad\qquad N(0) = 1.$$

In this example the time discretization does not introduce an approximation error, because we are in the special situation of Section 13.1.2.4. If $\omega_1, \ldots, \omega_{n_{\text{paths}}}$ are paths of a Monte Carlo simulation, then we have for the price V of a European option with maturity t_k and strike K

$$V(0) \approx N(0)\frac{1}{n_{\text{paths}}} \sum_{j=1}^{n_{\text{paths}}} \frac{\max\left(S(t_k, \omega_j) - K, 0\right)}{N(t_k)}.$$

We can extend the object-oriented design from Figure 13.3 to derive the class BLACKSCHOLESMODEL from the abstract base class LOGNORMALPROCESS. The class BLACKSCHOLESMODEL implements the methods providing the initial value (returning $S(0)$), the drift (returning r), and the factor loading (returning σ). In this context the factor loading is identical to the volatility.[2] In addition the class implements a method that returns the corresponding numéraire.

13.2.3.2 Separation of Product and Model

The evaluation of a derivative product, in our case a simple European option, is realized in its own class STOCKOPTION. This class does not communicate directly with the BLACKSCHOLESMODEL. Instead it expects an *interface* MONTECARLOSTOCKPROCESS-MODEL and the model implements this *interface*.

The interface MONTECARLOSTOCKPROCESSMODEL means that the stock model makes the stock process and the numéraire available to the stock product as a Monte Carlo simulation. All corresponding Monte Carlo evaluations of stock products expect this interface only. All corresponding Monte Carlo stock models implement this interface. This produces a separation of product and model. The model used to evaluate the products may be exchanged for another, as long as the interface is respected.

We will use this principle in the object-oriented design of the LIBOR market model, a multidimensional interest rate model. There we will reuse the classes BROWNIANMOTION and LOGNORMALPROCESS; see Section 19.6.

---

[2] In a multi factor model the factor loading is given by the square root of the covariance matrix.

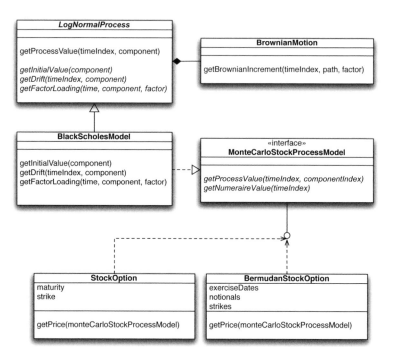

Figure 13.4. *UML Diagram: Evaluation under a Black-Scholes model via Monte Carlo simulation.*

13.2.3.3 Model-Product Communication Protocol

In designing a Monte Carlo simulation, one of the first steps is the design of the core data objects representing the realizations of the stochastic process. These are the data objects that are sent from the model to the product (via call to a method of the model). They define the *communication protocol* between model and product.

If we think of a more complex model, e.g., a model for a family of forward rates $\{L(T_i, T_{i+1}) \mid i = 0, 1, 2, \ldots, n - 1\}$, as given by the LIBOR market model, then the product needs access to a whole family of stochastic processes to process it.

The question we would like to briefly address here is how these data should be stored and in what order it is usually accessed and processed. Of course, the favored solution may vary with the specific application, so we shall remain on a fairly general level. More specifically we would like to consider the order in which data objects are aggregated. For example: Is it better to store a family of stochastic processes, each being a family of one-dimensional random variables (parametrized by simulation time), each being a set of realizations on a path, i.e.,

$$\{\{\{X_i(t_j, \omega_k) \mid k = 0, 1, 2, \ldots\} \mid j = 0, 1, 2, \ldots\} \mid i = 0, 1, 2, \ldots\},$$

or would we rather store a family of sample paths (with path index k) of a function of t_j (with time index j) each being vector values (with component index i), i.e.,

$$\{\{\{X_i(t_j, \omega_k) \mid i = 0, 1, 2, \ldots\} \mid j = 0, 1, 2, \ldots\} \mid k = 0, 1, 2, \ldots\}. \tag{13.7}$$

Examples of such data objects are a family of forward rates $X_i := L(T_i, T_{i+1})$, e.g., as modeled by a LIBOR market model, or a set of stock price processes X_i, for example, as underlyings of a basket.

One is tempted to believe that for a Monte Carlo simulation the paths ω_k are totally independent objects and thus that it would be reasonable to have the index k on the top-most level of aggregation/parametrization, as is the case in (13.7). Indeed, this would make it easy to parallelize the processing since the algorithm could be called with different subsets of paths. However, this is only possible for simple (say European) options.

The solution we recommend is to aggregate the data (from outer to inner objects) as a family of random variables parametrized by t_j, each random variable being a vector consisting of one-dimensional random variables parametrized by i, each one-dimensional random variable being represented by a vector of evaluations on sample paths parametrized by k, i.e.,

$$\{\{\{L(T_i, T_{i+1}; t_j, \omega_k) \mid k = 0, 1, 2, \ldots\} \mid i = 0, 1, 2, \ldots\} \mid j = 0, 1, 2, \ldots\}.$$

In other words: We build our data object or array (from inner to outer) as follows:

Core Object: Random Variable   The core object is a one-dimensional random variable evaluated on given sample paths,

$$x_k := X(\omega_k),$$

defining a vector of realizations $\hat{X} := (x_1, \ldots, x_m)^\top$. There are two major reasons for using this vector as a basic object:

- Product payoffs are functions of the underlyings, i.e., functions operating on random variables. This makes it possible to define the functions as functions acting on vectors (vector arithmetic), which greatly increases the readability of the code. Loops over the paths are hidden inside the methods acting on the random variable objects. For example, the payoff of an option $\max(S_T - X, 0)$ would appear as such in the code while the pathwise evaluation is hidden in the implementation of the max-function.

- For Bermudan options it is necessary to calculate conditional expectations. To do so, one needs access to the realizations on other paths, e.g., when using the regression method (see Section 15.10.1). To some extent, the pricing of Bermudan options breaks the naive parallelization of pricings through subsets of sample paths. See the discussion of the foresight bias in Section 15.9.

Aggregation 1: Vector of Random Variables of Same Simulation Time
When simulating multiple stochastic processes, like, for example, a basket of underlyings or a family of forward rates, access to the whole family for a fixed simulation time t is normally required. So on the next aggregation level the basic object thus is a vector of random variables sharing the same measurability property,

$$X_i \qquad \text{where } X_i \text{ is } \mathcal{F}_t\text{-measurable.}$$

This makes sense since we often build a new stochastic process by defining its time t value as a function of the time t value of other stochastic processes. We give two examples:

- At time t, the value of a basket of stocks is the sum of the underlyings S_T^i.

- At time t the swap rate is a function of forward rates $L_i(t)$.

Aggregation 2: Time-Discrete Stochastic Process   Aggregating these vectors of random variables over all simulation times is finally the complete description of the Monte Carlo simulation.

Storage, Access, and Processing   To summarize, we say that for most applications the storage should be allocated as a three-dimensional array

```
double[][][]    process;
```

where

```
double[] randomVariable = process[simulationTimeIndex][processIndex];
```

results in a reference (!) to a vector containing the sample paths of a random variable $X_j(t_i)$. If basic functionalities are implemented as methods acting on random variables, then we may work directly with this reference. It may be convenient to define this as a class, eventually endowing it with additional information, e.g.,

```
class RandomVariable {
    double      time;   // Filtration time (if used for stochastic processes)
    double[]    realizations;
}
```

Remark 181 (Counterexample):   There are applications where other storage layouts are advantageous. Path-dependent products such as Asian or lookback options usually require the application of a function to a path (parametrized by t_j). In such cases it may be convenient to work with (13.7).

Remark 182 (Performance and Readability of the Code):   The storage layout has an impact on the performance. Usually a large number of paths (10,000–100,000) for a few stochastic processes (1–100) given at a modest number of time discretization points (50–500) are considered. Therefore, it is sensible to use the random variable as the core object so that one needs to allocate only a few objects containing large continuous blocks of memory (which is more efficient than doing it vice versa) and one can optimize core methods which iterate often.[3]

Last but not least, the entire mathematical theory is built on random variables as the central modeling entity. Thus, using random variables as core objects will improve the readability of the code. Whenever code is developed as a collaborative effort, this should be considered as a top priority.

13.2.4 Review

Through the Monte Carlo simulation we can evaluate simple and pure path-dependent products. The Monte Carlo evaluation of derivatives where an expectation has to be

---

[3] Consideration of a large *single* (one-dimensional) array and working on it might be the most efficient implementation, but it will make it difficult to comply with basic principles of object-oriented design (like data hiding) and will most likely make the code difficult to maintain and extend.

calculated that is conditional to a future time t_i is nontrivial, e.g.,

$$\mathrm{E}\left(X(t_j) \mid \mathcal{F}_{t_i}\right)$$

for $t_j > t_i > t_0$.[4] Why this is nontrivial becomes apparent when we consider the filtration:

In a path simulation in general no two paths will have a common past. The reason is that the number of possible states $X(t_i)$ is much higher than the number of paths that are simulated. If no two paths have a common past, then we have the following filtration:

$$\tilde{\mathcal{F}}_{t_i} = \tilde{\mathcal{F}} \quad \text{for all } t_i > t_0$$

and

$$\tilde{\mathcal{F}}_{t_0} = \tilde{\Omega}.$$

The calculation of a condition expectation in a Monte Carlo simulation is not straightforward, since no suitable time discretization of the filtration $\mathcal{F}_{t_i}$ of X is given.[5] The filtration of the Monte Carlo simulation depicted in Figure 13.2 is given by

$$\tilde{\mathcal{F}}_{t_0} = \{\emptyset, \tilde{\Omega}\},$$
$$\tilde{\mathcal{F}}_{t_1} = \sigma(\{\{\omega_1\}, \{\omega_2\}, \{\omega_3\}, \{\omega_4\}\}),$$
$$\tilde{\mathcal{F}}_{t_2} = \sigma(\{\{\omega_1\}, \{\omega_2\}, \{\omega_3\}, \{\omega_4\}\}),$$
$$\tilde{\mathcal{F}}_{t_3} = \sigma(\{\{\omega_1\}, \{\omega_2\}, \{\omega_3\}, \{\omega_4\}\}).$$

For the path simulation depicted in Figure 13.5 the filtrations are not all the same. The filtration is given by

$$\tilde{\mathcal{F}}_{t_0} = \{\emptyset, \Omega\},$$
$$\tilde{\mathcal{F}}_{t_1} = \sigma(\{\{\omega_1, \omega_2\}, \{\omega_3\}, \{\omega_4, \omega_5, \omega_6\}\}),$$
$$\tilde{\mathcal{F}}_{t_2} = \sigma(\{\{\omega_1\}, \{\omega_2\}, \{\omega_3\}, \{\omega_4\}, \{\omega_5, \omega_6\}\}),$$
$$\tilde{\mathcal{F}}_{t_3} = \sigma(\{\{\omega_1\}, \{\omega_2\}, \{\omega_3\}, \{\omega_4\}, \{\omega_5\}, \{\omega_6\}\}),.$$

To achieve a time discretization of the filtration, we restrict the possible values of X in each simulation time step. We thus assume a discretization of state space.

---

[4] Such a conditional expectation would be necessary for the evaluation of a Bermudan option.
[5] In Chapter 15 we will present special methods for the evaluation of conditional expectation in a Monte Carlo simulation.

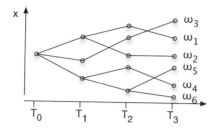

Figure 13.5. *Monte Carlo simulation.*

13.3 Discretization of State Space

13.3.1 Definitions

Instead of the time-discrete process

$$X(t_{i+1}) = X(t_i) + \mu(t_i, X(t_i))\, \Delta t_i + \sigma(t_i, X(t_i))\, \Delta W(t_i)$$

we will consider the time-discrete process

$$X(t_{i+1}) = X(t_i) + \mu(t_i, X(t_i))\, \Delta t_i + \sigma(t_i, X(t_i))\, \Delta B(t_i), \qquad (13.8)$$

where the increments $\Delta B(t_i)$ are random variables that take only a finite number of values and are mutually independent. According to Theorem 31 we can choose the $\Delta B(t_i)$ such that in the limit $\Delta t_i \to 0$ we recover a Brownian motion, i.e., for X we recover the original time-continuous process.

Using the increments $\Delta B(t_i)$ the process X from (13.8) can take on only a finite number of values too:

$$X(t_i) \in \{x_{t_i}^j \mid j = 0, \dots, n_{t_i} \; ; \; i = 0, \dots, n_{\text{times}}\}.$$

We denote this set as a *lattice*. Let

$$\{p_i^{j_1, j_2} \mid j_1 = 0, \dots, n_i \; ; \; j_2 = 0, \dots, n_{i+1} \; ; \; i = 0, \dots, n_{\text{times}} - 1\},$$

denote the transition probabilities of the $\Delta B(t_i)$'s, i.e., $p_i^{j_1, j_2} := P(X(t_{i+1}) = x_{t_{i+1}}^{j_2} \mid X(t_i) = x_{t_i}^{j_1})$. Depending on the probability distribution assumed for the $\Delta B(t_i)$, the matrix of transitional probabilities $(p_i^{j_1, j_2})_{j_1, j_2}$ from t_i to t_{i+1} is sparse. For a binomial tree, i.e., binomial distributed $\Delta B(t_i)$'s, in each row only two entries are nonzero.

195

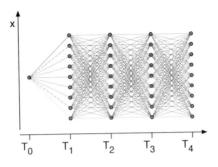

Figure 13.6. *Lattice.*

In the given situation of a discretized state space, it is far more efficient to store transition probabilities than to simulate all possible paths. For example, the lattice in Figure 13.6 exhibits 9 states at each of the 4 time steps. In this lattice there exist $9^4 = 6561$ paths; however, there are only $9 + 3 \times 9^2 = 252$ transition probabilities. The amount of memory needed to store all paths grows exponentially with time, whereas the amount of memory needed to store all states and transition probabilities shows only linear growth over time.

If we use binomial distributed increments $\Delta B(t_i)$, then we obtain a binomial tree. Figure 13.8 shows the paths that can be distinguished in the binomial tree depicted in Figure 13.7.

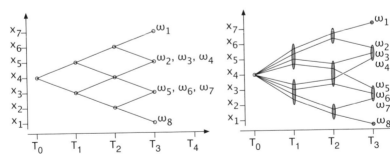

Figure 13.7. *Binomial tree*

Figure 13.8. *Paths of the binomial tree*

For the binomial tree the filtration is no longer trivial. It exhibits a hierarchy of refinements. For the binomial tree in Figure 13.7 the filtration is given by

$$\mathcal{F}_{t_0} = \{\emptyset, \Omega\},$$

$$\mathcal{F}_{t_1} = \sigma(\{\{\omega_1, \omega_2, \omega_3, \omega_5\}, \{\omega_4, \omega_6, \omega_7, \omega_8\}\}),$$

$$\mathcal{F}_{t_1} = \sigma(\{\{\omega_1, \omega_2\}, \{\omega_3, \omega_5\}, \{\omega_4, \omega_6\}, \{\omega_7, \omega_8\}\}),$$

$$\mathcal{F}_{t_3} = \sigma(\{\{\omega_1\}, \{\omega_2\}, \{\omega_3\}, \{\omega_4\}, \{\omega_5\}, \{\omega_6\}, \{\omega_7\}, \{\omega_8\}\}),$$

i.e., in addition we have a suitable time discretization of the filtration. This allows for a calculation of conditional expectations.

13.3.2 Backward Algorithm

Given a lattice we may calculate the conditional expectation of a function f of $X(t_{i+1})$, condition to $X(t_i) = x_{t_i}^{j_1}$ as

$$E^{\mathbb{P}}(f(X(t_{i+1})) \mid \{X(t_i) = x_{t_i}^{j_1}\}) \approx \sum_{j_2=1}^{n_{t_{i+1}}} p_i^{j_1, j_2} \, f(x_{t_{i+1}}^{j_2}).$$

A step-by-step application finally gives the expectation $E^{\mathbb{P}}(f(X(t_{i+1})) \mid \{X(t_0) = x_{t_0}^1\})$. Assuming that $X(t_0) = \text{const.} = x_{t_0}^1$, i.e., that the lattice has a single state in t_0, this corresponds to

$$E^{\mathbb{P}}(f(X(t_{i+1})) \mid \mathcal{F}(t_0)).$$

This procedure is called *backward algorithm*. If we consider the case of a numéraire N and a derivative product V given at time t_{i+1} as a function of the states $X(t_{i+1})$, then we find

$$\frac{V(t_i, x_{t_i}^{j_1})}{N(t_i, x_{t_i}^{j_1})} \approx \sum_{j_2=1}^{n_{t_{i+1}}} p_i^{j_1, j_2} \, \frac{V(t_{i+1}, x_{t_{i+1}}^{j_2})}{N(t_{i+1}, x_{t_{i+1}}^{j_2})}. \tag{13.9}$$

Thus, financial products which are functions of the states $X(t_i)$ are evaluated in a lattice by storing the numéraire-relative prices at the nodes $x_{t_{i+1}}^{j_2}$ and calculating the N-relative value in node $x_{t_i}^{j_1}$ via (13.9). The transition (13.9) is also called *rollback*.

13.3.3 Review

13.3.3.1 Path Dependencies

If a lattice with states $x_{t_i}^j$ and transition probabilities $p_i^{j_1, j_2}$ is set up, we are able to calculate (certain) conditional expectations. However, it is nontrivial to calculate path-dependent products, i.e., financial products that not only depend on the current state of the underlying but also on the history (the paths). The backward algorithm carries information from the future backward in time but cannot consider information from the past. In contrast the Monte Carlo simulation is a *forward algorithm* that carries information forward in time.

13.3.3.2 Course of Dimension

A further problem of lattices becomes apparent for vector values processes X. The number of possible transitions (and thus the amount of transition probabilities that must be stored) grows exponentially with the dimension of the vector X, and already for dimensions like 3 or higher the numerical calculation of the rollback (13.9) is critical with respect to the resources required (CPU time and memory).

13.4 Path Simulation through a Lattice: Two Layers

To calculate a path-dependent product in a lattice we may create a Monte Carlo simulation according to the state-discretized process (13.8). This may be depicted as a "Monte Carlo simulation through a lattice" or a second Monte Carlo layer laid over the lattice. See Figure 13.9.

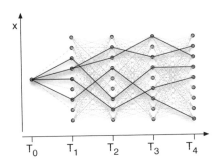

Figure 13.9. *Lattice with overlain Monte Carlo simulation*

 Further Reading: An in-depth discussion of numerical methods to approximate stochastic processes is to be found in [21]. ◁|

CHAPTER 14

Numerical Methods for Partial Differential Equations

The Feynman-Kač Theorem creates the link to partial differential equations (PDEs). The calculation of an expectation, i.e., the pricing of a derivative, becomes equivalent to the solution of the PDE:

$$
\frac{\partial u}{\partial t} + \sum_{i=1}^{d} \mu_i(t, x)\frac{\partial u}{\partial x_i} + \sum_{i,j=1}^{d} \gamma_{i,j}(t, x)\frac{\partial^2 u}{\partial x_i \partial x_j} = 0 \qquad \text{auf } I \times \Omega
$$

$$
u(T, \cdot) = \phi(\cdot) \qquad \text{on } \mathbb{R}^d.
$$

Endowed with the *pricing PDE* we can apply the proven numerical methods for the solution of partial differential equations, like *finite differences* and *finite elements*. In the context of PDEs, the binomial or trinomial tree is just a special variant of a finite difference method (namely an explicit Euler scheme). On the other hand, PDE implementations are just a special version of lattices.

The field of numerical methods for partial differential equations is huge. In this book we focus more on Monte Carlo methods, which, to some extent, have a much broader range of application. Here is a brief reference to some literature.

 Further Reading: Numerical methods for partial differential equations in the context of mathematical finance may be found in Günther and Jüngel [17], Seydel [32], and Wilmott [40]. A discussion of the implementation of the Cheyette model's PDE is given in Kohl-Landgraf [84].

◁|

CHAPTER 15

Pricing Bermudan Options in a Monte Carlo Simulation

15.1 Introduction

Let us first consider the simple case of a Bermudan option $V_{\text{berm}}(T_1, T_2)$ with two exercise dates only (Figure 15.1): The option holder has the right to receive an underlying $V_{\text{underl},1}$ in T_1 or wait and retain the right to either receive an underlying $V_{\text{underl},2}$ in T_2 or receive nothing. Put differently, the option holder has the choice of receiving the underlying value $V_{\text{underl},1}(T_1)$ or the value of an option $V_{\text{option}}(T_1)$ on $V_{\text{underl},2}(T_2)$. The Bermudan may be interpreted as an option on an option. In T_1 the optimal exercise is given by choosing the maximum value

$$\max(V_{\text{underl}}(T_1), V_{\text{option}}(T_1)), \tag{15.1}$$

where (having chosen a numéraire N)

$$\begin{aligned}
V_{\text{option}}(T_1) &= N(T_1) \, \text{E}^{\mathbb{Q}^N}\left(\frac{V_{\text{option}}(T_2)}{N(T_2)} \,\Big|\, \mathcal{F}_{T_1}\right) \\
&= N(T_1) \, \text{E}^{\mathbb{Q}^N}\left(\frac{\max(V_{\text{underl},2}(T_2), 0)}{N(T_2)} \,\Big|\, \mathcal{F}_{T_1}\right). \tag{15.2}
\end{aligned}$$

is the value of the option with exercise in T_2, evaluated in T_1.

Thus, to evaluate the exercise criterion (15.1) it is necessary to calculate a conditional expectation. The calculation of a conditional expectation within a Monte Carlo simulation is a nontrivial problem. The two main issues are *complexity* and *foresight bias*, which we will illustrate in Section 15.5 and 15.6. In the following section we will present methods to efficiently estimate conditional expectations and/or

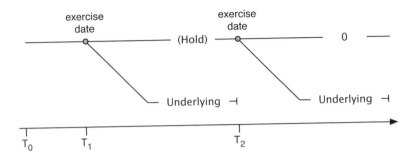

Figure 15.1. *A simple Bermudan option with two exercise dates.*

the Bermudan exercise criterion within Monte Carlo simulation. The application to the pricing of Bermudan options is exemplary. The methods presented are not limited to Bermudan option pricing.

15.2 Bermudan Options: Notation

Reconsider the general definition of a Bermudan option (see Definition 146). Let $\{T_i\}_{i=1,\ldots,n}$ denote a set of exercise dates and $\{V_{\mathrm{underl},i}\}_{i=1,\ldots,n}$ a corresponding set of underlyings. The Bermudan option is the right to receive at one and only one time T_i the corresponding underlying $V_{\mathrm{underl},i}$ (with $i = 1,\ldots,n$) or receive nothing.

At each exercise date T_i, the optimal strategy compares the value of the product upon exercise with the value of the product upon nonexercise and chooses the larger one. Thus the value of the Bermudan is given recursively

$$\underbrace{V_{\mathrm{berm}}(T_i,\ldots,T_n;T_i)}_{\substack{\text{Bermudan with}\\ \text{exercise dates}\\ T_i,\ldots,T_n}} := \max \big(\underbrace{V_{\mathrm{berm}}(T_{i+1},\ldots,T_n;T_i)}_{\substack{\text{Bermudan with}\\ \text{exercise dates}\\ T_{i+1},\ldots,T_n}} , \underbrace{V_{\mathrm{underl},i}(T_i)}_{\substack{\text{Product}\\ \text{received upon}\\ \text{exercise in } T_i}} \big), \qquad (15.3)$$

where $V_{\mathrm{berm}}(T_n;T_n) := 0$ and $V_{\mathrm{underl},i}(T_i)$ denotes the value of the underlying $V_{\mathrm{underl},i}$ at exercise date T_i.

15.2.1 Bermudan Callable

The most common Bermudan option is the Bermudan callable[1]. For a Bermudan Callable the underlyings consist of periodic payments X_k and differ only by the start of the periodic payments. The value of the underlying then becomes

$$V_{\text{underl},i}(T_i) = V_{\text{underl}}(T_i, \ldots, T_n; T_i) = N(T_i) \, \mathrm{E}^{\mathbb{Q}^N} \left(\sum_{k=i}^{n-1} \frac{X_k}{N(T_{k+1})} \,\Big|\, \mathcal{F}_{T_i} \right)$$

Here X_k denotes a payment fixed in T_k (i.e., $\mathcal{F}_{T_k}$-measurable) and paid in T_{k+1}. This is the usual setup for interest rate Bermudan callables. Other payment dates are a minor modification; they simply change the time argument of the numéraire. For the value upon nonexercise we have as before

$$V_{\text{berm}}(T_{i+1}, \ldots, T_n; T_i) = N(T_i) \mathrm{E}^{\mathbb{Q}^N} \left(\frac{V_{\text{berm}}(T_{i+1}, \ldots, T_n; T_{i+1})}{N(T_{i+1})} \,\Big|\, \mathcal{F}_{T_i} \right).$$

If the value of the underlying cannot be expressed by means of an analytical formula, *two* conditional expectations have to be evaluated to calculate the exercise strategy (15.3).

15.2.2 Relative Prices

Since the conditional expectation of a numéraire-relative price is a numéraire-relative price, the presentation will be simplified by considering the numéraire-relative quantities. We will therefore define

$$\tilde{V}_{\text{underl},i}(T_j) := \frac{V_{\text{underl},i}(T_j)}{N(T_j)} \quad \text{and} \quad \tilde{V}_{\text{berm},i}(T_j) := \frac{V_{\text{berm}}(T_i, \ldots, T_n; T_j)}{N(T_j)},$$

thus we have

$$\tilde{V}_{\text{berm},n} \equiv 0,$$
$$\tilde{V}_{\text{berm},i+1}(T_i) = \mathrm{E}^{\mathbb{Q}^N} (\tilde{V}_{\text{berm},i+1}(T_{i+1}) \mid \mathcal{F}_{T_i}),$$
$$\tilde{V}_{\text{berm},i}(T_i) = \max \left(\tilde{V}_{\text{berm},i+1}(T_i), \; \tilde{V}_{\text{underl},i}(T_i) \right),$$

and in the case of a Bermudan callable

$$\tilde{V}_{\text{underl},i}(T_j) = \mathrm{E}^{\mathbb{Q}^N} \left(\sum_{k=i}^{n-1} \tilde{X}_k(T_{k+1}) \,\Big|\, \mathcal{F}_{T_j} \right) \quad \text{where} \quad \tilde{X}_i(T_{i+1}) := \frac{X_i}{N(T_{i+1})}.$$

---

[1] See Remark 154 on the naming *Bermudan callable*.

The relative prices are marked by a tilde.

Remark 183 (Notation): The processes $t \mapsto \tilde{V}_{\mathrm{underl},i}(t)$ and $t \mapsto \tilde{V}_{\mathrm{berm},i}(t)$ are $\mathcal{F}_t$ conditional expectations of $\tilde{V}_{\mathrm{underl},i}(T_i)$ and $\tilde{V}_{\mathrm{berm},i}(T_i)$, respectively, and thus martingales by definition. The time-discrete processes $i \mapsto \tilde{V}_{\mathrm{underl},i}(T_i)$, $i \mapsto \tilde{V}_{\mathrm{berm},i}(T_i)$ consist of different products at different times and are thus not normally time-discrete martingales.

15.3 Bermudan Option as Optimal Exercise Problem

A Bermudan option consists of the right to receive one (and only one) of the underlyings $V_{\mathrm{underl},i}$ at the corresponding exercise date T_i. The recursive definition (15.3) represents the optimal exercise strategy in each exercise time. We formalize this optimal exercise strategy:

For a given path $\omega \in \Omega$ let

$$T(\omega) := \min\{T_i : V_{\mathrm{berm},i+1}(T_i, \omega) < V_{\mathrm{underl},i}(T_i, \omega)\},$$

and

$$\eta : \{1, \ldots, n-1\} \times \Omega \to \{0, 1\}, \qquad \eta(i, \omega) := \begin{cases} 1 & \text{if } T_i \geq T(\omega) \\ 0 & \text{else.} \end{cases}$$

The definitions of T and η give equivalent descriptions of the exercise strategy: $T(\omega)$ is the optimal exercise time on a given path ω; $\eta(\cdot, \omega)$ is an indicator function which changes from 0 to 1 at the time index i corresponding to $T_i = T(\omega)$. The boundary $\partial\{\eta = 1\}$ of the set $\{\eta = 1\}$ is termed the exercise boundary. It should be noted that $\eta(k)$ is $\mathcal{F}_{T_k}$-measurable.

15.3.1 Bermudan Option Value as Single (Unconditioned) Expectation: The Optimal Exercise Value

With the definition of the optimal exercise strategy T (or η) it is possible to define a random variable which allows the Bermudan option value to be expressed as a single (unconditioned) expectation. With

$$\tilde{U}(T_i) := \tilde{V}_{\mathrm{underl},i}(T_i) \quad i = 1, \ldots, n$$

denoting the relative price of the i-th underlying; upon its exercise date T_i we have for the Bermudan value

$$\tilde{V}_{\mathrm{berm}}(T_0) = \mathrm{E}^{\mathbb{Q}}\left(\tilde{U}(T) \mid \mathcal{F}_{T_0}\right).$$

For the Bermudan callable we may alternatively write

$$\tilde{V}_{\text{berm}}(T_0) = \mathrm{E}^{\mathbb{Q}}\left(\sum_{k=1}^{n-1} \tilde{X}_k(T_{k+1})\, \eta(k)\,\Big|\, \mathcal{F}_{T_0}\right).$$

The random variable $\tilde{U}(T)$ can be calculated directly using the *backward algorithm*. We will look at this in the next section and conclude by giving $\tilde{U}(T)$ a name.

Definition 184 (Option Value upon Optimal Exercise):
Let $\tilde{U}$ be the stochastic process whose time t value $U(t)$ is the (numéraire-relative) option value received upon exercise in t. Let T be the optimal exercise strategy. The random variable $\tilde{U}(T)$, where

$$\tilde{U}(T)[\omega] := U(T(\omega), \omega)$$

is the (numéraire-relative) *option value received upon optimal exercise.* The (numéraire-relative) Bermudan option value is given by $\mathrm{E}^{\mathbb{Q}}(\tilde{U}(T)\,|\,\mathcal{F}_{T_0})$.

Thus the value of $V_{\text{berm}}(T_1, \ldots, T_n)$ can be expressed through a single expectation conditioned to T_0 and does not need an expectation conditional to a later time to be calculated, *if* we have the optimal exercise date $T(\omega)$ (and thus $\eta(\cdot, \omega)$) for any path ω.

Remark 185 (Stopped Process): The random variable $\tilde{U}(T)$ is termed a *stopped process*. $\tilde{U}$ is a stochastic process and T is a random variable with the interpretation of a (stochastic) time. Furthermore T is a stopping time; see Definition 197. Here the stochastic process $\tilde{U}$ is the family of underlyings received upon exercise, parametrized by exercise time, and T is the optimal exercise time. Thus $\tilde{U}(T)$ is the underlying received upon optimal exercise. All quantities are stochastic.

15.4 Bermudan Option Pricing—The Backward Algorithm

The random variable $\tilde{U}(T)$ can be derived in a Monte Carlo simulation through the backward algorithm, *given* the exercise criterion (15.3), i.e., the conditional expectation. The algorithm consists of the application of the recursive definition of the Bermudan value in (15.3) with a slight modification. Let:

Induction start:

$$\tilde{U}_{n+1} \equiv 0.$$

Induction step $i + 1 \to i$ **for** $i = n, \ldots, 1$**:**

$$\tilde{U}_i = \begin{cases} \tilde{U}_{i+1} & \text{if } \tilde{V}_{\text{underl},i}(T_i) < \mathrm{E}^{\mathbb{Q}}(\tilde{U}_{i+1} | \mathcal{F}_{T_i}) \\ \tilde{V}_{\text{underl},i}(T_i) & \text{else.} \end{cases}$$

From the Tower law[2] $\mathrm{E}^{\mathbb{Q}}(\tilde{U}_{i+1} | \mathcal{F}_{T_i}) = \mathrm{E}^{\mathbb{Q}}(\tilde{V}_{\text{berm},i+1}(T_i) | \mathcal{F}_{T_i})$ by induction and thus

$$\tilde{V}_{\text{berm}}(T_1, \ldots, T_n, T_0) = \mathrm{E}^{\mathbb{Q}}(\tilde{U}_1 | \mathcal{F}_{T_0}) \tag{15.4}$$

and $\tilde{U}_1 = \tilde{U}(T)$ with the notation from the previous section.

 Interpretation: The recursive definition of $\tilde{U}_i$ differs from the recursive definition of $\tilde{V}_{\text{berm},i}(T_i)$. We have

$$\tilde{U}_i = \begin{cases} \tilde{U}_{i+1} & \text{if } \tilde{V}_{\text{underl},i}(T_i) < \mathrm{E}^{\mathbb{Q}}(\tilde{U}_{i+1} | \mathcal{F}_{T_i}) \\ \tilde{V}_{\text{underl},i}(T_i) & \text{else,} \end{cases}$$

and

$$\tilde{V}_{\text{berm},i}(T_i) = \begin{cases} \mathrm{E}^{\mathbb{Q}}(\tilde{V}_{\text{berm},i+1}(T_{i+1}) | \mathcal{F}_{T_i}) & \text{if } \tilde{V}_{\text{underl},i}(T_i) < \mathrm{E}^{\mathbb{Q}}(\tilde{U}_{i+1} | \mathcal{F}_{T_i}) \\ \tilde{V}_{\text{underl},i}(T_i) & \text{else.} \end{cases}$$

There is a subtle but crucial difference. While both definitions give the Bermudan option value (through application of (15.4)), the definition of $\tilde{U}_i$ requires the conditional expectation operator only to calculate the exercise criterion.

Since a Monte Carlo simulation requires advanced methods to obtain an (often not very accurate) estimate for the conditional expectation, it is important to reduce their use.

Note that $\tilde{V}_{\text{berm},i}(T_i)$ is $\mathcal{F}_{T_i}$-measurable by definition as a $\mathcal{F}_{T_i}$ conditional expectation, while all $\tilde{U}_i$ are at most $\mathcal{F}_{T_n}$-measurable since they are defined pathwise from $\mathcal{F}_{T_k}$-measurable random variables $\tilde{V}_{\text{underl},k}(T_k)$ for $i \leq k \leq n$. ◁|

The pricing of a Bermudan option may thus be reduced to either the calculation of conditional expectations or to the calculation of the optimal exercise strategy T.

As a motivation, in Sections 15.5 and 15.6 we will look at two methods which are not suitable for calculating conditional expectations.

---

[2] See Exercise 2 on page 479.

15.5 Resimulation

Let us consider the simplified example of a Bermudan option as given in Section 15.1. If no analytical calculation of the conditional expectation (15.2) is possible and if Monte Carlo is the numerical tool for calculating expectations, the straightforward way to calculate the conditional expectation is to create in T_1 a new Monte Carlo simulation (conditioned) on each path—see Figure 15.2. This leads to a much higher number of total simulation paths needed.

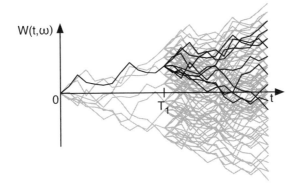

Figure 15.2. *Brute-force calculation of the conditional expectation by (pathwise) resimulation.*

If one considers more than one exercise date (option on option on option...), this method becomes particularly impractical. The required number of paths, i.e., the complexity of the algorithm and thus the calculation time, grows exponentially with the number of exercise dates. This creates the need for efficient alternatives.

 Interpretation: The calculation of conditional expectation in a path simulation requires further measures since the path simulation does not offer a suitable discretization of the filtration. ◁

15.6 Perfect Foresight

If one refuses to use a full resimulation and sticks to the paths generated in the original simulation, then one effectively estimates the conditional expectation by a single path, namely by

$$\mathrm{E}^{Q}\left(\frac{V(T_2)}{N(T_2)} \mid \mathcal{F}_{T_1}\right) \approx \frac{V(T_2)}{N(T_2)}.$$

Basically, this is a limit case of the resimulation where each resimulation consists of a single path only, namely the one of the original simulation. If this estimate is used in the exercise criterion, the exercise will be *superoptimal* since it is based on future information that would be unknown otherwise.

The exercise criterion at time T_1 may only depend on information available in T_1, i.e., on $\mathcal{F}_{T_1}$-measurable random variables. The estimate $\frac{V(T_2)}{N(T_2)}$ is *not* $\mathcal{F}_{T_1}$-measurable.

For an illustration of the superoptimality, consider the simulation consisting of two paths; see Figure 15.3. Both paths are identical on $[0, T_1]$, i.e., $\mathcal{F}_{T_1} = \{\emptyset, \Omega\} = \{\emptyset, \{\omega_1, \omega_2\}\}$. We consider the option V to receive either $S(T_1) = 2$ at time T_1 or $S(T_2) \in \{1, 4\}$ at a later time T_2. The random variable $\eta : \Omega \to \{0, 1\}$ denotes the exercise strategy for T_1: It is 1 on paths that exercise in T_1, otherwise 0. With *perfect foresight* the superoptimal exercise strategy is $T(\omega_1) = T_2$, $T(\omega_2) = T_1$, i.e., $\eta(\omega_1) = 0$, $\eta(\omega_2) = 1$, and an average value of $V(T_0) = \frac{1}{2}(4 + 2) = \frac{6}{2}$ will be received. Note that then η is not $\mathcal{F}_{T_1}$-measurable. The exercise decision is made in T_1 with knowledge of the future outcome. If we restrict the exercise strategy to the set of $\mathcal{F}_{T_1}$-measurable random variables, we either get $\frac{1}{2}(4 + 1) = \frac{5}{2}$ using $\eta \equiv 0$ or $\frac{1}{2}(2 + 2) = \frac{4}{2}$ using $\eta \equiv 1$. Thus the optimal, $\mathcal{F}_{T_1}$-measurable (and thus admissible) exercise strategy is $\eta(\omega_1) = \eta(\omega_2) = 0$.

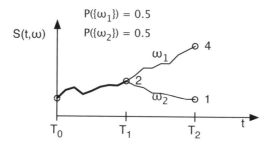

Figure 15.3. *Illustration of perfect foresight.*

Perfect foresight is not a suitable method for estimating conditional expectation and calculating the exercise criterion.

15.7 Conditional Expectation as Functional Dependence

Let us reconsider the calculation of the conditional expectation through brute-force resimulation as described in Section 15.5 and depicted in Figure 15.2. On each path of the original simulation a resimulation has to be created. These resimulations differ in their initial conditions (e.g., the value $S(T_1)$ in a simulation of a stock price following a Black-Scholes model, or the values $L_i(T_1)$ in a simulation of forward rates following a LIBOR market model). The initial conditions are $\mathcal{F}_{T_1}$-measurable random variables (known as of T_1). Thus the conditional expectation is a function of these initial conditions (and possibly other model parameters known in T_1). If it is known that the conditional expectation is a function of an $\mathcal{F}_{T_1}$-measurable random variable Z (we assume here that $Z : \Omega \to \mathbb{R}^d$ with some d), we have

$$\mathrm{E}^{\mathbb{Q}^N}\left(\frac{V(T_2)}{N(T_2)} \mid \mathcal{F}_{T_1}\right) = \mathrm{E}^{\mathbb{Q}^N}\left(\frac{V(T_2)}{N(T_2)} \mid Z\right); \qquad (15.5)$$

see Figure 15.4.

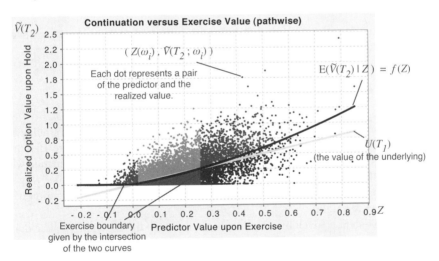

Figure 15.4. *Predictor variable versus realized value (continuation value): A diagram showing the path value of the predictor variable $Z(\omega_i)$ and the path value of $\tilde{V}(T_2; \omega_i) = \frac{V(T_2;\omega_i)}{N(T_2;\omega_i)}$. The conditional expectation is a function of Z dividing the cloud of dots.*

Interpretation: If the random variable Z is such that $\mathcal{F}_{T_1}$ is the smallest σ field with respect to which Z is measurable (i.e., we have $Z^{-1}(\mathcal{B}(\mathbb{R}^q)) = \mathcal{F}_{T_1}$), then Equation (15.5) is merely the definition of an expectation conditioned on a random variable. If, however, the conditional expectation on the left hand side (i.e., $\frac{V(T_1)}{N(T_1)}$) is known to be measurable with respect to a smaller σ field (e.g., because its functional depends on a smaller set of random variables), then it might be advantageous to use the right-hand side representation. This representation is also useful for deriving an approximation, e.g., if the functional dependence with respect to one component of Z is known to be weak and thus neglectable.

Example: Consider a LIBOR market model with stochastic processes for the forward rates $L_1, L_2, \ldots L_n$. In T_1 we wish to calculate the conditional expectation of a derivative with a numéraire-relative payoff that depends on $L_2, \ldots, L_k$ only (e.g., on a swap rate). While the filtration $\mathcal{F}_{T_1}$ is generated by the full set of forward rates $L_1(T_1), L_2(T_1), \ldots L_n(T_1)$ it is sufficient to know $L_2(T_1), \ldots, L_k(T_1)$ to describe the conditional expectation (i.e., the conditional value of the product).   ◁|

We will now describe methods that derive the functional dependence of the conditional expectation from a given set of random variables.

15.8 Binning

In a path simulation the approximation of $E^Q\left(\frac{V(T_2)}{N(T_2)} \mid Z\right)$ will be given by averaging all paths for which Z attains the same value. For the simple example in Figure 15.3 this would remove the *perfect foresight* since $S(T_1)^{-1}(2) = \Omega$. In general the situation will be such that there are no two or more paths for which Z attains the same value—apart from the construction of the unfeasible resimulation. Thus this approximation will show a perfect foresight.

An improvement is given by a *binning* method, where the averaging will be done over those paths for which Z lies in a neighborhood (*bin*). If the quantities are continuous, we have

$$E^Q\left(\frac{V(T_2)}{N(T_2)} \mid Z\right)[\omega] \approx E^Q\left(\frac{V(T_2)}{N(T_2)} \mid Z \in U_\epsilon(Z(\omega))\right),$$

where $U_\epsilon(Z(\omega)) := \{y \mid \|Z(\omega) - y\| < \epsilon\}$.

Instead of defining a bin $U_\epsilon(Z(\omega))$ for each path ω, it is more efficient to start with a partition of $Z(\Omega)$ into a finite set of disjoint bins $U_i \subset Z(\Omega)$. The approximation of the conditional expectation

$$\mathrm{E}^{Q}\left(\frac{V(T_2)}{N(T_2)} \,\Big|\, Z(\omega)\right)$$

will then be given by

$$H_i := \mathrm{E}^{Q}\left(\frac{V(T_2)}{N(T_2)} \,\Big|\, Z \in U_i\right)$$

where U_i denotes the set with $Z(\omega) \in U_i$; see Figure 15.5.

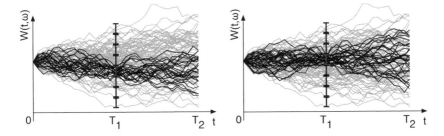

Figure 15.5. *Calculation of the conditional expectation by binning: Neighboring paths, i.e., paths which belong to the same bin, are bundled. The bins are defined by means of the $\mathcal{F}_{T_1}$-measurable predictor variable Z. The figure shows the special case $Z = W(T_1)$.*

Example: Pricing of a Simple Bermudan Option on a Stock

We illustrate the method in a simple Black-Scholes model for a stock S. In T_1 we wish to evaluate the option of receiving $N_1\ (S(T_1) - K_1)$ in T_1 or receiving $N_2\ \max(S(T_2) - K_2, 0)$ at later time T_2 (where N_1, N_2 (notional), K_1, K_2 (strike) are given). The optimal exercise in T_1 compares the exercise value with the value of the T_2 option, i.e.,

$$\mathrm{E}^{Q}\left(\frac{N_2\ \max(S(T_2) - K_2, 0)}{N(T_2)} \,\Big|\, \mathcal{F}_{T_1}\right).$$

From the model specification, e.g., here a Black-Scholes model

$$\mathrm{d}S(t) = r\,S(t)\,\mathrm{d}t + \sigma S(t)\,\mathrm{d}W^{Q}(t), \qquad N(t) = \exp(r\,t),$$

it is obvious that the price of the T_2 option seen in T_1 is a given function $S(T_1)$ and the given model parameters (r, σ). Thus it is sufficient to calculate

$$\mathrm{E}^{Q}\left(\frac{N_2\ \max(S(T_2) - K_2, 0)}{N(T_2)} \,\Big|\, S(T_1)\right).$$

211

In this example the functional dependence is known analytically. It is given by the Black-Scholes formula (4.3). Nevertheless we use the binning to calculate an approximation to the conditional expectation. If we plot

$$\underbrace{\frac{N_2 \ \max(S(T_2, \omega_i) - K_2, 0)}{N(T_2)}}_{\text{Continuation Value}} \quad \text{as a function of} \quad \underbrace{S(T_1, \omega_i)}_{\text{Underlying}}$$

we obtain the scatter plot in Figure 15.6, left. For a given $S(T_1)$ none or very few values of the *continuation values* exist. An estimate is not possible or else exhibits a foresight bias. For an interval $[S_1 - \epsilon, S_1 + \epsilon]$ with sufficiently large ϵ we have enough values to estimate

$$E^Q \left(\frac{N_2 \ \max(S(T_2) - K_2, 0)}{N(T_2)} \ \Big| \ S(T_1) \in [S_1 - \epsilon, S_1 + \epsilon] \right)$$

which in turn may be used as an estimate of

$$E^Q \left(\frac{N_2 \ \max(S(T_2) - K_2, 0)}{N(T_2)} \ \Big| \ S(T_1) = S_1 \right).$$

In Figure 15.6, right, we calculate this estimate for $S_1 = 1$ and $\epsilon = 0.05$.

15.8.1 Binning as a Least-Square Regression

Consider the binning again: As an estimate of the conditional expectation $E^Q \left(\frac{V(T_2)}{N(T_2)} \mid Z(\omega) \right)$ we calculated the conditional expectation

$$H_i := E^Q \left(\frac{V(T_2)}{N(T_2)} \ \Big| \ Z \in U_i \right) \tag{15.6}$$

given a *bin* U_i with $Z(\omega) \in U_i$.

For the expectation operator E^Q an alternative characterization may be used:

Lemma 186 (Characterization of the Expectation as Least-Square Approximation): The expectation of a random variable X is the number h for which $X - h$ has the smallest variance (i.e., $L_2(\Omega)$ norm).

Proof: Let X be a real-valued random variable. Then we have for any $h \in \mathbb{R}$

$$E((X - h)^2) = E(X^2) - 2 \, E(X) \, h + h^2 =: f(h).$$

Since $f' = 0 \Leftrightarrow h = E(X)$ and $f'' = 1 \geq 0$, we have that f attains its minimum in $h = E(X)$. For vector-valued random variables this follows componentwise. The same result holds for conditional expectations. $\qquad \Box$

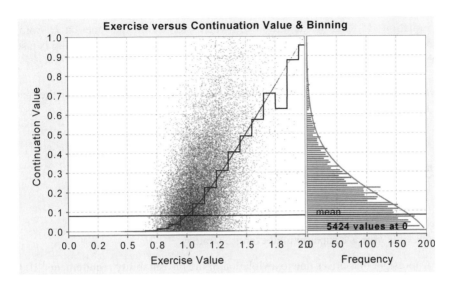

Figure 15.6. *The continuation value as a function of the underlying (spot value) and the calculation of the conditional expectation by a binning*

Using Lemma 186 we can write (15.6) as a minimization problem:

$$E^Q\left(\left(\frac{V(T_2)}{N(T_2)} - H_i\right)^2 \Big| Z \in U_i\right) = \min_{G \in \mathbb{R}} E^Q\left(\left(\frac{V(T_2)}{N(T_2)} - G\right)^2 \Big| Z \in U_i\right).$$

For disjoint bins U_i this may be written in a single minimization problem for the vector $(H_i)_{i=1,\dots}$:

$$\sum_i E^Q\left(\left(\frac{V(T_2)}{N(T_2)} - H_i\right)^2 \Big| Z \in U_i\right) = \min_{G_i \in \mathbb{R}} \sum_i E^Q\left(\left(\frac{V(T_2)}{N(T_2)} - G\right)^2 \Big| Z \in U_i\right). \quad (15.7)$$

This condition admits an alternative interpretation: H_i represents the piecewise constant function (constant on U_i) with the minimal distance from $\frac{V(T_2)}{N(T_2)}$ in the least-square sense.

Let $\mathcal{H}$ be the space of functions $H : \Omega \to \mathbb{R}$ being constant on the bins $Z^{-1}(U_i)$.[3] Let $H \in \mathcal{H}$ with $H(\omega) := H_i$ for $\omega \in Z^{-1}(U_i)$. Then (15.7) is equivalent to

$$E^Q\left(\left(\frac{V(T_2)}{N(T_2)} - H\right)^2 \Big| Z\right) = \min_{G \in \mathcal{H}} E^Q\left(\left(\frac{V(T_2)}{N(T_2)} - G\right)^2 \Big| Z\right). \quad (15.8)$$

---

[3] Note that the bins U_i were defined as subsets of $Z(\Omega)$, whereas here we consider H as a function on Ω.

Equation (15.8) is the definition of a regression: Find the function H from a function space $\mathcal{H}$ with minimum distance to $\frac{V(T_2)}{N(T_2)}$ in the L_2 norm. Binning is just a special choice of functional space:

Lemma 187 (Binning as L_2 Regression):   Binning is an L_2 regression on the space of functions being piecewise constant on U_i.

15.9 Foresight Bias

Definition 188 (Foresight Bias (Definition 1)):
A *foresight bias* is a superoptimal exercise strategy.

A foresight bias arises due to a violation of the measurability requirements: If the exercise decision in T_i is based on a random variable which is not $\mathcal{F}_{T_i}$-measurable, the exercise may be superoptimal, i.e., better than if based on the information theoretically available ($\mathcal{F}_{T_i}$). If we use the same Monte Carlo simulation to first estimate the exercise criterion and then use this criterion to price the derivative, we will definitely generate a foresight bias. In this case the foresight bias is created by the Monte Carlo error of the estimate, which is in general not $\mathcal{F}_{T_i}$-measurable. The existence of this problem becomes obvious if we consider a limit case of binning where each bin contains a single path only. Here we would have perfect foresight.

If our exercise criterion at time T_i uses only $\mathcal{F}_{T_i}$-measurable random variables, then there is—in theory—no foresight bias. If, however, the exercise criterion is calculated within a Monte Carlo simulation, the Monte Carlo error of the calculation represents a non-$\mathcal{F}_{T_i}$-measurable random variable; thus it induces a foresight bias. In this case we can give an alternative definition for the foresight bias:

Definition 189 (Foresight Bias (Definition 2)):
The *foresight bias* is the value of the option on the Monte Carlo error.

As the number of paths increases the foresight bias introduced by binning converges to zero since the Monte Carlo error with respect to a bin converges to zero.

A general solution to the problem of a foresight bias is given by using two independent Monte Carlo simulations: One to estimate the exercise criterion (for binning this is given by the H_i corresponding to the U_i's), the other to apply the criterion in pricing. This is a numerical removal of the foresight bias. In [67] an analytic formula for the (Monte Carlo error-induced) foresight bias is derived. It can be used to correct the foresight bias analytically.

15.10 Regression Methods—Least-Square Monte Carlo

 Motivation (Disadvantage of Binning): The partition of the state space $Z(\Omega)$ into a finite number of bins results in a piecewise constant approximation of the conditional expectation. An obvious improvement would be to approximate the conditional expectation by some smooth function of the state variable Z.

The considerations in Section 15.8.1 suggest a simple yet powerful improvement to the binning: The function giving our estimate for the conditional expectation is defined by a least-square approximation (regression). ◁|

15.10.1 Least-Square Approximation of the Conditional Expectation

Let us start with a fairly general definition of the *least-square approximation* of the conditional expectation of random variable U.

Definition 190 (Least-Square Approximation of the Conditional Expectation): ⌐
Let $(\Omega, \mathcal{F}, \mathbb{Q}, \{\mathcal{F}_t\})$ be a filtered probability space and V an $\mathcal{F}_{T_1}$-measurable random variable defined as the conditional expectation of U:

$$V = E^{\mathbb{Q}}(U \mid \mathcal{F}_{T_1}),$$

where U is at least $\mathcal{F}$-measurable. Furthermore let $Y := (Y_1, \ldots, Y_p)$ be a given $\mathcal{F}_{T_1}$-measurable random variable and $f : \mathbb{R}^p \times \mathbb{R}^q$ a given function. Let $\Omega^* = \{\omega_1, \ldots \omega_m\}$ be a drawing from Ω (e.g., a Monte Carlo simulation corresponding to $\mathbb{Q}$) and $\alpha^* := (\alpha_1, \ldots, \alpha_q)$ such that

$$\|U - f(Y, \alpha^*)\|_{L_2(\Omega^*)} = \min_{\alpha} \|U - f(Y, \alpha)\|_{L_2(\Omega^*)}$$

where $\|U - f(Y, \alpha^*)\|_{L_2(\Omega^*)}^2 = \sum_{j=1}^{m} (U(\omega_j) - f(Y(\omega_j), \alpha^*))^2$. We set

$$V^{\mathrm{LS}} := f(Y, \alpha^*).$$

The random variable V^{LS} is $\mathcal{F}_{T_1}$-measurable. It is defined over Ω and a *least-square* approximation of V on Ω^*. ⌐

The approach of Carriere [59], Longstaff and Schwartz [86] uses a function f with $q = p$ and

$$f(y_1, \ldots, y_p, \alpha_1, \ldots, \alpha_p) := \sum_{i=1}^{p} \alpha_i \, y_i,$$

such that α^* may be calculated analytically as a linear regression.

Lemma 191 (Linear Regression): Let $\Omega^* = \{\omega_1, \ldots, \omega_m\}$ be a given sample space, $V : \Omega^* \to \mathbb{R}$ and $Y := (Y_1, \ldots, Y_p) : \Omega^* \to \mathbb{R}^p$ given random variables. Furthermore let

$$f(y_1, \ldots, y_p, \alpha_1, \ldots, \alpha_p) := \sum \alpha_i y_i.$$

Then we have for any α^* with $X^T X \alpha^* = X^T v$

$$\|V - f(Y, \alpha^*)\|_{L_2(\Omega^*)} = \min_\alpha \|V - f(Y, \alpha)\|_{L_2(\Omega^*)},$$

where

$$X := \begin{pmatrix} Y_1(\omega_1) & \ldots & Y_p(\omega_1) \\ \vdots & & \vdots \\ Y_1(\omega_m) & \ldots & Y_p(\omega_m) \end{pmatrix}, \qquad v := \begin{pmatrix} V(\omega_1) \\ \vdots \\ V(\omega_m) \end{pmatrix}.$$

If $(X^T X)^{-1}$ exists, then $\alpha^* := (X^T X)^{-1} X^T v$.

Proof: See Appendix B.5. □|

Definition 192 (Basis Functions): ⌐

The random variables $Y_1, \ldots, Y_p$ of Lemma 191 are called *basis functions* (*explanatory variables*). ⌐

15.10.2 Example: Evaluation of a Bermudan Option on a Stock (Backward Algorithm with Conditional Expectation Estimator)

Consider a simple Bermudan option on a stock. The Bermudan should allow exercise at times $T_1 < T_2 < \ldots T_n$. Upon exercise in T_i the holder of the option will receive

$$N_i \left(S(T_i) - K_i\right)$$

once, but nothing if no exercise is made.

We will apply the *backward algorithm* to derive the optimal exercise strategy. All payments will be considered in their numéraire-relative form. Thus the exercise criterion is given by a comparison of the conditional expectation of the payments received upon nonexercise with the payments received upon exercise.

Induction Start: $t > T_n$. After the last exercise time we have

- The value of the (future) payments is $\tilde{U}_{n+1} = 0$

Induction Step: $t = T_i, i = n, n - 1, n - 2, \ldots 1.$ In T_i we have

- In the case of exercise in T_i the value is

$$\tilde{V}_{\text{underl},i}(T_i) := \frac{N_i(S(T_i) - K_i)}{N(T_i)}. \tag{15.9}$$

- In the case of nonexercise in T_i the value is $\tilde{V}_{\text{hold},i}(T_i) = \mathrm{E}^{\mathbb{Q}}(\tilde{U}_{i+1} \mid \mathcal{F}_{T_i})$. This value is estimated through a regression for given paths $\omega_1, \ldots, \omega_m$:

  - Let B_j be given ($\mathcal{F}_{T_i}$-measurable) basis functions.[4] Let the matrix X consist of the column vectors $B_j(\omega_k)$, $k = 1, \ldots, m$. Then we have

$$\begin{pmatrix} \tilde{V}_{\text{hold},i}(T_i, \omega_1) \\ \vdots \\ \tilde{V}_{\text{hold},i}(T_i, \omega_m) \end{pmatrix} \approx X \cdot (X^{\mathsf{T}} \cdot X)^{-1} \cdot X^{\mathsf{T}} \cdot \begin{pmatrix} \tilde{U}_{i+1}(\omega_1) \\ \vdots \\ \tilde{U}_{i+1}(\omega_m) \end{pmatrix}. \tag{15.10}$$

- The value of the payments of the product in T_i under optimal exercise is given by

$$\tilde{U}_i := \begin{cases} \tilde{V}_{\text{underl},i}(T_i) & \text{if } \tilde{V}_{\text{hold},i}(T_i) < \tilde{V}_{\text{underl},i}(T_i) \\ \tilde{U}_{i+1} & \text{else.} \end{cases}$$

Remark 193:   Our example is of course just the backward algorithm with an explicit specification of an underlying (15.9) and an explicit specification of an exercise criterion, here given by the estimator of the conditional expectation (15.10).

15.10.3 Example: Evaluation of a Bermudan Callable

Consider a Bermudan callable. The Bermudan should allow exercise at times $T_1 < T_2 < \ldots T_n$. Upon exercise in T_i the holder of the option will receive a payment of X_i in T_{i+1}, i.e., the relative value $\tilde{X}_i(T_{i+1}) := \frac{X_i}{N(T_{i+1})}$.

We will apply the backward algorithm to derive the optimal exercise strategy. All payments will be considered in their numéraire-relative form.

Induction Start: $t > T_n.$ After the last exercise time we have

- The value of the (future) payments is $\tilde{U}_{n+1} = 0$.

---

[4] Suitable basis functions for this example are 1 (constant), $S(T_i)$, $S(T_i)^2$, $S(T_i)^3$, etc., such that the regression function f will be a polynomial in $S(T_i)$.

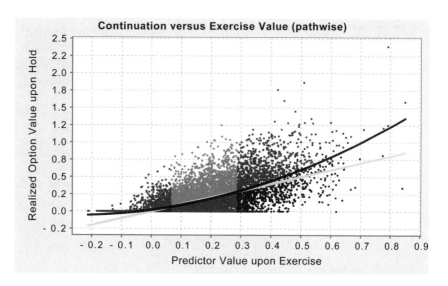

Figure 15.7. *Regression of the conditional expectation estimator without restriction of the regression domain: We consider a Bermudan option with two exercise dates $T_1 = 1.0$, $T_1 = 2.0$. Notional and strike are as follows: $N_1 = 0.7$, $N_2 = 1.0$, $K_1 = 0.82$, $K_2 = 1.0$. The model for the underlying S is a Black-Scholes model with $r = 0.05$ and $\sigma = 20\%$. The plot shows the values received upon exercise depending on the values received upon nonexercise in T_1. Each dot corresponds to a path. The regression polynomial gives the estimator for the expectation of the value upon nonexercise. It is optimal to exercise if this estimate lies above the value received upon exercise. The regression polynomial is a second-order polynomial in $S(T_1)$.*

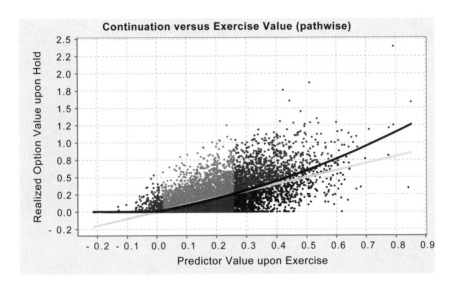

Figure 15.8. *Regression of the conditional expectation estimator with restriction of the regression domain: Parameters as in Figure 15.7. The regression polynomial is a second-order polynomial in* $\max(S(T_1) - K_1, 0)$. *Thus, values where* $S(T_1) - K_1 \leq 0$ *are aggregated into a single point. For the product under consideration this is advantageous since for* $S(T_1) - K_1 \leq 0$ *exercise is not optimal with probability* 1. *This restriction of the regression domain increases the regression accuracy over the remaining regression domain. Compare with Figure 15.7.*

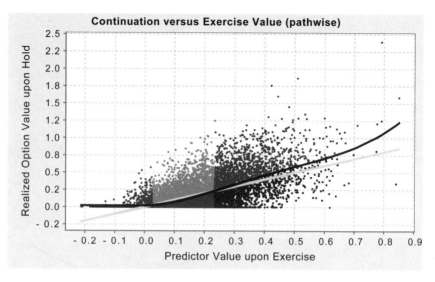

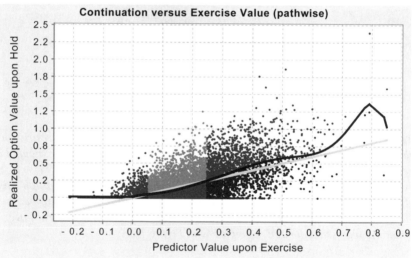

Figure 15.9. *Regression of the conditional expectation estimator using a polynomial of fourth (above) and eighth (below) order in* $\max(S(T_1) - K_1, 0)$. *Parameters as in Figure 15.7. A polynomial of higher order shows wiggles at the boundary of the regression domain. However, only a few paths are affected by the wrong estimate. Restricting the regression domain may reduce the errors (compare the left end of the regression domain with the right end).*

Induction Step: $t = T_i, i = n, n-1, n-2, \dots 1.$ In T_i we have

- In the case of exercise in T_i the value is

$$
\tilde{V}_{\text{underl},i}(T_i) := \mathrm{E}^{\mathbb{Q}^N}\left(\sum_{k=i}^{n-1} \frac{X_k}{N(T_{k+1})} \,\Big|\, \mathcal{F}_{T_i} \right).
$$

This value is estimated by a regression for given paths $\omega_1, \dots, \omega_m$:

Let B_j^1 be given ($\mathcal{F}_{T_i}$-measurable) basis functions. Let the matrix X^1 consist of the column vectors $B_j^1(\omega_k)$, $k = 1, \dots, m$. Then we have

$$
\begin{pmatrix} \tilde{V}_{\text{underl},i}(T_i, \omega_1) \\ \vdots \\ \tilde{V}_{\text{underl},i}(T_i, \omega_m) \end{pmatrix} \approx X^1 \cdot (X^{1,\mathsf{T}} \cdot X^1)^{-1} \cdot X^{1,\mathsf{T}} \cdot \begin{pmatrix} \displaystyle\sum_{k=i}^{n-1} \frac{X_k(\omega_1)}{N(T_{k+1}, \omega_1)} \\ \vdots \\ \displaystyle\sum_{k=i}^{n-1} \frac{X_k(\omega_m)}{N(T_{k+1}, \omega_m)} \end{pmatrix}.
$$

$$(15.11)$$

- In the case of nonexercise in T_i the value is $\tilde{V}_{\text{hold},i}(T_i) = \mathrm{E}^{\mathbb{Q}}(\tilde{U}_{i+1} \mid \mathcal{F}_{T_i})$. This value is estimated by a regression for given paths $\omega_1, \dots, \omega_m$:

Let B_j^0 be given ($\mathcal{F}_{T_i}$-measurable) basis functions. Let the matrix X^0 consist of the column vectors $B_j^0(\omega_k)$, $k = 1, \dots, m$. Then we have

$$
\begin{pmatrix} \tilde{V}_{\text{hold},i}(T_i, \omega_1) \\ \vdots \\ \tilde{V}_{\text{hold},i}(T_i, \omega_m) \end{pmatrix} \approx X^0 \cdot (X^{0,\mathsf{T}} \cdot X^0)^{-1} \cdot X^{0,\mathsf{T}} \cdot \begin{pmatrix} \tilde{U}_{i+1}(\omega_1) \\ \vdots \\ \tilde{U}_{i+1}(\omega_m) \end{pmatrix}.
$$

- The value of the payments of the product in T_i under optimal exercise is given by

$$
\tilde{U}_i := \begin{cases} \displaystyle\sum_{k=i}^{n-1} \frac{X_k}{N(T_{k+1})} & \text{if } \tilde{V}_{\text{hold},i}(T_i) < \tilde{V}_{\text{underl},i}(T_i) \\ \tilde{U}_{i+1} & \text{else.} \end{cases}
$$

Remark 194 (Bermudan Callable): The modification to the backward algorithm to price a Bermudan callable consists of the use of two conditional expectation estimators: one for the continuation value and (additionally) one for the underlying. As before, the conditional expectation estimators are used only for the exercise criterion (and not for the payment).

Remark 195 (Longstaff-Schwartz):

- The estimator of the conditional expectation is used in the estimation of the exercise strategy only.

- The choice of basis functions is crucial to the quality of the estimate.

Clément, Lamberton, and Protter [60] showed convergence of the Longstaff-Schwartz regression method to the exact solution.

15.10.4 Implementation

MonteCarloConditionalExpectationLongstaffSchwartz
setBasisFunctionsEstimator(RandomVariable[] basisFunctionsEstimator) setBasisFunctionsPredictor(RandomVariable[] basisFunctionsPredictor)
getConditionalExpectation(RandomVariable randomVariable)

Figure 15.10. *UML Diagram: Conditional expectation estimator. The method* `setBasisFunctionsEstimator` *sets the basis functions which form the matrix X. The method* `setBasisFunctionsPredictor` *sets the basis functions which form the matrix X^*. These are the same basis functions as for X, but possibly evaluated in an independent Monte Carlo simulation (to avoid foresight bias). The method* `getConditionalExpectation` *calculates the regression parameter $\alpha^* = (X^{\mathsf{T}} \cdot X)^{-1} \cdot X^{\mathsf{T}} \cdot v$ from a given vector v and returns the conditional expectation estimator $X^* \cdot \alpha^*$ of v. See Lemma 191.*

The Longstaff-Schwartz conditional expectation estimator may easily be implemented in a corresponding class, independent of the given model or Monte Carlo simulation—see Figure 15.10. This class contains nothing more than a linear regression, but the methodology may be replaced by alternative algorithms (e.g., nonparametric regressions).

As pointed out in the discussion of the backward algorithm, it is not normally necessary to explicitly calculate the exercise strategy in the form of T or η. It is sufficient to calculate the random variables $\tilde{U}_i$ in a backward recursion. Since finally only $\tilde{U}_1$ is needed to calculate the price of the Bermudan option, the $\tilde{U}_i$'s may be stored (updated) in the same vector of Monte Carlo realizations.

15.10.5 Binning as Linear Least-Square Regression

We return once again to the binning. In Section 15.8.1 it turned out that binning may be interpreted as least-square regression with a specific set of basis functions: The indicator variables of the bins U_j, which we denote by

$$h_j(\omega) := \begin{cases} 1 & \text{for } \omega \in U_j \\ 0 & \text{else.} \end{cases} \tag{15.12}$$

We now give an explicit calculation using the linear regression algorithm with the bin indicator variables as basis functions.

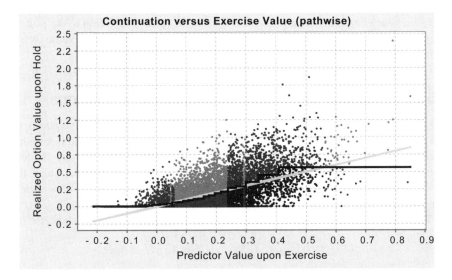

Figure 15.11. *Binning using the linear regression algorithm with piecewise constant basis functions: We use 20 bins (basis functions). Each bin consist of approximately the same number of paths. Model and product parameters are as in Figure 15.7.*

Let ω_k denote the paths of a Monte Carlo simulation and X the matrix $(h_j(\omega_k))$, j column index, k row index. Since the U_j's are disjoint, we have $X^{\mathsf{T}}X = \operatorname{diag}(m_1, \dots, m_p)$, where m_j is the number of paths for which $h_j(\omega_k) = 1$. Thus we have for the regression parameter

$$\alpha^* = (X^{\mathsf{T}} \cdot X)^{-1} \cdot X^{\mathsf{T}} \cdot v = \operatorname{diag}\left(\frac{1}{m_1}, \dots, \frac{1}{m_p}\right) \cdot X^{\mathsf{T}} \cdot v.$$

It follows that the regression parameter gives the expectation on the corresponding bin:

$$\alpha_j^* = \frac{1}{m_j} \sum_{v_k \in U_j} v_k \qquad \text{for } j = 1, \dots, p.$$

15.11 Optimization Methods

 Motivation: In the discussion of the backward algorithm it has become obvious that the conditional expectation estimator is needed to derive the optimal exercise strategy only. Since a suboptimal exercise will lead to a lower Bermudan price, the optimal exercise has an alternative characterization: It maximizes the Bermudan value. A solution to the pricing problem of the Bermudan thus consists of maximizing the Bermudan value over a suitable, sufficiently large space of admissible[5] exercise strategies. ◁|

15.11.1 Andersen Algorithm for Bermudan Swaptions

The following method was proposed for the valuation of Bermudan swaptions by Andersen [44]. We thus restrict our presentation to the evaluation of the Bermudan callable and use the notation of Section 15.10.3. In [44] the method appears less generic than the Longstaff-Schwartz regression. However, one might reformulate the optimization method in a fairly generic way. As the optimization is a high dimensional one, the method then becomes less useful in practice.

The exercise strategy is given by a parametrized function of the underlyings

$$I_i(\lambda) := f(V_{\text{underl},i}(T_i, \omega), \dots, V_{\text{underl},n-1}(T_i, \omega), \lambda),$$

where we replace the optimal exercise

$$\tilde{V}_{\text{underl},i}(T_i) \;<\; \mathrm{E}^{\mathbb{Q}}(\tilde{U}_{i+1} \mid \mathscr{F}_{T_i})$$

by

$$I_i(\lambda) \;>\; 0.$$

Here the function f may represent a variety of exercise criteria, e.g.,

$$I_i(\omega, \lambda) = V_{\text{underl},i}(T_i, \omega) - \lambda. \tag{15.13}$$

---

[5] By an admissible exercise strategy we denote one that respects the measurability requirements. As we noted, a violation of measurability requirements, i.e., a foresight bias or even perfect foresight, will result in a superoptimal strategy. The Bermudan value with a superoptimal strategy is higher than the Bermudan value with the optimal strategy; however, superoptimal exercise is impossible.

We assume that I_i is such that it may be calculated without resimulation, i.e., we assume that the underlyings $V_{\text{underl},j}(T_i)$ are either given by an analytic formula or a suitable approximation. For example, this is the case for a swap within a LIBOR market model. If we use the optimization method within the backward algorithm, it now looks as follows:

Induction Start: $t > T_n$.   After the last exercise time we have

- The value of the (future) payments is $\tilde{U}_{n+1} = 0$.

Induction Step: $t = T_i, i = n, n-1, n-2, \ldots 1$.   In T_i we have

- In the case of exercise in T_i the value is $\tilde{V}_{\text{underl},i}(T_i)$

- In the case of nonexercise in T_i the value is $\tilde{V}_{\text{hold},i}(T_i) = \mathrm{E}^{\mathbb{Q}}(\tilde{U}_{i+1} \mid \mathcal{F}_{T_i})$. This value is estimated through an optimization for given paths $\omega_1, \ldots, \omega_m$:

  - $I_i(\lambda, \omega) = f(V_{\text{underl},i}(T_i, \omega), \ldots, V_{\text{underl},n-1}(T_i, \omega), \lambda)$.

  - $\tilde{U}_i(\lambda, \omega) := \begin{cases} \tilde{V}_{\text{underl},i}(T_i) & \text{if } I_i(\lambda, \omega) > 0 \\ \tilde{U}_{i+1} & \text{else.} \end{cases}$

  - $\tilde{V}_{\text{berm},i}(T_0, \lambda) = \mathrm{E}^{\mathbb{Q}}(\tilde{U}_i(\lambda) \mid \mathcal{F}_{T_0}) \approx \frac{1}{m} \sum_k \tilde{U}_i(\lambda, \omega_k)$

  - $\lambda^* = \arg \max_\lambda \left(\frac{1}{m} \sum_k \tilde{U}_i(\lambda, \omega_k) \right)$

- The value of the payments of the product in T_i under optimal exercise is given by

$$\tilde{U}_i(\omega) := \left. \begin{cases} \tilde{V}_{\text{underl},i}(T_i) & \text{if } I_i(\lambda^*, \omega) > 0 \\ \tilde{U}_{i+1} & \text{else} \end{cases} \right\} = \tilde{U}_i(\lambda^*, \omega).$$

The exercise strategy is estimated in T_i by choosing the λ^* for which I_i gives the maximal Bermudan option value. This is done by going backward in time, from exercise date to exercise date.

15.11.2 Review of the Threshold Optimization Method

15.11.2.1 Fitting the Exercise Strategy to the Product

Let us apply the optimization method to the pricing of a simple Bermudan option on a stock following a Black-Scholes model. This shows that a too simple choice of the exercise strategy will give surprisingly unreliable results.

The simple strategy (15.13) fails for the simplest type of Bermudan option. Consider the option to receive

$$N_1(S(T_1) - K_1)$$

in T_1 or receive

$$\max(N_2(S(T_2) - K_2), 0)$$

in T_2, where—as before—N_i and K_i denote notional and strike and S follows a Black-Scholes model. This gives us an analytic formula for the option in T_2 and thus the true optimal exercise. Figure 15.12 shows an example where the optimization of the simple strategy (15.13) gives the value of the Bermudan option.

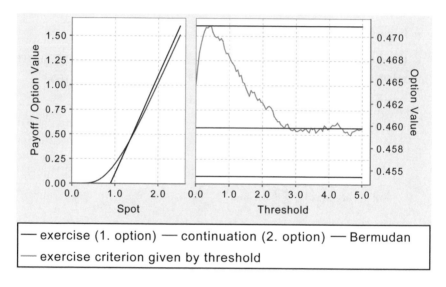

Figure 15.12. *Example of the successful optimization of the exercise criterion (intersection of the two price curves, left). The graph on the right shows the Bermudan option value as a function of the exercise threshold λ.*

A small change in notional N_1 and strike K_1 changes the picture. If both are smaller than N_2 and K_2, respectively, we obtain two intersection points of the exercise and continuation value. In T_1 it is optimal to exercise in between these two intersection points. Our simple exercise criterion cannot render this case. Optimizing the threshold parameter λ shows two maxima: the value of the two European options "exercise never" and "exercise always". Both values are below the true Bermudan option value; see Figure 15.13, right.

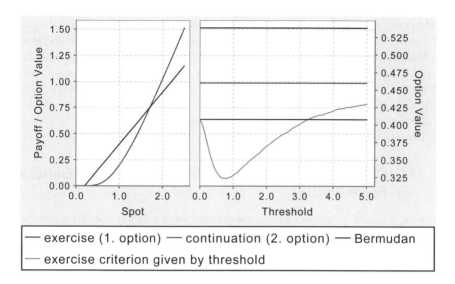

Figure 15.13. *Example of a failing optimization of the exercise criterion (intersection of the two price curves, left). The graph on the right shows the Bermudan option value as a function of the exercise threshold λ.*

The conclusion of this example is that the choice of the exercise strategy has to be made carefully in accordance with the product. But this remark applies to some extent to any method.

15.11.2.2 Disturbance of the Optimizer through Discontinuities and Local Minima

The Monte Carlo Bermudan price calculated from the backward algorithm is a discontinuous function of the exercise criterion. The Bermudan price jumps if the exercise criterion $I(\omega, \lambda)$ changes sign for a given path ω. The price jumps by the difference of *exercise value* and *continuation value*. Even in the case of an optimal exercise criterion (i.e., $\lambda = \lambda_{max}$) we see a jump in price since even then exercise value and continuation value will generally be different (at optimal exercise, only the expected (!) continuation value equals the exercise value).

As a function of λ, the price will not only exhibit discontinuities, but also small local maxima induced by them; see Figure 15.12, right. These may prevent the optimizing algorithm from finding the global maxima.

However, if there are a sufficient number of paths, the local maxima appear only on a small scale. The jumps in price will be of the order $O(\frac{1}{n})$, where n denotes the number of paths. Thus with a robust minimizer one would rarely encounter this problem. For example, consider the case that the limit function (for $n \to \infty$) would satisfy an estimate of the form $V_{\max} - V(\lambda) > C(\lambda - \lambda_{\max})^2$, i.e., without the Monte Carlo discontinuities no other local maxima or saddle point would exist. Then a bisection search on the disturbed function will miss the true maxima only by the order of a jump $O(\frac{1}{n})$.

15.11.3 Optimization of Exercise Strategy: A More General Formulation

There is a trivial generalization of the optimization method considered in Section 15.11.1:

- The exercise criterion will be given as a function of arbitrary $\mathcal{F}_{T_i}$-measurable random variables.

- The exercise criterion will be given as a function of a parameter vector $\lambda \in \mathbb{R}^k$.

Thus we replace the "true" exercise criterion used in the backward algorithm

$$\tilde{V}_{\text{underl},i}(T_i) \; < \; E^Q(\tilde{U}_{i+1} \,|\, \mathcal{F}_{T_i})$$

by a function

$$I_i(\lambda, \omega) \; := \; f(B_{i,1}(\omega), \dots, B_{i,m}(\omega), \lambda),$$

where $B_{i,j}$ is a set of $\mathcal{F}_{T_i}$-measurable random variables and $\lambda \in \mathbb{R}^k$.

15.11.4 Comparison of Optimization Method and Regression Method

The difference between the optimization method and the regression method becomes apparent in Figures 15.7 to 15.9. While the regression method requires the regression functions to give a good fit to the conditional expectations across the whole domain of the independent variable, the optimization method only requires that the functional $I_i(\lambda)$ captures the exercise boundary. In Figures 15.7 to 15.9 the conditional expectation estimator is a curve, but the exercise boundary is given by two points only (the intersection of the conditional expectation estimator with the bisector).

Thus the optimization method can cope with far fewer parameters than the regression. On the other hand, as noted in the example above, it is far more important that the functional is adapted to the Bermudan product under consideration.

It is trivial to choose the map $I_i(\lambda)$ such that the optimal λ^* give the same or a better value than the least-square regression. If $B_{i,j}$ denote the basis functions used in the least-square regression for exercise date T_i, we set

$$I_i(\lambda_1, \ldots, \lambda_n) = (\lambda_1 B_{i,1} + \cdots + \lambda_1 B_{i,m}) - \tilde{V}_{\text{underl},i}(T_i).$$

Then for $\lambda = \alpha^*$, where α^* is the regression parameter from the least-square regression, the exercise criterion agrees with the one from the least-square regression. This result however is rarely an advantage of the optimization method since in practice a high dimensional optimization does not represent an alternative.

15.12 Duality Method: Upper Bound for Bermudan Option Prices

 Motivation: So far, the ansatz to evaluate a Bermudan option has been to estimate the exercise strategy, i.e., the stopping time τ. If this estimate is itself a stopping time, i.e., no foresight bias is present, then the evaluation with the estimated exercise strategy gives a lower bound for the true value of the Bermudan option. The stopping time τ maximizes the option price; all other stopping times give lower prices because they are suboptimal. In order to decide if the given lower bound is sufficiently close to the optimal value, a corresponding upper bound would come in handy. ◁|

In this section we present a method that gives, in the limit of a vanishing Monte Carlo error, an upper bound for the Bermundan option value. The upper bound itself converges to the true option value as a given space of test functions converges to the whole space. Putting aside the Monte Carlo error, together with the lower bound methods, Sections 15.8 to 15.11.1, we then obtain an interval for the true value of the Bermudan price.

The method was introduced by Rogers [93], Haugh and Kogan [77], and originally by Davis and Karatzas [62].

15.12.1 Foundations

Let us review some foundations.

Definition 196 (Supermartingale): ⌐
The stochastic process $\{X_t, \mathcal{F}_t \; ; \; 0 \le t < \infty\}$ is called a *supermartingale* with respect to the filtration $\mathcal{F}_t$ and the measure $\mathbb{P}$, if

$$X_s \ge \mathrm{E}(X_t \mid \mathcal{F}_s), \quad \mathbb{P}\text{-almost surely}, \quad \forall\, 0 \le s < t < \infty.$$

⌟

Definition 197 (Stopping Time (Time-Discrete Process)): ⌐
Let $T : \Omega \to \{T_0, \dots, T_n\}$ denote a random variable. T is called a *stopping time* if

$$\{T = T_i\} \in \mathcal{F}_{T_i}. \tag{15.14}$$

⌟

Remark 198 (Stopping Time): Equivalent to (15.14) is

$$\{T \le T_i\} \in \mathcal{F}_{T_i}, \tag{15.15}$$

since $\mathcal{F}_{T_j} \subset \mathcal{F}_{T_i}$ for $T_j < T_i$.

Interpretation: A stopping time may be viewed as the mathematical representation of an exercise strategy of a Bermudan option. If for a given path ω, $T(\omega) = T_i$ denotes the time at which an event occurs (e.g., the decision to exercise an option), then (15.15) constitutes the requirement that the decision is based on information that is available on or before T_i—expressed through the $\mathcal{F}_{T_i}$-measurability.

The condition (15.15) precludes a *foresight bias*. Compare Section 15.6[6].　◁|

Given a stochastic process and a stopping time, we may construct the *stopped process*:

Definition 199 (Stopped Process):　⌐

Let X denote a stochastic process and T a stopping time. The process X^T given through X *stopped in T*, is defined by

$$X^T(t, \omega) = \begin{cases} X(t, \omega) & t < T(\omega) \\ X(T(\omega), \omega) & t \geq T(\omega). \end{cases}$$

⌐

Interpretation: On every path ω the stopped process X^T coincides with X for times before the stopping time. For times on or after the stopping time the stopped process takes the constant value of X at the stopping time. It is stopped. This is similar to what happens for the (relative) price of a Bermudan option. Before the exercise date, the option value is stochastic. Upon exercise the value is frozen to the value upon exercise.　◁|

Definition 200 (Snell Envelope):　⌐

Let Z denote a time-discrete adapted process and U defined by

$$U(T_n) = Z(T_n),$$
$$U(T_i) = \max(Z(T_i), \mathrm{E}(U(T_{i+1}) \mid \mathcal{F}_{T_i})).$$

The process U is called *Snell envelope* of Z.　⌐

Lemma 201 (Snell Envelope): The Snell envelope of Z is the smallest supermartingale which dominates Z.

Theorem 202 (Doob-Meyer Decomposition): Let U denote a (time-discrete) supermartingale. Then there exists a (unique) decomposition

$$U(T_i) = M(T_i) - A(T_i),$$

---

[6] For the example discussed in Section 15.6 $T(\omega_1) = T(\omega_2) = T_1$ and $T(\omega_1) = T(\omega_2) = T_2$ are both stopping times (check!), however, the superoptimal exercise strategy $T(\omega_1) = T_1$, $T(\omega_2) = T_2$ is not a stopping time.

where M is a (time-discrete) martingale and A is a (time-discrete) previsible nondecreasing process (i.e., $A(T_{i-1}, \omega) \le A(T_i, \omega)$) where $A(T_0) = 0$.

Proof:

$$M(T_i) := M(T_{i-1}) + U_i - \mathrm{E}(U_i \mid \mathcal{F}_{T_{i-1}}), \qquad M(T_0) = U(T_0),$$
$$A(T_i) := A(T_{i-1}) + U_{i-1} - \mathrm{E}(U_i \mid \mathcal{F}_{T_{i-1}}), \qquad A(T_0) = 0.$$

$\Box$

15.12.2 American Option Evaluation as Optimal Stopping Problem

We repeat the Bermudan option evaluation as optimal stopping: Let $0 = T_0 < T_1 < \cdots < T_n$. Let $V_{\mathrm{underl}}(T_i)$, $i = 1, \ldots, n-1$ denote a sequence of $\mathcal{F}_{T_i}$-measurable random variables, giving the value of financial products at time T_i. We define recursively

$$V_{\mathrm{berm}}(T_i) := \max(V_{\mathrm{underl}}(T_i),\ \mathrm{E}^{\mathbb{Q}^N}(V_{\mathrm{berm}}(T_{i+1}) \frac{N(T_i)}{N(T_{i+1})} \mid \mathcal{F}_{T_i})),$$

where $(N, \mathbb{Q}^N)$ denotes a given numéraire-martingale measure pair. Furthermore let

$$\tilde{V}_{\mathrm{U}}(T_i) := \frac{V_{\mathrm{underl}}(T_i)}{N(T_i)}, \qquad \tilde{V}_{\mathrm{B}}(T_i) := \frac{V_{\mathrm{berm}}(T_i)}{N(T_i)}$$

denote the corresponding N-relative prices, i.e.,

$$\tilde{V}_{\mathrm{B}}(T_i) := \max(\tilde{V}_{\mathrm{U}}(T_i),\ \mathrm{E}^{\mathbb{Q}^N}(\tilde{V}_{\mathrm{B}}(T_{i+1}) \mid \mathcal{F}_{T_i})).$$

Theorem 203 (American Option Price—Dual Formulation (Time-Continuous Version)): Let
$$V_{\mathrm{amer}}(0) := \sup_{T \text{ stopping time}} \mathrm{E}(\tilde{V}_{\mathrm{U}}(T) \mid \mathcal{F}_0).$$

Then
$$V_{\mathrm{amer}}(0) = \inf_{M \in H_0^1} \mathrm{E}(\sup_{0 \le t \le T} (\tilde{V}_{\mathrm{U}}(t) - M(t)) \mid \mathcal{F}_0),$$

where H_0^1 denotes the set of all martingales M with $\sup_{0 \le t \le T} |M(t)| \in \mathrm{L}_1(\Omega)$ and $M(0) = 0$.

Since we consider only the evaluation of Bermudan options, i.e., options a finite number of exercise dates, we give a time-discrete version of Theorem 203:

Theorem 204 (Bermudan Option Price—Dual Formulation (Time-Discrete Version)): Let

$$\tilde{V}_B(T_i) := \max(\tilde{V}_U(T_i), \, E(\tilde{V}_B(T_{i+1}) \mid \mathcal{F}_{T_i})).$$

$$\tilde{V}_B(T_0) = \inf_{M \in H_0^1} E(\max_{T_i}(\tilde{V}_U(T_i) - M(T_i)) \mid \mathcal{F}_0), \qquad (15.16)$$

where H_0^1 denotes the set of all (time-discrete) martingales M with

$$M(T_0) = 0$$
$$M(T_i) = E(M(T_{i+1}) \mid \mathcal{F}_{T_i}) \qquad\qquad M(T_i) \in L_1(\Omega)$$

and $\max_{T_i}(\tilde{V}_U(T_i) - M(T_i))$ has to be understood pathwise, i.e.,

$$\max_{T_i}(\tilde{V}_U(T_i) - M(T_i))[\omega] = \max_{T_i}(\tilde{V}_U(T_i, \omega) - M(T_i, \omega))$$

(and *not* as a maximum over all stopping-times).

Remark 205 (On Theorem 204):   First, it is worth noting that in Theorem 204 the maximum is applied pathwise. If the pathwise maximum is applied to V_{underl} then $\max_{T_i}(V_{\text{underl}}(T_i))$ gives the value of a perfect foresight. As we will show in Lemma 206, a foresight error is precluded by the martingale M.

The expectation in (15.16) is conditioned to $\mathcal{F}_{T_0} = \mathcal{F}_0$ only, i.e., it is not required to calculate a conditional expectation at a later time. However, the expectation has to be minimized over all martingales M. This is as complex as the calculation of the optimal exercise strategy (by maximizing over all stopping times τ). But Theorem 204 indeed gives an upper bound for the value of the Bermudan option:

$$V_{\text{berm}}(T_0) \leq E^{\mathbb{Q}^N}\Big(\max_{T_i}(V_{\text{underl}}(T_i) \frac{N(0)}{N(T_i)} - M(T_i) \mid \mathcal{F}_0)\Big), \qquad M \in H_0^1,$$

and this upper bound may be arbitrarily close to the option price $V_{\text{berm}}(T_0)$ if M is suitably chosen.

Lemma 206 (Eliminating *Foresight*):
Let $\tilde{V}_U(T_i)$, $\tilde{V}_B(T_i)$ be as before, i.e.,

$$\tilde{V}_B(T_i) := \max_{T_i}(\tilde{V}_U(T_i), \, E(\tilde{V}_B(T_{i+1}) \mid \mathcal{F}_{T_i})), \qquad (15.17)$$

and M as in the Doob-Meyer decomposition of $\tilde{V}_B(T_i)$

$$M(T_0) := 0,$$
$$M(T_i) := M(T_{i-1}) + \tilde{V}_B(T_i) - E(\tilde{V}_B(T_{i+1}) \mid \mathcal{F}_{T_i}). \qquad (15.18)$$

Let $T^{\text{opt}} : \Omega \to \{T_1, \ldots, T_n\}$ denote the (optimal) stopping time given by

$$T^{\text{opt}}(\omega) := \min\{T_j : \tilde{V}_U(T_j, \omega) \geq \mathrm{E}(\tilde{V}_B(T_{j+1}) \mid \mathcal{F}_{T_j})[\omega]\}$$

(this implies $\tilde{V}_B(T_0) = \mathrm{E}(\tilde{V}_U(T^{\text{opt}}) \mid \mathcal{F}_{T_0})$).

Then we have (pathwise!)

$$\max_{T_j}(\tilde{V}_U(T_j, \omega) - M(T_j, \omega)) = \tilde{V}_U(T^{\text{opt}}(\omega), \omega) - M(T^{\text{opt}}, \omega).$$

Proof: We give the proof only with (15.17) and (15.18). Let $\omega \in \Omega$. By definition we have

$$\tilde{V}_B(T_j, \omega) = \mathrm{E}(\tilde{V}_B(T_{j+1}) \mid \mathcal{F}_{T_j})[\omega] \qquad \text{for } T_j < T^{\text{opt}}(\omega) \qquad (15.19)$$

$$\tilde{V}_B(T_j, \omega) \geq \mathrm{E}(\tilde{V}_B(T_{j+1}) \mid \mathcal{F}_{T_j})[\omega] \qquad \text{for } T_j \geq T^{\text{opt}}(\omega) \qquad (15.20)$$

and

$$\tilde{V}_B(T_i, \omega) = \tilde{V}_U(T_i, \omega) \qquad \text{for } T_i = T^{\text{opt}}(\omega). \qquad (15.21)$$

1. For $T_i = T^{\text{opt}}(\omega)$ and $T_j < T_i$ we have

- $\tilde{V}_U(T_i, \omega) - M(T_i, \omega) = \tilde{V}_B(T_i, \omega) - M(T_i, \omega)$

- $\tilde{V}_B(T_{j+1}, \omega) - M(T_{j+1}, \omega)$
  $= \tilde{V}_B(\cancel{T_{j+1}, \omega}) - M(T_j, \omega) - \tilde{V}_B(\cancel{T_{j+1}, \omega}) + \mathrm{E}(\tilde{V}_B(T_{j+1}, \omega) \mid \mathcal{F}_{T_j})[\omega]$
  $= \tilde{V}_B(T_j, \omega) - M(T_j, \omega)$

- $\tilde{V}_B(T_j, \omega) - M(T_j, \omega) \geq \tilde{V}_U(T_j, \omega) - M(T_j, \omega)$

and thus for $T_j \leq T_i$:

$$\tilde{V}_U(T_j, \omega) - M(T_j, \omega) \leq \tilde{V}_U(T_i, \omega) - M(T_i, \omega).$$

2. For $T_i = T^{\text{opt}}(\omega)$ and $T_j > T_i$ we have

- $\tilde{V}_U(T_i, \omega) - M(T_i, \omega)$
  $= \tilde{V}_U(T_i, \omega) - M(T_{i+1}, \omega) + \tilde{V}_B(T_{j+1}, \omega) - \mathrm{E}(\tilde{V}_B(T_{j+1}) \mid \mathcal{F}_{T_j})[\omega]$
  $= \tilde{V}_B(T_i, \omega) - M(T_{i+1}, \omega) + \tilde{V}_B(T_{j+1}, \omega) - \mathrm{E}(\tilde{V}_B(T_{j+1}) \mid \mathcal{F}_{T_j})[\omega]$
  $\geq \tilde{V}_B(T_{i+1}, \omega) - M(T_{i+1}, \omega)$

- $\tilde{V}_B(T_j, \omega) - M(T_j, \omega)$
  $= \tilde{V}_B(T_j, \omega) - M(T_{j+1}, \omega) + \tilde{V}_B(T_{j+1}, \omega) - \mathrm{E}(\tilde{V}_B(T_{j+1},) \mid \mathcal{F}_{T_j})[\omega]$
  $\geq \tilde{V}_B(T_{j+1}, \omega) - M(T_{j+1}, \omega)$

- $\tilde{V}_B(T_{j+1}, \omega) - M(T_{j+1}, \omega) \geq \tilde{V}_U(T_{j+1}, \omega) - M(T_{j+1}, \omega)$

and thus for $T_j \geq T_i$

$$\tilde{V}_U(T_i, \omega) - M(T_i, \omega) \geq \tilde{V}_U(T_j, \omega) - M(T_j, \omega).$$

Thus, for all T_j

$$\tilde{V}_U(T_i, \omega) - M(T_i, \omega) \geq \tilde{V}_U(T_j, \omega) - M(T_j, \omega).$$

$\square$|

Remark 207 (Foresight Bias): Lemma 206 shows that the martingale M eliminates the foresight bias. We have

$$\tilde{V}_U(T_i) - M(T_i) \geq \tilde{V}_U(T_j) - M(T_j) \qquad\qquad \forall\, T_j,$$

however, in general

$$\tilde{V}_U(T_i) \not\geq \tilde{V}_U(T_j) \qquad\qquad \forall\, T_j.$$

Lemma 208 (Optimal Stopping): It is

$$E(\tilde{V}_U(T^{\mathrm{opt}}) \mid \mathcal{F}_{T_0}) \;=\; E(\max_{T_j}(\tilde{V}_U(T_j) - M(T_j)) \mid \mathcal{F}_{T_0}).$$

Proof: It is

$$E(\tilde{V}_U(T^{\mathrm{opt}}) \mid \mathcal{F}_{T_0}) \overset{\mathrm{Def.}\ T^{\mathrm{opt}}}{=} E(\sup_{T\ \mathrm{stopping-time}} \tilde{V}_U(T) \mid \mathcal{F}_{T_0})$$

$$= E(\sup_{T\ \mathrm{stopping-time}} \tilde{V}_U(T) - M(T) \mid \mathcal{F}_{T_0})$$

$$\overset{\mathrm{La.}\ 206}{=} E(\max_{T_i} \tilde{V}_U(T_i) - M(T_i) \mid \mathcal{F}_{T_0})$$

$\square$|

15.13 Primal-Dual Method: Upper and Lower Bound

The calculation of the exercise strategy and from it the price of the Bermudan option via an (approximation) of the conditional expectation is called *primal*. The calculation of the exercise strategy via the stopping time through Section 15.12 is called *dual*.

With the result from Lemma 206 we can combine both methods. From Theorem 204 and Lemma 206 we have

$$\tilde{V}_B(T_0) = E(\max_{T_i}(\tilde{V}_U(T_i) - M(T_i)) \mid \mathcal{F}_0), \qquad (15.22)$$

with

$$M(T_i) := M(T_{i-1}) + U_i - E(U_i \mid \mathcal{F}_{T_{i-1}}) \qquad M(T_0) = U(T_0). \qquad (15.23)$$

The conditional expectation in (15.23) can be estimated with a *primal* method. If the conditional expectation has already been estimated in a calculation of the lower bound or the Bermudan option price, then (15.22) immediately gives a corresponding upper bound.

Compare also [45].

Further Reading: The discussions in [16], [18], and [53] consider Monte Carlo methods for derivative pricing in general. In [60] the convergence of the regression methods is proven. The optimization method for Bermudan swaptions is given in [44], and a primal-dual method is given in [45]. The discussion of the foresight bias is found in [67]. ◁|

Experiment: At http://www.christian-fries.de/finmath/applets/BermudanStockOptionPricing.html the evaluation of a Bermudan option on a stock following a Black-Scholes model may be studied. There is a choice of different evaluation methods. ◁|

CHAPTER 16

Pricing Path-Dependent Options in a Backward Algorithm

A backward algorithm, e.g., as given by a model implemented as a lattice, allows the calculation of the conditional expectation

$$V(T_{i-1}) = N(T_{i-1}) \, \mathrm{E}^{\mathbb{Q}^N} \left(\frac{V(T_i)}{N(T_i)} \,\Big|\, \mathcal{F}_{T_{i-1}} \right),$$

and thus defines induction steps $T_i \rightarrow T_{i-1}$ backward in time. Path-dependent quantities cannot be considered directly. One way of allowing for path-dependent quantities in a backward algorithm is to eliminate the path dependency by extending the state space.

16.1 State Space Extension

Let V denote a product whose time T_i value depends on a quantity C_i given by an update rule

$$C_i = f(T_i, C_{i-1}, X_i), \qquad C_0 = \text{constant}, \tag{16.1}$$

were X_i is a random variable that is a function of the time T_i values of the model primitives, i.e., non-path dependent. Thus X_i and hence C_i are $\mathcal{F}_{T_i}$ measurable. Equation (16.1) constitutes the path-dependency of C_i; it may not be written as a function of the time T_i values of the model primitives. It depends on the past since it depends on the previous value C_{i-1}.

To remove the path-depency in V we add C_i as an additional state. We consider the time T_i value of V as a function of C_i

$$V(T_i) = V(T_i, C_i), \qquad i = 0, 1, \ldots, n.$$

Then backward algorithm is:

- Given $V(T_i, C_i)$.

- Apply the update rule to define

$$\tilde{V}(T_i, C_{i-1}, X_i) \; := \; V(T_i, f(C_{i-1}, X_i)) \qquad (16.2)$$

- Define

$$V(T_{i-1}, C_{i-1}) \; := \; N(T_{i-1}) \, \mathrm{E}^{\mathcal{Q}^N} \left(\frac{\tilde{V}(T_i, C_{i-1}, X_i)}{N(T_i)} \, \Big| \, \mathcal{F}_{T_{i-1}} \right). \qquad (16.3)$$

Note that conditional to $\mathcal{F}_{T_{i-1}}$ the state C_{i-1} is a constant.

 Interpretation (State Space Extension): The method is called state space extension because the discrete stochastic process $T_i \mapsto C(T_i) := C_i$ can be interpreted as an additional state of the model and (16.1) defines the evolution of this process. Seen over this extended space the product is non-path dependent. ◁|

16.2 Implementation

In order to implement the state space extension we discretize the additional state random variables C_i into k_i state values

$$C_i \in \{c_{i,1}, \ldots, c_{i,k_i}\}.$$

For the implementation of the update rule (16.2) an interpolation has to be used, e.g., a linear interpolation

$$\tilde{V}(T_i, c_{i-1,j}, X_i) \approx$$
$$\frac{f(c_{i-1,j}, X_i) - c_{i,l_j}}{c_{i,l_j+1} - c_{i,l_j}} V(T_i, c_{i,l_j+1}) + \frac{c_{i,l_j+1} - f(c_{i-1,j}, X_{i+1})}{c_{i,l_j+1} - c_{i,l_j}} V(T_i, c_{i,l_j}),$$

where l_j is such that $c_{i,l_j} \le f(c_{i-1,j}, X_i) < c_{i,l_j+1}$.

Then the conditional expectation (16.3) is calculated for each state realization $c_{i-1,j}$ giving $V(T_{i-1}, c_{i-1,1}), \ldots, V(T_{i-1}, c_{i-1,k_i})$.

Remark 209: For some products the value $V(T_i, c)$ is linear in c. In such cases two states are sufficient and the approximation of the update rule by the linear interpolation is exact. Examples are zero structures, where the additional state is the accrued notional; the value of the future cashflow is linear in the notional.

16.3 Path-Dependent Bermudan Options

The state space extension can be used in any pricing code that uses a backward algorithm. It is not limited to models implemented on a lattice, although this might have been suggested by the word "extension". A state space extension can also be used to consider path-dependent quantities in a backward algorithm in a Monte-Carlo simulation, e.g., for the pricing of path-dependet bermudan options in Monte-Carlo.

Combining the Bermudan's optimal exercise (15.3) with the state space extension the backward algorithm is

- Given $V_{berm,i+1}(T_i, C_i)$, $V_{underl,i}(T_i, C_i)$.

- Apply the update rule to define

$$\tilde{V}_{berm,i+1}(T_i, C_{i-1}, X_i) := V_{berm,i+1}(T_i, f(C_{i-1}, X_i)) \qquad (16.4)$$
$$\tilde{V}_{underl,i}(T_i, C_{i-1}, X_i) := V_{underl,i}(T_i, f(C_{i-1}, X_i)).$$

- Apply the optimal exercise for exercise date T_i to define

$$\underbrace{\tilde{V}_{berm,i}(T_i, C_{i-1}, X_i)}_{\substack{\text{Bermudan with} \\ \text{exercise dates} \\ T_i, \ldots, T_n}} := \max\,(\,\underbrace{\tilde{V}_{berm,i+1}(T_i, C_{i-1}, X_i)}_{\substack{\text{Bermudan with} \\ \text{exercise dates} \\ T_{i+1}, \ldots, T_n}} ,\; \underbrace{\tilde{V}_{underl,i}(T_i, C_{i-1}, X_i)}_{\substack{\text{Product} \\ \text{received upon} \\ \text{exercise in } T_i}}\,).$$

- Define

$$V_{berm,i}(T_{i-1}, C_{i-1}) := N(T_{i-1})\, E^{Q^N}\!\left(\frac{\tilde{V}_{berm,i}(T_i, C_{i-1}, X_i)}{N(T_i)}\,\middle|\, \mathcal{F}_{T_{i-1}}\right).$$

Tip: Here, the application of the update rule is performed before the application of the optimal exercise. However, the two steps may be interchanged. If one adds a state space extension to the Bermudan pricing it may appear more natural to apply the update rule *after* the optimal exercise has been applied. However, this will likely introduce numerical problems. If we consider

$$V_{berm,i}(T_i, C_i) := \max\,(V_{berm,i+1}(T_i, C_i) ,\; V_{underl,i}(T_i, C_i))$$

and then apply the update rule

$$\tilde{V}_{berm,i}(T_i, C_{i-1}, X_i) := V_{berm,i}(T_i, f(C_{i-1}, X_i)),$$

it is difficult to implement an accurate interpolation of the update rule since $c \mapsto V_{\text{berm},i}(T_i, f(c, X_i))$ is not a smooth function in c (the max() will introduce kinks).

On the other hand, in (16.4) the update rule is applied to the conditional expectation of the previous backward induction step, which is usually a smooth function of the states. ◁|

16.4 Examples

We illustrate the method of state space extension for the valuation of a snowball/memory (Definition 164) and for the evaluation of a flexi-cap (Definition 169).

16.4.1 Evaluation of a Snowball in a Backward Algorithm

A snowball/memory pays a coupon C_i in T_{i+1} which depends on the previous coupon. The coupon C_i is given by an update rule

$$C_i = f(T_i, C_{i-1}, X_i)$$

with an $\mathcal{F}_{T_i}$-measurable X_i (the index), i.e., C_i is path-dependent.

We add the value of the coupon C_i as an additional state and write the product value as a function of this state.

The backward induction from 16.1 gives the product value $V(T_0, C_0)$ as a function of the initial (or past) coupon C_0, which is known.

Remark 210 (In Arrears Fixing): Note that we assumed that the index X_i is a function of the time T_i values of the model primitives and thus $\mathcal{F}_{T_i}$-measurable. If the index X_{i-1} is a function of the time T_i values of the model primitives, i.e., fixed in arrears, then the additional state variable is the value of the *previous* coupon.

16.4.2 Evaluation of a Autocap in a *Backward Algorithm*

An autoccap pays at time T_{i+1} the amount

$$X_i := N \max\left(L_i(T_i) - K_i, 0\right) (T_{i+1} - T_i)$$
$$\cdot \begin{cases} 1 & \text{if } |\{j : j < i \text{ and } L_j(T_j) - K_j > 0\}| < n_{\text{maxEx}} \\ 0 & \text{else,} \end{cases}$$

where $L_i(t) := L(T_i, T_{i+1}; t)$ denotes the forward rate for the period $[T_i, T_{i+1}]$ seen in $t \leq T_i$.

As a function of the processes L_i the payoff X_i is path-dependent since the payoff function is not given by the random variables $L_k(T_i)$ alone, but also by the past realizations of the processes L_k (entering though $L_j(T_j)$, $j < i$).

We extend the model by the stochastic process

$$\eta(t) : \Omega \to \{0, \dots, n-1\},$$
$$\omega \mapsto \big|\{L_j(T_j) - K > 0 \mid T_j \le t,\ j = 1, \dots, n-1\}\big|.$$

Given L_i, η the payoffs X_i are a function of the realizations $L_i(T_i)$, $\eta(T_i)$:

$$X_i := N \ \max\left(L_i(T_i) - K_i,\ 0\right)\ (T_{i+1} - T_i) \cdot \begin{cases} 1 & \text{if } \eta(T_j) < n_{\text{maxEx}} \\ 0 & \text{else.} \end{cases}$$

CHAPTER 17

Sensitivities (Partial Derivatives) of Monte Carlo Prices

17.1 Introduction

The technique of risk-neutral pricing, i.e., the change toward the martingale measure, allows us to calculate the cost of a (self-financing) replication portfolio, to be expressed as an expectation. The determination of the replication portfolio itself is not necessary. However, once a pricing formula or pricing algorithm (e.g., a Monte Carlo simulation) has been derived, the replication portfolio can be given in terms of the partial derivatives of the price with respect to current model parameters (like the initial values of the underlyings).[1] The partial derivatives of the price with respect to the model parameters are also called sensitivities, or *Greeks*. They are important to assess the risk of a financial product; see also Chapter 7.

For complex products, like Bermudan options, an analytic pricing formula is usually not available. The pricing has to be done numerically. Under a high-dimensional model, like the LIBOR market model, the numerical method of choice is usually a Monte Carlo simulation. Given that, we will investigate the numerical calculation of sensitivities (partial derivatives) of Monte Carlo prices.

The simplest way of calculating a derivative is by applying finite differences. Unfortunately, this can lead to a Monte Carlo algorithm giving unstable or inaccurate results.

---

[1] Note that all market parameters enter into model parameters via the calibration of the model.

17.2 Problem Description

Let us consider a pricing algorithm that uses Monte Carlo simulation to calculate the price of a financial product as the expectation of the numéraire-relative value under an equivalent martingale measure $\mathbb{Q}$:

$$V(t_0) \;=\; N(t_0)\,\mathrm{E}^{\mathbb{Q}}\!\left(\frac{V(T)}{N(T)}\,\Big|\,\mathcal{F}_{t_0}\right).$$

We are interested in the calculation of a partial derivative of $V(t_0)$ with respect to some model parameter, e.g., the initial values of the underlying ($\rightarrow$ delta), the volatility ($\rightarrow$ vega), etc.

Since we treat this problem as a general numerical problem, not necessarily related to derivative pricing, we do not adopt a specific model but use a notation that is slightly more general. To fix notation, let us restate Monte Carlo pricing first.

17.2.1 Pricing using Monte-Carlo Simulation

Assume that our model is given as a stochastic process X, for example an Itô process

$$\mathrm{d}X \;=\; \mu\,\mathrm{d}t + \sigma\cdot\mathrm{d}W(t)$$

modeling our model primitives like functions of the underlyings (e.g., financial products (stocks) or rates (forward rates, swap rates, FX rates)). For example, for the Black-Scholes model we would have $X = (\log(S), \log(B))$. Let $X^*(t_i)$ denote an approximation of $X(t_i)$ generated by some (time) discretization scheme, e.g., an Euler scheme:

$$X^*(t_{i+1}) \;=\; X^*(t_i) + \mu(t_i)\Delta t_i + \sigma(t_i)\cdot\Delta W(t_i)$$

or one of the more advanced schemes[2]. We assume that our financial product depends only on realizations of X at a finite number of time points, i.e., we assume that the risk-neutral pricing of the financial product may be expressed as the expectation (with respect to the pricing measure) of a function f of some realizations $Y :=$ $(X(t_0), X(t_1), \ldots, X(t_m))$. This is true for many products (e.g., Bermudan options). If these are approximated through the realizations of the numerical scheme, we have

$$\mathrm{E}(f(Y)\,|\,\mathcal{F}_{t_0}) \;\approx\; \mathrm{E}(f(Y^*)\,|\,\mathcal{F}_{t_0}) \;=\; \mathrm{E}(f((X^*(t_0), X^*(t_1), \ldots, X^*(t_m)))\,|\,\mathcal{F}_{t_0}).$$

Here f denotes the numéraire-relative payoff function.

---

[2] For alternative schemes see, e.g., [21, 70]

The Monte Carlo pricing consists of the averaging over some (often equidistributed) sample paths ω_i, $i = 1, \ldots, n$:

$$E(f(Y^*)|\mathcal{F}_{t_0}) \approx \hat{E}(f(Y^*)|\mathcal{F}_{t_0}) := \frac{1}{n} \sum_{i=1}^{n} f(Y^*(\omega_i)).$$

To summarize: We have two approximation steps involved: The first one approximates the time-continuous process by a time-discrete process. The second one approximates the expectation by a Monte Carlo simulation of n sample paths. This is the minimum requirement to have the pricing implemented as a Monte Carlo simulation.

17.2.2 Sensitivities from Monte Carlo Pricing

Assume that θ denotes some model parameter[3] or a parametrization of a generic market data movement and let Y_θ denote the model realizations dependent on that parameter. Let us further assume that ϕ_{Y_θ} denotes the probability density of Y_θ. Then the analytic calculation of the sensitivity is given by

$$\frac{\partial}{\partial \theta} E(f(Y_\theta) \mid \mathcal{F}_{t_0}) = \frac{\partial}{\partial \theta} \int_{\mathbb{R}^m} f(y) \phi_{Y_\theta}(y) \, dy.$$

While the payoff f may be discontinuous, the density in general is a smooth function of θ in which case the expectation $E(f(Y_\theta)|\mathcal{F}_{t_0})$ (the price) is a smooth function of θ, too. The price inherits the smoothness of ϕ_{Y_θ}.

The calculation of sensitivities using finite differences on a Monte-Carlo-based pricing algorithm is known to exhibit instabilities, if the payoff function is not smooth enough, e.g., if the payoff exhibits discontinuities as for a digital option. The difficulties arise when we consider the Monte Carlo approximation. It inherits the regularity of the payoff f, not that of the density ϕ:

$$\hat{E}(f(Y_\theta)|\mathcal{F}_{t_0}) = \frac{1}{n} \sum_{i=1}^{n} f(Y_\theta(\omega_i)).$$

So while $E(f(Y_\theta)|\mathcal{F}_{t_0})$ may be smooth in θ, the Monte Carlo approximation $\hat{E}(f(Y_\theta)|\mathcal{F}_{t_0})$ may have discontinuities. In this case a finite difference approximation of the derivative applied to the Monte Carlo pricing will perform poorly.

17.2.3 Example: The Linear and the Discontinuous Payout

The challenge in calculating Monte Carlo sensitivities becomes obvious if we consider two very simple examples:

---

[3] So for *delta* θ is an initial value $X(0)$, for *vega* θ denotes a volatility, etc.

17.2.3.1 Linear Payout

First consider a linear payout, say

$$f(X(T)) = a\,X(T) + b.$$

The (discounted) payout depends only on the time T realization of X (as one would have for a European option). Let $Y_\theta(\omega) := X(T, \omega, \theta)$, where θ denotes some model parameter. The partial derivative of the Monte Carlo value of the payout with respect to θ is

$$\frac{\partial}{\partial\theta}\hat{E}(f(Y_\theta)|\mathcal{F}_{t_0}) = \frac{1}{n}\sum_{i=1}^{n}\frac{\partial}{\partial\theta}f(Y_\theta(\omega_i)) = \frac{1}{n}\sum_{i=1}^{n}a\,\frac{\partial}{\partial\theta}Y_\theta(\omega_i).$$

Obviously the accuracy of the Monte Carlo approximation depends on the variance of $\frac{\partial Y_\theta}{\partial\theta}$ only. When $\frac{\partial}{\partial\theta}Y_\theta(\omega_i)$ does not depend on ω_i, then the Monte Carlo approximation gives the exact value of the partial derivative, even if we use only a single path.

17.2.3.2 Discontinuous Payout

Next, consider a discontinuous payout, say

$$f(X(T)) = \begin{cases} 1 & \text{if } X(T) > K \\ 0 & \text{else.} \end{cases}$$

Analytically we know from $Y_{\theta+h} = Y_\theta + \frac{\partial Y_\theta}{\partial\theta}h + O(h^2)$ and

$$E^Q(f(Y_{\theta+h}) \mid \mathcal{F}_{t_0}) = Q\left(\left\{Y_\theta > K - \frac{\partial Y_\theta}{\partial\theta}h - O(h^2)\right\}\right) = \int_{K - \frac{\partial Y_\theta}{\partial\theta}h - O(h^2)}^{\infty}\phi_{Y_\theta}(y)\,dy$$

that

$$\lim_{h\to 0}\frac{1}{2h}(E^Q(f(Y_{\theta+h}) \mid \mathcal{F}_{t_0}) - E^Q(f(Y_{\theta-h}) \mid \mathcal{F}_{t_0})) = \phi_{Y_\theta}(K)\frac{\partial Y_\theta}{\partial\theta}.$$

However, the partial derivative of the Monte Carlo value of the payout is

$$\frac{\partial}{\partial\theta}\hat{E}(f(Y_\theta)|\mathcal{F}_{t_0}) = \frac{1}{n}\sum_{i=1}^{n}\frac{\partial}{\partial\theta}f(Y_\theta(\omega_i)) = 0 \text{ assuming that } Y_\theta(\omega_i) \neq K \text{ for all } i.$$

Thus, here, the partial derivative of the Monte Carlo value is always wrong.

17.2.4 Example: Trigger Products

The two simple examples above suggest that a finite difference approximation of a Monte Carlo price works well if the payout is smooth, but fails if the payout exhibits discontinuities. The problem becomes a bit more subtle if we consider products where the discontinuous behavior is just one part of the payout which, in addition, may also be of more complex nature. Consider, for example, the autocap. For given times $T_1, \ldots, T_n$ the autocap pays at each payment date T_{i+1} the payout of a caplet $\max(L(T_i, T_{i+1}; T_i) - K_i, 0)(T_{i+1} - T_i)$, but does so only if the number of nonzero payments up to T_i is less than some $n_{\max Ex}$. This latter condition represents a trigger which makes the otherwise continuous payoff discontinuous; see Figures 17.1 and 17.2.

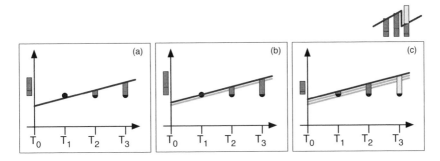

Figure 17.1. *The payoff of an autocap paying a maximum of two out of three caplets, considered under a parallel shift of the interest rate curve (black line). The strike rate is depicted by a dot, a positive payout is marked in dark gray: In scenarios (a) and (b) the first caplet does not lead to a positive payout while the second and third caplet do generate a positive payout. The shift of the interest rate curve from (a) to (b) changes the payout continuously. In scenario (c) the first caplet leads to a positive payout. Since the autocap is limited to two positive payouts the payout of the third caplet is lost as soon as the first caplet pays a positive amount. Thus, from scenario (b) to (c) the payout of the autocap changes discontinuously.*

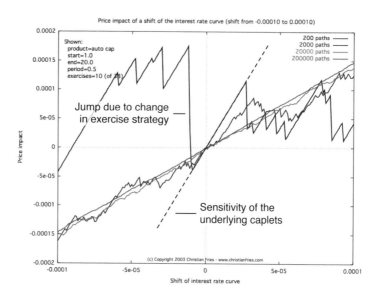

Figure 17.2. *The value of an autocap as a function of the shift size of a parallel shift of the interest rate curve. Using only a small number of paths, a small shift does not lead to a change of the exercise strategy. The price change is driven by the sensitivity of the underlying caplets. Thus, for small shifts one might be tempted to call the sensitivity stable. For a larger shift the exercise strategy changes on some paths, leading to a jump in payoff.*

17.3 Generic Sensitivities: Bumping the Model

The finite difference approximation calculates the sensitivity by

$$\frac{\partial}{\partial \theta} E(f(Y_\theta)|\mathcal{F}_{t_0}) \approx \frac{E(f(Y_{\theta+h})|\mathcal{F}_{t_0}) - E(f(Y_{\theta-h})|\mathcal{F}_{t_0})}{2h}.$$

This *brute-force finite difference* calculation of sensitivities is sometimes referred to as *bumping the model*. Bumping the model has a charming advantage: If you keep your model and your pricing code separated (a design pattern one should always consider), then you may implement a generic code for generating sensitivities by feeding the pricing code with differently bumped models. In other words:

Once the pricing code is written, all sensitivities are available.   (17.1)

It seems as if you get sensitivities almost for free (i.e., without any effort in modeling and implementation) and the only price you pay is a doubling of calculation time compared to pricing. However, it is known that applying such a finite difference approximation to a Monte Carlo implementation will often result in extremely large Monte Carlo errors. Especially if the payout function of the derivative is discontinuous, this Monte Carlo error tends to infinity as h tends to zero. And discontinuous payout is present whenever a trigger feature is present.

Sensitivities in Monte Carlo are known as a challenge. Numerous methods have been proposed for calculating sensitivities in Monte Carlo, among them the likelihood ratio [55] and the application of Malliavin calculus [22], which has attracted increased attention recently [65]. These methods improve the robustness of sensitivities but require more information.

It appears as if the measures you have to take to improve Monte Carlo sensitivities will lose the advantage (17.1) of bumping the model. Later, we will present a method (which is also an implementation design pattern) that makes it possible to calculate sensitivities through bumping the model while providing the accuracy and robustness achieved by the likelihood ratio or Malliavin calculus approach. The method is essentially a likelihood ratio reconsidered on the level of the numerical scheme.

There are basically two different methods for calculating sensitivities in Monte Carlo:

- The pathwise method, which differentiates the payout on every simulation path; see Section 17.5

- The likelihood ratio method, which differentiates the probability density; see Section 17.6.

Numerically, these two methods may be realized as:

- (Traditional) finite differences; see Section 17.4.

- Finite differences applied to a proxy simulation scheme.

However, a proxy simulation scheme is a much more powerful design; see Chapter 18. It is also possible to mix the two approaches by considering a partial proxy simulation scheme; see Section 18.2.

In the following we will present the different methods for calculating sensitivities in Monte Carlo simulations. Each section starts with a short description of the approximating formula and gives the method requirements and properties as bullet points. We assume that a Monte Carlo pricing algorithm has been implemented and we mention only requirements additional to the pricing.

17.4 Sensitivities by Finite Differences

The finite difference approximation is given by

$$\frac{\partial}{\partial \theta} E^Q(f(Y_\theta) \mid \mathcal{F}_{t_0}) \approx \frac{1}{2h}(E^Q(f(Y_{\theta+h}) \mid \mathcal{F}_{t_0}) - E^Q(f(Y_{\theta-h}) \mid \mathcal{F}_{t_0}))$$

$$\approx \frac{1}{2h}(\hat{E}^Q(f(Y_{\theta+h}) \mid \mathcal{F}_{t_0}) - \hat{E}^Q(f(Y_{\theta-h}) \mid \mathcal{F}_{t_0}))$$

$$= \frac{1}{n}\sum_{i=1}^{n}\frac{1}{2h}(f(Y_{\theta+h}(\omega_i)) - f(Y_{\theta-h}(\omega_i)))$$

Requirements

⊕ No additional information from the model SDE[4] X
⊕ No additional information from the simulation scheme $X^*(t_{i+1})$
⊕ No additional information from the payout f
⊕ No additional information on the nature of θ ($\Rightarrow$ generic sensitivities)

Properties

⊖ Biased derivative for *large h* due to finite difference of order h
⊖ Extremely large variance for discontinuous payouts and *small h* (order h^{-1})

The most important feature of finite differences is their genericity. Once the pricing code has been written, all kinds of sensitivities may be calculated.

For smooth payouts, the finite difference approximation converges to the derivative for $h \to 0$. Thus, if the payout is smooth, small shift sizes h are favorable. Using large h the approximation of the derivative is biased.

For discontinuous payouts, as $h \to 0$ the finite difference of the Monte Carlo price does not converge to the derivative of the Monte Carlo price. The reason is that for discontinuous payouts the Monte Carlo approximation ($n \to \infty$) and the approximation of the derivative ($h \to 0$) are not interchangeable.

For discontinuous payouts finite differences with a fixed, small shift size h perform poorly. The contribution of a discontinuity to the sensitivity may be calculated analytically. It is the jump size multiplied by the probability density at the discontinuity. Finite differences resolve this contribution only through those sample paths which fall into a neighborhood around the discontinuity, having the width of the shift size. Thus, if the shift size is small, the discontinuity is resolved by a few points, ultimately resulting in a large Monte Carlo error. For discontinuous payouts large shift sizes

---

[4] SDE: stochastic differential equation.

are preferable. However, if the shift size is large, the derivative becomes biased by second order effects (if present).

Since finite difference does not require anything more than a given pricing algorithm, we are tempted to apply it to any product for which a Monte Carlo pricing may be calculated. If the product exhibits discontinuities in the payout, the finite difference approximation tends to be unreliable, and a careful analysis of the Monte Carlo error for a given shift size h has to be performed.

17.4.1 Example: Finite Differences Applied to Smooth and Discontinuous Payout

Let us consider a finite difference approximation of the partial derivative for the case of the linear payout $f(X(T)) = a\,X(T) + b$ from Section 17.2.3.1. We have

$$
\begin{aligned}
\frac{\partial}{\partial \theta} \mathrm{E}^{Q}(f(Y_\theta) \mid \mathcal{F}_{t_0}) &\approx \frac{1}{2h}\left(\mathrm{E}^{Q}(f(Y_{\theta+h}) \mid \mathcal{F}_{t_0}) - \mathrm{E}^{Q}(f(Y_{\theta-h}) \mid \mathcal{F}_{t_0})\right) \\
&\approx \frac{1}{2h}\left(\hat{\mathrm{E}}^{Q}(f(Y_{\theta+h}) \mid \mathcal{F}_{t_0}) - \hat{\mathrm{E}}^{Q}(f(Y_{\theta-h}) \mid \mathcal{F}_{t_0})\right) \\
&= \frac{1}{n}\sum_{i=1}^{n}\frac{1}{2h}\left(f(Y_{\theta+h}(\omega_i)) - f(Y_{\theta-h}(\omega_i))\right) \\
&= \frac{1}{n}\sum_{i=1}^{n} a\,\frac{1}{2h}\left(Y_{\theta+h}(\omega_i) - Y_{\theta-h}(\omega_i)\right),
\end{aligned}
$$

which is a good approximation, if $\frac{\partial}{\partial \theta}Y_\theta(\omega_i) \approx \frac{1}{2h}(Y_{\theta+h}(\omega_i) - Y_{\theta-h}(\omega_i))$. This is usually the case, and throughout this chapter we assume that the model is such that its realizations $Y_\theta(\omega_i)$ are smooth in the model parameters θ.

For the discontinuous payout $f(X(T)) = 1$ if $X(T) > K$ and $f(X(T)) = 0$ else, considered in Section 17.2.3.2, we have

$$
\begin{aligned}
\frac{\partial}{\partial \theta} \mathrm{E}^{Q}(f(Y_\theta) \mid \mathcal{F}_{t_0}) &\approx \frac{1}{2h}\left(\mathrm{E}^{Q}(f(Y_{\theta+h}) \mid \mathcal{F}_{t_0}) - \mathrm{E}^{Q}(f(Y_{\theta-h}) \mid \mathcal{F}_{t_0})\right) \\
&\approx \frac{1}{2h}\left(\hat{\mathrm{E}}^{Q}(f(Y_{\theta+h}) \mid \mathcal{F}_{t_0}) - \hat{\mathrm{E}}^{Q}(f(Y_{\theta-h}) \mid \mathcal{F}_{t_0})\right) \\
&= \frac{1}{n}\sum_{i=1}^{n}\frac{1}{2h}\left(f(Y_{\theta+h}(\omega_i)) - f(Y_{\theta-h}(\omega_i))\right) \\
&= \frac{1}{n}\sum_{i=1}^{n}\frac{1}{2h}
\begin{cases}
1 & \text{if } Y_{\theta-h}(\omega_i) < K < Y_{\theta+h}(\omega_i) \\
-1 & \text{if } Y_{\theta-h}(\omega_i) > K > Y_{\theta+h}(\omega_i) \\
0 & \text{else.}
\end{cases}
\end{aligned}
$$

This is a valid approximation, but it has a large Monte Carlo variance, since the true value is sampled by 0 and $\frac{1}{2h}$ occurring in the appropriate frequency. If h gets smaller, then we have to represent the true value by a sampling of 0 and a very large constant.

Simplified Example: Assume for simplicity that Y_θ is linear in θ, i.e., we have $Y_{\theta+h} = Y_\theta + \frac{\partial Y_\theta}{\partial \theta} h$ and thus

$$
\frac{f(Y_{\theta+h}(\omega_i)) - f(Y_{\theta-h}(\omega_i))}{2h} =
\begin{cases}
\dfrac{1}{2h} & \text{if } Y_{\theta-h}(\omega_i) < K < Y_{\theta+h}(\omega_i) \\[2mm]
\dfrac{-1}{2h} & \text{if } Y_{\theta-h}(\omega_i) > K > Y_{\theta+h}(\omega_i) \\[2mm]
0 & \text{else.}
\end{cases}
$$

$$
=
\begin{cases}
\dfrac{\text{sign}\left(\frac{\partial Y_\theta}{\partial \theta}\right)}{2h} & \text{if } Y_\theta(\omega_i) \in [K - \epsilon, K + \epsilon] \\[2mm]
0 & \text{else.}
\end{cases}
$$

where $\epsilon := \left|\frac{\partial Y_\theta}{\partial \theta}\right| h$. For the probability we have

$$
q := \mathbb{Q}(Y_\theta \in [K - \epsilon, K + \epsilon]) \approx \phi_{Y_\theta}(K)\, 2\epsilon = \phi_{Y_\theta}(K)\left|\frac{\partial Y_\theta}{\partial \theta}\right| 2h.
$$

In other words: We are sampling the partial derivative of the expectation by a binomial experiment:

$$
\frac{\text{sign}\left(\frac{\partial Y_\theta}{\partial \theta}\right)}{2h} \quad \text{with probability } q \qquad \text{and} \qquad 0 \quad \text{with probability } 1 - q.
$$

The expectation of this binomial experiment is

$$
\frac{\text{sign}\left(\frac{\partial Y_\theta}{\partial \theta}\right)}{2h} q + 0\,(1 - q) \approx \phi_{Y_\theta}(K)\frac{\partial Y_\theta}{\partial \theta},
$$

which is the desired analytic value for the finite difference approximation as $h \to 0$. The variance of the binomial experiment is

$$
\left(\frac{1}{2h}\right)^2 q\,(1 - q) \approx \phi_{Y_\theta}(K)\frac{\partial Y_\theta}{\partial \theta}\,(1 - q)\frac{1}{2h} = O\left(\frac{1}{2h}\right),
$$

which explodes as $h \to 0$.

17.5 Sensitivities by Pathwise Differentiation

The pathwise differentiation method is given by

$$
\frac{\partial}{\partial \theta} E^Q(f(Y(\theta)) \mid \mathcal{F}_{t_0}) = \frac{\partial}{\partial \theta} \int_\Omega f(Y(\omega, \theta)) \, dQ(\omega) = \int_\Omega \frac{\partial}{\partial \theta} f(Y(\omega, \theta)) \, dQ(\omega)
$$

$$
= \int_\Omega f'(Y(\omega, \theta)) \cdot \frac{\partial Y(\omega, \theta)}{\partial \theta} \, dQ(\omega) = E^Q\left(f'(Y(\theta)) \cdot \frac{\partial Y(\theta)}{\partial \theta} \mid \mathcal{F}_{t_0} \right)
$$

$$
\overset{f \text{ smooth}}{\approx} \hat{E}^Q\left(f'(Y(\theta)) \cdot \frac{\partial Y(\theta)}{\partial \theta} \mid \mathcal{F}_{t_0} \right) = \frac{1}{n} \sum_{i=1}^n f'(Y(\omega_i, \theta)) \cdot \frac{\partial Y(\omega_i, \theta)}{\partial \theta}
$$

Requirements

- $\ominus$ Additional information on the model SDE X
- $\oplus$ No additional information on the simulation scheme $X(t_{i+1})$
- $\ominus$ Additional information on the payout f (derivative of f must be known)
- $\ominus$ Additional information on the nature of θ ($\Rightarrow$ no generic sensitivities)

Properties

- $\oplus$ Unbiased derivative.
- $\oplus$ Discontinuous payouts may be dealt with (interpret f' as distribution; see below).

The pathwise method requires the knowledge of the derivative of the payout f' and the derivative of the process realizations with respect to the parameter θ, i.e., $\frac{\partial Y(\omega_i, \theta)}{\partial \theta}$. It is thus only applicable for a restricted class of models and model parameters, where $\frac{\partial Y(\omega_i, \theta)}{\partial \theta}$ may be calculated analytically.

It seems as if a discontinuity in the payout cannot be dealt with, since we require f' to exist. However, the impact of a discontinuity can be calculated analytically; see Section 17.5.2.

It is a major disadvantage of the method that it requires special knowledge of the payout function and of model realizations.

17.5.1 Example: Delta of a European Option under a Black-Scholes Model

We consider a Black-Scholes Model:

$$
S(t) = S(0) \exp(\bar{r}\, t - \frac{1}{2}\bar{\sigma}^2 t + \bar{\sigma}\, W(t)), \qquad S(0) = S_0
$$

$$
B(t) = B(0) \exp(\bar{r}\, t).
$$

(17.2)

In this case we have, e.g.,

$$\frac{\partial}{\partial S_0} S(T) = \frac{S(T)}{S_0}.$$

Using the notation above, our model primitive is $X = (S, B)$. We assume that the payout of our derivative depends on $Y = X(T) = (S(T), B(T))$ only, i.e., we are considering a European option. Then we have

$$\frac{\partial}{\partial S_0} E^Q(f(S(T)) \mid \mathcal{F}_{t_0}) = E^Q\left(f'(S(T)) \cdot \frac{S(T)}{S_0} \mid \mathcal{F}_{t_0} \right)$$

$$\approx \hat{E}^Q\left(f'(S(T)) \cdot \frac{S(T)}{S_0} \mid \mathcal{F}_{t_0} \right)$$

$$= \frac{1}{n} \sum_{i=1}^{n} f'(S(T, \omega_i)) \cdot \frac{S(T, \omega_i)}{S_0}.$$

17.5.2 Pathwise Differentiation for Discontinuous Payouts

In case that the payout f exhibits discontinuities the pathwise method may be applied, provided that f allows for a decomposition

$$f = g + \sum_i \alpha_i \mathbf{1}(\{Y(\theta) > y_i\}),$$

with g being smooth. In this case we have

$$\frac{\partial}{\partial \theta} E^Q(f(Y(\theta)) \mid \mathcal{F}_{T_0}) = \frac{\partial}{\partial \theta} \int_\Omega f(Y(\omega, \theta)) \, d\mathbb{Q}(\omega) = \int_\Omega \frac{\partial}{\partial \theta} f(Y(\omega, \theta)) \, d\mathbb{Q}(\omega)$$

$$= \int_\Omega f'(Y(\omega, \theta)) \cdot \frac{\partial Y(\omega, \theta)}{\partial \theta} \, d\mathbb{Q}(\omega) = E^Q\left(f'(Y(\theta)) \cdot \frac{\partial Y(\theta)}{\partial \theta} \mid \mathcal{F}_{T_0} \right)$$

$$\overset{g \text{ smooth}}{\approx} \hat{E}^Q\left(g'(Y(\theta)) \cdot \frac{\partial Y(\theta)}{\partial \theta} \mid \mathcal{F}_{T_0} \right) + \sum_i \alpha_i \cdot \phi(y_i) \cdot \frac{\partial Y(\theta)}{\partial \theta} \Big|_{Y(\theta)=y_i}.$$

See [83, 94] for examples of how to use pathwise differentiation with discontinuous payouts (there in the context of nth to default swaps, CDOs[5]).

---

[5] CDO: Credit Default Obligation.

17.6 Sensitivities by Likelihood Ratio Weighting

The pathwise method differentiates the path value $Y(\theta)$ of the underlying process realizations Y. Provided there is a probability density $\phi_{Y(\theta)}$ of $Y(\theta)$ we may write the expectation as a convolution with the density. The likelihood ratio weighting [16, 53, 55] is then given by

$$
\frac{\partial}{\partial \theta} \mathrm{E}^Q(f(Y(\theta)) \mid \mathcal{F}_{t_0}) = \frac{\partial}{\partial \theta} \int_\Omega f(Y(\omega, \theta)) \, \mathrm{d}Q(\omega) = \frac{\partial}{\partial \theta} \int_{\mathbb{R}^m} f(y) \, \phi_{Y(\theta)}(y) \, \mathrm{d}y
$$

$$
= \int_{\mathbb{R}^m} f(y) \, \frac{\frac{\partial}{\partial \theta}\phi_{Y(\theta)}(y)}{\phi_{Y(\theta)}(y)} \, \phi_{Y(\theta)}(y) \, \mathrm{d}y = \mathrm{E}^Q(f(Y) \, w(\theta) \mid \mathcal{F}_{t_0})
$$

$$
\approx \hat{\mathrm{E}}^Q(f(Y) \, w(\theta) \mid \mathcal{F}_{t_0}) = \frac{1}{n} \sum_{i=1}^n f(Y(\omega_i)) \, w(\theta, \omega_i),
$$

where

$$
w(\theta) := \frac{\frac{\partial}{\partial \theta}\phi_{Y(\theta)}(Y(\theta))}{\phi_{Y(\theta)}(Y(\theta))}.
$$

Requirements

- $\ominus$ Additional information on the model SDE X $(\to \phi_{Y(\theta)})$
- $\oplus$ No additional information on the simulation scheme $X(t_{i+1})$
- $\oplus$ No additional information on the payout f
- $\ominus$ Additional information on the nature of θ $(\Rightarrow$ no generic sensitivities$)$

Properties

- $\oplus$ Unbiased derivative.
- $\oplus$ Discontinuous payouts may be dealt with.

The likelihood ratio method requires no additional information on the payout function. This is an advantage compared to the pathwise differentiation. However, it requires that the density of the model SDE's realizations $X(t)$ is known and, furthermore, that its derivative is known analytically with respect to the parameter θ. This is rarely the case and thus a major drawback of the method.

The likelihood ratio method does not require the payout to be smooth. The method works very well for calculating the impact of a discontinuity in the payout. However, the method has its problems with smooth payouts: The Monte Carlo error of the approximation using likelihood ratio is larger than the Monte Carlo error of the finite difference approximation. We give a simple example of this effect next.

17.6.1 Example: Delta of a European Option under a Black-Scholes Model Using Pathwise Derivative

Let us look again at a European option using the Black-Schloes model (17.2). Since B is deterministic, we need to consider the probability density of S. Since $\log(S(T))$ is normally distributed (see Chapter 4), we have for the density of $S(T)$

$$\phi_{S(T)}(s) = \frac{\phi_{\text{std.norm.}}\left(\frac{1}{\bar{\sigma}\sqrt{T}}(\log(s) + \bar{r}(T) - \frac{1}{2}\bar{\sigma}(T)^2 - \log(S_0))\right)}{s},$$

where $\phi_{\text{std.norm.}}(x) = \frac{1}{\sqrt{2\pi}}\exp(-x^2/2)$ is the density of the standard normal distribution.

Thus, the delta of a European option with (numéraire-rebased) payout $f(S(T), B(T))$, calculated by the likelihood ratio method, is given by

$$E^Q\left(f(S(T), B(T))\frac{\frac{\partial}{\partial S_0}\phi_{S(T)}(S(T))}{\phi_{S(T)}(S(T))}\ \middle|\ \mathcal{F}_0\right).$$

17.6.2 Example: Variance Increase of the Sensitivity when using Likelihood Ratio Method for Smooth Payouts

For some smooth payouts, the likelihood ratio method may perform less accurately than the pathwise method (Section 17.5) or its finite difference approximation (Section 17.4). A simple example illustrates this effect: Consider constant payout $f(S(T), B(T)) = b$. Then, the likelihood ratio method gives the delta of this option as

$$E^Q\left(f(S(T), B(T))\frac{\frac{\partial}{\partial S_0}\phi_{S(T)}(S(T))}{\phi_{S(T)}(S(T))}\ \middle|\ \mathcal{F}_0\right) = \int_{\mathbb{R}} f(s, B(T))\frac{\partial}{\partial S_0}\phi_{S(T)}(s)\,\mathrm{d}s$$

$$\overset{f=b=\text{const.}}{=} b\int_{\mathbb{R}}\frac{\partial}{\partial S_0}\phi_{S(T)}(s)\,\mathrm{d}s$$

and indeed (using substituation $y = \log(s)$, $\mathrm{d}y = \frac{1}{s}\mathrm{d}s$) we see that the delta is zero:

$$= \int_{\mathbb{R}}\frac{\partial}{\partial S_0}\phi_{\text{std.norm.}}\left(\frac{1}{\bar{\sigma}\sqrt{T}}(y + \bar{r}(T) - \frac{1}{2}\bar{\sigma}(T)^2 - \log(S_0))\right)\mathrm{d}y = 0.$$

The Monte Carlo approximation is

$$\mathrm{E}^{\mathbb{Q}}\left(f(S(T), B(T)) \, \frac{\frac{\partial}{\partial S_0}\phi_{S(T)}(S(T))}{\phi_{S(T)}(S(T))} \, \middle| \, \mathcal{F}_0 \right)$$

$$\approx \frac{1}{n}\sum_{i=1}^{n} f(S(T,\omega_i), B(T)) \cdot \frac{\frac{\partial}{\partial S_0}\phi(S(T,\omega_i))}{\phi(S(T,\omega_i))} = \frac{1}{n}\sum_{i=1}^{n} b \cdot \frac{\frac{\partial}{\partial S_0}\phi(S(T,\omega_i))}{\phi(S(T,\omega_i))},$$

which is in general nonzero. It is an approximation of zero, having some variance.

On the other hand, note that the pathwise method and even a finite difference approximation thereof would give a delta of zero with zero Monte Carlo variance.

17.7 Sensitivities by Malliavin Weighting

The Malliavin weighting [22, 65] is similar to the likelihood ratio method: The sensitivity is expressed as the expectation of a weighted payout function:

$$\frac{\partial}{\partial\theta}\mathrm{E}^{\mathbb{Q}}(f(Y(\theta)) \,|\, \mathcal{F}_{t_0}) = \mathrm{E}^{\mathbb{Q}}(f(Y(\theta)) \, w(\theta) \,|\, \mathcal{F}_{t_0})$$

$$\approx \hat{\mathrm{E}}^{\mathbb{Q}}(f(Y(\theta)) \, w(\theta) \,|\, \mathcal{F}_{t_0}) = \frac{1}{n}\sum_{i=1}^{n} f(Y(\theta,\omega_i)) \, w(\theta,\omega_i).$$

Requirements

⊖ Additional information on the model SDE X ($\rightarrow w$)

⊕ No additional information on the simulation scheme $X(t_{i+1})$

⊕ No additional information on the payout f

⊖ Additional information on the nature of θ ($\Rightarrow$ no generic sensitivities)

Properties

⊕ Unbiased derivative.

⊕ Discontinuous payouts may be dealt with.

Benhamou [47] showed that the likelihood ratio corresponds to the Malliavin weights with minimal variance and may be expressed as a conditional expectation of all corresponding Malliavin weights (we thus view the likelihood ratio as an example of the Malliavin weighting method).

However, here the weights are derived directly through Malliavin calculus, which makes this method more general and applicable even if the density is not known. The derivation of the Malliavin weights requires in-depth knowledge of the underlying continuous process X and it is heavily dependent on the nature of θ.

17.8 Proxy Simulation Scheme

The proxy simulation scheme defines a design of a Monte Carlo pricing engine that has the remarkable properties that the application of finite differences to the pricing will result in likelihood ratio weighted sensitivities without actually the need to know the density ϕ analytically. Thus it combines the robustness of likelihood ratio or Malliavin weighting with the genericity of finite differences.

Since the proxy simulation scheme method is not solely devoted to the calculation of sensitivities, it will be discussed in Chapter 18. Here, we will summarize the key properties.

The Monte Carlo sensitivity under a proxy simulation scheme is given by

$$
\begin{aligned}
\frac{\partial}{\partial \theta} \mathrm{E}^Q(f(Y^*(\theta)) \mid \mathcal{F}_{t_0}) &\approx \frac{1}{2h}(\mathrm{E}^Q(f(Y^*(\theta+h)) \mid \mathcal{F}_{t_0}) - \mathrm{E}^Q(f(Y^*(\theta-h)) \mid \mathcal{F}_{t_0})) \\
&= \frac{\partial}{\partial \theta} \int_{\mathbb{R}^m} f(y) \frac{1}{2h}(\phi_{Y^*(\theta+h)}(y) - \phi_{Y^*(\theta-h)}(y)) \, dy \\
&= \int_{\mathbb{R}^m} f(y) \frac{\frac{1}{2h}(\phi_{Y^*(\theta+h)}(y) - \phi_{Y^*(\theta-h)}(y))}{\phi_{Y^\circ}(y)} \phi_{Y^\circ}(y) \, dy \\
&\approx \frac{1}{n} \sum_{i=1}^{n} f(Y^\circ(\omega_i)) \frac{1}{2h}(w(\theta+h, \omega_i) - w(\theta-h, \omega_i)).
\end{aligned}
$$

—see Section 18.1 for a definition of Y° and Y^*.

Requirements

$\oplus$ No additional information on the model SDE X
$\ominus$ Additional information on the simulation scheme $X^*(t_{i+1})$, $X^\circ(t_{i+1})$
$\oplus$ No additional information on the payout f
$\oplus$ No additional information on the nature of θ ($\Rightarrow$ generic sensitivities)

Properties

$\odot$ Biased derivative (but small shift h possible!).
$\oplus$ Discontinuous payouts may be dealt with.

CHAPTER 18

Proxy Simulation Schemes for Monte Carlo Sensitivities and Importance Sampling

In this chapter we describe the proxy simulation scheme technique as it is given in [66, 69, 70].

18.1 Full Proxy Simulation Scheme

We take the notation of the previous chapter (see Section 17.2 and 17.3) and consider *two* time-discrete schemes for the stochastic process X:

X^* $t_i \mapsto X^*(t_i)$ $i = 0, 1, 2, \ldots$ time discretization scheme of X
                                          $\rightarrow$ *target scheme*

X° $t_i \mapsto X^\circ(t_i)$ $i = 0, 1, 2, \ldots$ any other time-discrete stochastic process
                                          (assumed to be *close* to X^*)
                                          $\rightarrow$ *proxy scheme*

Let $Y = (X(t_1), \ldots, X(t_m))$, $Y^* = (X^*(t_1), \ldots, X^*(t_m))$, $Y^\circ = (X^\circ(t_1), \ldots, X^\circ(t_m))$. Let $\phi_{Y^\circ}(y)$ denote the *density* of Y° and $\phi_{Y^*}(y)$ the *density* of Y^*. We require

$$\forall y : \phi^{Y^\circ}(y) = 0 \Rightarrow \phi^{Y^*}(y) = 0. \tag{18.1}$$

261

18.1.1 Pricing under a Proxy Simulation Scheme

Using the additional scheme X° the pricing of a payout function f is now performed in the following way: We have $\mathrm{E}^{\mathbb{Q}}(f(Y(\theta)) \,|\, \mathcal{F}_{t_0}) \approx \mathrm{E}^{\mathbb{Q}}(f(Y^*(\theta)) \,|\, \mathcal{F}_{t_0})$ and furthermore

$$
\begin{aligned}
\mathrm{E}^{\mathbb{Q}}(f(Y^*(\theta)) \,|\, \mathcal{F}_{t_0}) &= \int_\Omega f(Y^*(\omega, \theta)) \, \mathrm{d}\mathbb{Q}(\omega) = \int_{\mathbb{R}^m} f(y) \, \phi_{Y^*(\theta)}(y) \, \mathrm{d}y \\
&= \int_{\mathbb{R}^m} f(y) \, \frac{\phi_{Y^*(\theta)}(y)}{\phi_{Y^\circ}(y)} \, \phi_{Y^\circ}(y) \, \mathrm{d}y = \mathrm{E}^{\mathbb{Q}}(f(Y^\circ) \, w(\theta) \,|\, \mathcal{F}_{t_0}),
\end{aligned}
$$

where $w(\theta) = \frac{\phi_{Y^*(\theta)}(y)}{\phi_{Y^\circ}(y)}$.

For the Monte Carlo approximation this implies that the sample paths are generated from the scheme X° while the probability densities are corrected toward the target scheme X^*.

18.1.1.1 Basic Properties of a Proxy Simulation Scheme

- For $X^\circ = X^*$ we have $w(\theta) = 1$, and in this case the proxy simulation scheme corresponds to the ordinary Monte Carlo simulation of X^*.

- The proxy scheme X° and thus its realization vector Y° are seen as being independent of θ. This has important implications on the calculation of sensitivities; see Section 18.1.3.

- The requirement $\forall y : \phi^{Y^\circ}(y) = 0 \Rightarrow \phi^{Y^*}(y) = 0$ corresponds to the non-degeneracy condition of the diffusion matrix as it appears in the application of the likelihood ratio and Malliavin weights. However, here this requirement is far less restrictive since we are free to choose the proxy scheme X°.

18.1.2 Calculation of Monte Carlo Weights

For the most common numerical schemes the densities ϕ^{Y°, ϕ^{Y^*} and thus the Monte Carlo weights may be calculated numerically. Consider, for example, the schemes

Target scheme: $\quad X^*(t_{i+1}) = X^*(t_i) + \mu^{X^*}(t_i) \, \Delta t_i + \Sigma(t_i) \cdot \Gamma(t_i) \cdot \Delta U(t_i),$

Proxy scheme: $\quad X^\circ(t_{i+1}) = X^\circ(t_i) + \mu^{X^\circ}(t_i) \, \Delta t_i + \Sigma^\circ(t_i) \cdot \Gamma^\circ(t_i) \cdot \Delta U(t_i),$

where Σ denotes an invertible volatility matrix and Γ denotes a projection matrix, the factor matrix which defines the correlation structure $R = \Gamma\Gamma^\mathsf{T}$.

Assume for simplicity that $\mu^{X^*}(t_i)$ depends on $X^*(t_i)$ only (and similar for $\mu^{X^\circ}(t_i)$) (this holds for the Euler scheme), then we have for the transition probability densities:

$$\phi^{X^*}(t_i, X_i^*; t_{i+1}, X_{i+1}^*)$$
$$= \frac{1}{(2\Pi\,\Delta t_i)^{n/2}} \exp\left(-\frac{1}{2\,\Delta t_i}\,(\Lambda^{-1/2}F^\top\Sigma^{-1}(X_{i+1}^* - X_i^* - \mu^{X^*}(t_i)\,\Delta t_i))^2\right)$$

$$\phi^{X^\circ}(t_i, X_i^\circ; t_{i+1}, X_{i+1}^\circ)$$
$$= \frac{1}{(2\Pi\,\Delta t_i)^{n/2}} \exp\left(-\frac{1}{2\,\Delta t_i}\,(\Lambda^{\circ-1/2}F^{\circ\top}\Sigma^{\circ-1}(X_{i+1}^\circ - X_i^\circ - \mu^{X^\circ}(t_i)\,\Delta t_i))^2\right),$$

where we used the factor decomposition $(\text{PCA})^1$ $\Gamma = F \cdot \sqrt{\Lambda}$ where $\Lambda = \text{diag}(\lambda_1, \ldots, \lambda_m)$ are the nonzero eigenvalues of $\Gamma \cdot \Gamma^\top$.

Then the proxy scheme weights are given by

$$w(t_{i+1})\,|_{\mathcal{F}_{t_k}} = \prod_{j=k}^{i} \frac{\phi^{X^*}(t_j, X_j^\circ; t_{j+1}, X_{j+1}^\circ)}{\phi^{X^\circ}(t_j, X_j^\circ; t_{j+1}, X_{j+1}^\circ)}.$$

18.1.3 Sensitivities by Finite Differences on a Proxy Simulation Scheme

Applying a partial derivative with respect to some model parameter θ to a pricing under a proxy simulation scheme gives

$$\frac{\partial}{\partial\theta}E^Q(f(Y^*(\theta))\,|\,\mathcal{F}_{t_0}) \approx \frac{1}{2h}(E^Q(f(Y^*(\theta+h))\,|\,\mathcal{F}_{t_0}) - E^Q(f(Y^*(\theta-h))\,|\,\mathcal{F}_{t_0}))$$

$$= \frac{\partial}{\partial\theta}\int_{\mathbb{R}^m} f(y)\,\frac{1}{2h}(\phi_{Y^*(\theta+h)}(y) - \phi_{Y^*(\theta-h)}(y))\,\mathrm{d}y$$

$$= \int_{\mathbb{R}^m} f(y)\,\frac{\frac{1}{2h}(\phi_{Y^*(\theta+h)}(y) - \phi_{Y^*(\theta-h)}(y))}{\phi_{Y^\circ}(y)}\,\phi_{Y^\circ}(y)\,\mathrm{d}y$$

$$\approx \frac{1}{n}\sum_{i=1}^{n} f(Y^\circ(\omega_i))\,\frac{1}{2h}(w(\theta+h, \omega_i) - w(\theta-h, \omega_i)).$$

In other words, setting up the pricing using a proxy simulation scheme, apply finite differences to the pricing will result in an approximation of the likelihood ratio rather than an approximation of the pathwise differentiation.

---

^{1} See also Section 19.4.3.3 and Appendix B.3.

Requirements

⊕ No additional information on the model SDE X

⊖ Additional information on the simulation scheme $X^*(t_{i+1})$, $X^\circ(t_{i+1})$

⊕ No additional information on the payout f

⊕ No additional information on the nature of θ ($\Rightarrow$ generic sensitivities)

Properties

⊙ Biased derivative (but small shift h possible!).

⊕ Discontinuous payouts may be dealt with.

We noted above that additional information on the simulation scheme is required, that is, the densities of the two schemes. Note, however, that we require these densities to set up the pricing algorithm. For the sensitivity calculation no additional information is needed. Note also that the required densities are densities of numerical schemes, which can usually be calculated from known transition probability densities (see Section 18.1.2).

18.1.4 Localization

If the payout function f is smooth, then ordinary finite differences perform better than the weighting techniques. The latter shows an increase in Monte Carlo variance of the sensitivity. This effect is not only visible for smooth payouts f, but also for large finite difference shifts.

A solution that has been proposed in [65] is localization. Here the weighting is applied only to a region where the payoff is discontinuous.

Let g denote the localization function, i.e., a smooth function $0 \leq g \leq 1$ such that $g = 1$ at discontinuities of f. Consider the decomposition

$$f = (1 - g) f + g f.$$

We define the pricing of the payout f as

$$\mathrm{E}(f(Y^*)|\mathcal{F}_{t_0}) = \mathrm{E}((1 - g(Y^*)) f(Y^*)|\mathcal{F}_{t_0}) + \mathrm{E}\left(g(Y^\circ) f(Y^\circ) \frac{\phi_{Y^*}}{\phi_{Y^\circ}} \mid \mathcal{F}_{t_0}\right).$$

In other words; we use a pricing based on a proxy simulation scheme for $g f$ and a pricing based on direct simulation for $(1 - g) f$.

It should be noted that localization is carried out by a redefinition of the payout. The product is split into two parts, where one is priced by a direct simulation scheme and the other is priced by a proxy simulation scheme method. This allows us to implement localization on the product level, completely independent of the actual simulation properties. In addition, localization does not reduce the ability to calculate generic sensitivities.

In Section 18.3 we will consider a slightly different variant of localization, which uses information of the payout to modify the numerical scheme.

18.1.5 Object-Oriented Design

The proxy scheme simulation method may in part also be viewed as an implementation design. In Figure 18.1(a) we depict the object-oriented design of a standard Monte Carlo simulation where a change in market data results in a change of simulation path. In Figure 18.1(b) we contrast the proxy scheme simulation method where a change in market data results in a change of Monte Carlo weights.

In practice, we propose that the model driving the generation of the proxy schemes paths is calibrated to market data used for pricing while a market data scenario used for sensitivity calculation, i.e., by bumping the model, only impacts the Monte Carlo weights. A method should be offered to reset the proxy simulation's market data to the target simulation's market data.

18.1.6 Importance Sampling

The key idea of importance sampling is to generate the paths according to their importance to the application, not according to their probability law, and in doing so, adjust toward their probability by a suitable Monte Carlo weight (the change of measure).

Using a proxy simulation scheme, the paths are generated according to the proxy scheme while a Monte Carlo weight adjusts their probability toward the target scheme. Actually, once the proxy simulation scheme framework has been established, the Monte Carlo weights are calculated automatically from the two numerical schemes.

Thus, choosing the proxy scheme such that it creates paths according to their importance to the application is a form of importance sampling. It has the advantage that specifying a suitable process might come easier than calculating the optimal sampling and the corresponding Monte Carlo weights.

18.1.6.1 Example

Let us look at the pricing of an out-of-the-money (OTM) option under a lognormal model (like the Black-Scholes model or the LIBOR market model):

Log Euler scheme: $\log(X)(t_{j+1}) = \log(X)(t_j) + \mu(t)\,\Delta t_j + \sigma\,\Delta W(t_j)$

OTM option: $\max(X(T) - K, 0)$,

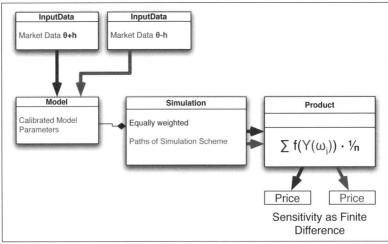

(a) Standard Monte Carlo Simulation

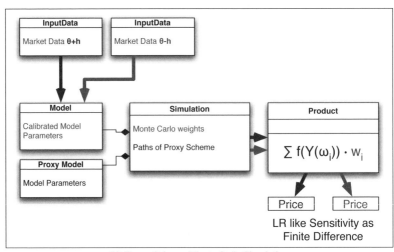

(b) Proxy Scheme Monte Carlo Simulation

Figure 18.1. *Object-oriented design of the Monte Carlo pricing engine: We depict the impact of a change of different market data scenarios $\theta + h$ and $\theta - h$ on the pricing code of a standard Monte Carlo simulation and a proxy scheme simulation.*

where $X(0) = X_0$ and $K \gg X_0$. The drift of the model is determined by the specific pricing measure. However, in our application we would prefer that the mean of $X(T)$ be close to the option strike K rather than being close to $\exp(\log(X_0) + \int_0^T \mu(t)\,dt)$. To achieve this, simply use a proxy scheme with artificial drift:

$$\text{Proxy scheme:} \quad \log(X)(t_{j+1}) = \log(X)(t_j) + \frac{\log(K) - \log(X_0)}{T}\,\Delta t_j + \sigma\,\Delta W(t_j)$$

$$\text{Target scheme:} \quad \log(X)(t_{j+1}) = \log(X)(t_j) + \mu(t)\,\Delta t_j + \sigma\,\Delta W(t_j)$$

This will bring the paths to the region that is important for the pricing of the option, while the proxy simulation scheme framework automatically adjusts probabilities accordingly. Figure 18.2 shows a comparison of the distribution of Monte Carlo prices obtained from direct simulation compared to the prices obtained from importance-adjusted proxy scheme simulation.

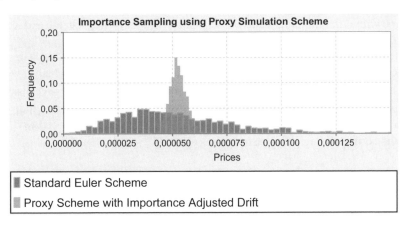

Figure 18.2. *Importance sampling using a drift-adjusted proxy scheme. The example was created using a LIBOR market model to price a caplet with strike $K = 0.3$, the initial forward rate being $X_0 = L_i(0) = 0.1$.*

267

18.2 Partial Proxy Simulation Schemes

The (full) proxy simulation scheme method requires the density of the target scheme realization to be zero if the density of the proxy scheme is zero; see Equation (18.1). In other words, it is required that the paths simulated under the proxy scheme comprise all paths possible under the target scheme. If the property is violated, then the Monte Carlo expectation using the weighted paths of the proxy scheme will leave out some mass. This limits the application of the full proxy simulation scheme. For the calculation of sensitivities the limitation means that we cannot calculate the sensitivity with respect to all possible perturbations.

However, in order to improve the calculation of sensitivities of trigger products it is not necessary to keep all underlying quantities rigid (as for a full proxy simulation); it is sufficient to keep the quantity that induces the discontinuity rigid. This gives rise to the notion of a partial proxy simulation scheme [69].

Let K^0 denote the unperturbed scheme and K^* some perturbation of K^0, e.g., a scheme with different initial data. We will call K^0 the reference scheme and K^* the target scheme.

The usual procedure of bump-and-revalue for computing Greeks would simulate paths of K^* having Monte Carlo weight $\frac{1}{n}$. The proxy simulation schemes would simulate paths of K^0 using Monte Carlo weights $\frac{1}{n} \cdot \frac{\phi^*}{\phi^0}$. Instead, here we consider a third scheme K^1, the *(partial) proxy simulation scheme* where paths are such that the pathwise values of some (but not all) components of K^1 (or a function thereof) agree with the corresponding pathwise quantities under K^0.

18.2.1 Linear Proxy Constraint

Let $\Pi(t_i)$ denote a projection operator of rank k. Let $v(t_i)$ be defined as

$$v(t_i) := (\Pi \cdot \Gamma(t_i))^{-1} \cdot (\Pi \cdot K^*(t_{i+1}) - \Pi \cdot K^0(t_{i+1})), \qquad (18.2)$$

where $(\Pi \cdot \Gamma(t_i))^{-1}$ is the quasi-inverse of $\Pi \cdot \Gamma(t_i)$, i.e., v is the solution of

$$\|\Pi \cdot K^0(t_{i+1}) - \Pi \cdot (K^*(t_{i+1}) - \Pi\Gamma(t_i)v(t_i))\|_{L_2} \to \min. \qquad (18.3)$$

We define the *k-dimensional partial proxy scheme K^1* as:

$$\begin{aligned} K^1(t_0) &:= K^*(t_0), \\ K^1(t_{i+1}) &:= K^*(t_{i+1}) - \Gamma(t_i) \cdot v(t_i). \end{aligned} \qquad (18.4)$$

The scheme K^1 has the following properties:

- It coincides with K^0 on the k-dimensional submanifold defined by Π, i.e., $\Pi \cdot K^1(t_i) = \Pi \cdot K^0(t_i)$.

- It is given through a mean shift $v(t_i)$ on the Brownian increment $\Delta W(t_i)$ of the target scheme K^*.

Consequently, the Monte Carlo weight of the partial proxy scheme is given by

$$w(t_i) = \frac{\phi^{K^*}(t_i, K^1(t_i); t_{i+1}, K^1(t_{i+1}))}{\phi^{K^1}(t_i, K^1(t_i); t_{i+1}, K^1(t_{i+1}))}.$$

In the case of a linear proxy constraint, the mean shift $v(t_i)$ is $\mathcal{F}_{t_i}$-measurable.[2] Then, using simple Euler schemes, the transition probabilities are

$$\phi^{K^1}(t_i, K^1(t_i); t_{i+1}, K^1(t_{i+1})) = \phi^W(t_i, W(t_i), t_{i+1}, W(t_{i+1})),$$
$$\phi^{K^*}(t_i, K^1(t_i); t_{i+1}, K^1(t_{i+1})) = \phi^W(t_i, W(t_i), t_{i+1}, W(t_{i+1}) - v(t_i)).$$
$$(18.5)$$

From this we can derive $w(t_i)$ as a simple analytic formula; see Section 18.2.4.2.

We would like to note that in (18.3) we may replace the projection operator by a general nonlinear function, if necessary. We will discuss this case in Section 18.2.3 and we will consider this case in our example in Section 18.2.7.

18.2.2 Comparison to Full Proxy Scheme Method

The full proxy simulation scheme introduced in Section 18.1 corresponds to $K^1 = K^0$. Thus, it is a special case of Equations (18.2) and (18.4) if Π is the identity and if

$$\Gamma(t_i)v(t_i) := K^*(t_{i+1}) - K^0(t_{i+1}) \qquad (18.6)$$

has a solution $v(t_i)$ (not only in the sense of a closest approximation). If, however, (18.6) has no solution, $v(t_i)$ from (18.2) still defines a valid mean shift for the scheme K^*. The scheme K^1 will be the closest approximation to K^0 fulfilling the measure continuity condition with respect to K^*.

A major advantage of the partial proxy scheme is that the projection Π may be chosen such that (18.2) has an exact solution with respect to the submanifold defined by Π, so K^1 and K^0 coincide on a k-dimensional submanifold. We will make use of this in our example in Section 18.2.6.

18.2.3 Nonlinear Proxy Constraint

An obvious (and commonly required) generalization is to replace the linear projection operator Π by a general, possibly nonlinear function $f : \mathbb{R}^n \to \mathbb{R}^k$ and define $v(t_i)$ as the solution of

$$f(t_{i+1}, K^0(t_{i+1})) = f(t_{i+1}, K^*(t_{i+1}) - \Gamma(t_i) \cdot v(t_i)). \qquad (18.7)$$

---

[2] We will later consider the general case of nonlinear proxy constraints and $\mathcal{F}_{t_{i+1}}$-measurable mean shifts; see Sections 18.2.3 and 18.2.4.

Thus we have $f(t_{i+1}, K^0(t_{i+1})) = f(t_{i+1}, K^1(t_{i+1}))$. An example of an application of this generalization is a LIBOR market model, where f represents a certain swap rate or function of swap rates (e.g., a CMS spread[3]). The condition will then ensure that the path values of the swap rate(s) are the same under K^0 and K^1.

18.2.3.1 Linearization of the Proxy Constraint

While a constraint like (18.7) will be the general application, its numerical implementation may be expensive since one has to solve the nonlinear equation on every path in every time step. However, if $K^*(t_{i+1})$ is a small perturbation of $K^0(t_{i+1})$, we may linearize Equation (18.7). In other words we would set

$$\Pi := f'(K^0(t_{i+1})). \tag{18.8}$$

Note that the proxy simulation method is constructed such that a finite difference using small perturbation will remain stable, i.e., $K^*(t_{i+1})$ may be chosen to be arbitrarily close to $K^0(t_{i+1})$.

18.2.3.2 Finite Difference Approximation of the Nonlinear Proxy Constraint

The linearization (18.8) of f may still result in relatively large computational costs, because the projection operator has to be calculated on every path. Note that we linearize around $K^0(t_{i+1}, \omega)$. Thus the quasi-inverse of $\Pi\Gamma$ has to be calculated on every path in every timestep. If we want to implement a faster calculation of the mean shift $v(t_i, \omega_j)$, we can calculate an approximate solution of (18.7) by guessing the directional shift $\tilde{v}(t_i)$ and finite differences to determine the shift size.

Assume we knew that the directional shift $\tilde{v}(t_i)$ does not lie in $\mathrm{Kern} f'\Gamma$. Then for some $\epsilon > 0$ calculate

$$\Delta_{-\epsilon\Gamma\tilde{v}(t_i)} f := \frac{1}{\epsilon}(f(t_{i+1}, K^*(t_{i+1}) - \epsilon\,\Gamma(t_i) \cdot \tilde{v}(t_i)) - f(t_{i+1}, K^0(t_{i+1}))) \tag{18.9}$$

and set

$$\Gamma \cdot v(t_i) := (\Delta_{-\epsilon\tilde{v}(t_i)})^{-1} \cdot \tilde{v}(t_i) \tag{18.10}$$

in the definition of the partial proxy scheme K^1 (18.4).

This solution has the desirable property that its implementation allows the constraint function f to be specified exogenously by the user; this constraint function may vary with the application.

---

[3] CMS: constant maturity swap; see Section 12.2.6.1

Example: If K is the log of the forward rates under a LIBOR market model and f is a swap rate, i.e., we would like to keep a swap rate rigid, then we can achieve this by modifying the first factor. This corresponds to $\tilde{v}(t_i) = (1, 0, \ldots, 0)$. From (18.9) we can calculate the impact of a shift of the first factor on the swap rate; from (18.10) we can calculate the required magnitude of this shift (it is a scalar equation with a scalar unknown $v_1(t_i)$).

We will consider a constraint like (18.7) next. In our benchmark application, a trigger option on an index like a CMS swap rate is considered under the LIBOR market model.

18.2.4 Transition Probability from a Nonlinear Proxy Constraint

18.2.4.1 The Proxy Constraint Revisited

There is subtle but crucial detail in the definition of the mean shift $v(t_i)$: It is defined by comparing $K^*(t_{i+1})$ to $K^0(t_{i+1})$:

$$f(t_{i+1}, K^0(t_{i+1})) = f(t_{i+1}, K^*(t_{i+1}) - \Gamma(t_i) \cdot v(t_i)), \qquad (18.11)$$

not by comparing $K^*(t_i)$ to $K^0(t_i)$. Thus, in general, $v(t_i)$ is a $\mathcal{F}_{t_{i+1}}$-measurable random variable, but not $\mathcal{F}_{t_{i+1}}$-measurable.[4] If we would define $v(t_i)$ through

$$f(t_{i+1}, K^0(t_i)) = f(t_{i+1}, K^*(t_i) - \Gamma(t_i) \cdot v(t_i)),$$

then it is not guaranteed that

$$f(t_{i+1}, K^0(t_{i+1})) = f(t_{i+1}, K^*(t_{i+1}) - \Gamma(t_i) \cdot v(t_i))$$

holds, after the drift and the diffusion from t_i to t_{i+1} has been applied. To account for the drift we could define $v(t_i)$ through

$$f(t_{i+1}, K^0(t_i) + \mu^0(t_i)\Delta t_i) = f(t_{i+1}, K^*(t_i) + \mu^*(t_i)\Delta t_i - \Gamma(t_i) \cdot v(t_i)), \qquad (18.12)$$

which makes $v(t_i)$ a $\mathcal{F}_{t_i}$-measurable random variable, but there is still no guarantee that the proxy constraint holds after the diffusion has been applied. However, it will be the case for linear constraints.

From this consideration it becomes obvious that for the linearization of the proxy constraint, we would have to linearize around $K^0(t_{i+1})$ and not around $K^0(t_i)$. As a solution of this linearization $v(t_i)$ will be $\mathcal{F}_{t_{i+1}}$-measurable only.

---

[4] In the following we will say $v(t_i)$ is $\mathcal{F}_{t_{i+1}}$-*measurable only*, if it is $\mathcal{F}_{t_{i+1}}$-measurable but not $\mathcal{F}_{t_i}$-measurable.

If the mean shift $v(t_i)$ is defined by (18.11) as an $\mathcal{F}_{t_{i+1}}$-measurable random variable, it means—using Euler schemes—that $v(t_i)$ depends nonlinearily on the increment $\Delta W(t_i)$, and the formula for the corresponding transition probability involves inverting this dependence. Here are two examples.

18.2.4.2 Transition Probabilities for General Proxy Constraints

If the proxy constraint on time t_{i+1} is linear, then it may be realized by an $\mathcal{F}_{t_i}$-measurable mean shift $v(t_i)$. In this case the calculation of the transition probabilities that form the Monte Carlo weight leads to very simple formulas. From (18.5) we find that for an $\mathcal{F}_{t_i}$-measurable mean-shift

$$w(t_i) = \prod_{k=1}^{m} \exp\left(-\frac{(x_k - v_k(t_i))^2 + x_k^2}{2\,\Delta t_i}\right),\qquad (18.13)$$

where $x_k := \Delta W_k(t_i)$.

If the mean shift $v(t_i)$ is only $\mathcal{F}_{t_{i+1}}$-measurable, then it is still possible to obtain a simple analytic formula for the transition probability; however, this formula requires the differentiation of the functional dependence of $v(t_i)$ on the increment $\Delta W(t_i)$.

Consider the general case where the mean shift $v(t_i)$ depends on the Brownian increment $\Delta W(t_i)$, i.e.,

$$v(t_i) = v(t_i, \Delta W(t_i)).$$

Define $\tilde{x} = g(x) := x - v(t_i, x)$. Obviously we have

$$\phi(\tilde{x})\,d\tilde{x} \stackrel{\tilde{x}=g(x)}{=} \phi(g(x)) \det\left(\frac{\partial v(t_i, x)}{\partial x}\right) dx = \frac{\phi(g(x)) \det\left(\frac{\partial g(x)}{\partial x}\right)}{\phi(x)} \phi(x)\,dx. \qquad (18.14)$$

Here x denotes the (realization of the) Brownian increment ΔW and ϕ denotes its probability density. Evaluating functions of $\tilde{x} = g(x)$ corresponds to pricing under the partial proxy scheme K^1; evaluating functions of x corresponds to the pricing under the target scheme K^*. From (18.14) we can read off the Monte Carlo weights for the pricing under the scheme K^1 as

$$w(t_i) = \det\left(I - \frac{\partial v(t_i, x)}{\partial x}\right) \prod_{k=1}^{m} \exp\left(-\frac{(x_k - v_k(t_i))^2 + x_k^2}{2\,\Delta t_i}\right),\qquad (18.15)$$

where $x_k := \Delta W_k(t_i)$.

Obviously this result is not limited to the case of Euler schemes. The only requirement with respect to the scheme is that it is generated by the Brownian increments $\Delta W(t_i)$ (e.g., as for a Milstein scheme). We summarize our result in a theorem.

Lemma 211 (Partial Proxy Simulation Scheme): Let $K^*(t_i)$, $i = 0, 1, 2, \ldots$, denote a numerical scheme generated from the Brownian increments $\Delta W(t_i)$, $i = 0, 1, 2, \ldots$

(target scheme), i.e.,

$$K^*(t_{i+1}) = K^*(t_{i+1}, K^*(t_i), \Delta W(t_i) - v(t_i))$$

Let $K^0(t_i)$, $i = 0, 1, 2, \ldots$ denote another numerical scheme, also generated from the Brownian increments $\Delta W(t_i)$ and close to K^*.

For a given function f (the proxy constraint) let $v(t_i)$ denote a solution of

$$f(t_{i+1}, K^*(t_{i+1}, K^*(t_i), \Delta W(t_i) - v(t_i))) = f(t_{i+1}, K^1(t_{i+1}))$$

and—assuming a solution exists—define the scheme K^1 by

$$K^1(t_{i+1}) := K^*(t_{i+1}, K^*(t_i), \Delta W(t_i) - v(t_i)).$$

Then the Monte Carlo pricing under the scheme K^* is, in the Monte Carlo limit, equivalent to the pricing under the scheme K^1 using the Monte Carlo weights $\prod w_i$ with w_i given by (18.15).

We call the scheme K^1 the (partial) proxy scheme satisfying the proxy constraint $f(t_{i+1}, K^1(t_{i+1})) = f(t_{i+1}, K^0(t_{i+1}))$.

18.2.4.3 Example

Since we desire an implementation that is both generic and fast, we would like to discuss a special case, sufficiently general for all our applications and simple enough to give direct formulas for the transition probabilities:

Assume that $v(t_i)$ is linearly dependent on the increment $\Delta W(t_i)$, i.e.,

$$v(t_i) := A(t_i) \cdot \Delta W(t_i) + b(t_i),$$

with A and b being $\mathcal{F}_{t_i}$-measurable. Then we have for the mean-shifted diffusion

$$\Delta W(t_i) - v(t_i) = (1 - A(t_i)) \cdot (\Delta W(t_i) - b(t_i)).$$

Thus the corresponding transition probability is normally distributed with mean $b(t_i)$ and standard deviation $(1 - A(t_i)) \sqrt{\Delta t_i}$. Note that if the target scheme is a small perturbation of the reference scheme, then $A(t_i)$ is small and $(1 - A(t_i))$ is nonsingular.

So here, the $\mathcal{F}_{t_{i+1}}$-measurable mean shift is given by an $\mathcal{F}_{t_i}$-measurable mean shift b and a scaling of the "factor" ΔW. We will make use of this in our next example: A proxy constraint stabilizing the calculation of vega, the sensitivity with respect to a change in the diffusion coefficient.

18.2.4.4 Approximating an $\mathcal{F}_{t_{i+1}}$-measurable Proxy Constraint by an $\mathcal{F}_{t_i}$-measurable Proxy Constraint

To allow rapid calculation of the transition probability we propose to approximate the proxy constraint (18.11) by (18.12). Thus $v(t_i)$ is an $\mathcal{F}_{t_i}$-measurable mean shift and the ratio of the transition probabilities is given by (18.13).

In addition we propose to linearize this constraint around $K^0(t_i) + \mu^0(t_i)\,\Delta t_i$, defining the linear proxy constraint by $\Pi := f'(K^0(t_i) + \mu^0(t_i)\,\Delta t_i)$.

All of our benchmark examples are based on the approximative constraint (18.12) or its linearization.

18.2.5 Sensitivity with Respect to the Diffusion Coefficients—Vega

If we consider only an $\mathcal{F}_{t_i}$-measurable mean shift applied to the Brownian increment $\Delta W(t_i)$, then the method is not applicable to the calculation of a sensitivity with respect to the diffusion coefficient $\Gamma(t_i)$—a.k.a. *vega*. The reason is simple: There is no $\mathcal{F}_{t_i}$-measurable mean shift that will ensure that the proxy constraint holds at t_{i+1} after a *different* ($\mathcal{F}_{t_{i+1}}$-measurable) diffusion has been applied—not even if the proxy constraint is a linear equation. Neglecting the Brownian increment, as suggested in Section 18.2.4.4, is a step in the wrong direction, since we are interested in the sensitivity with respect to the diffusion coefficient.

Of course, in our general formulation (18.11), an $\mathcal{F}_{t_{i+1}}$-measurable mean shift applied to the diffusion $\Delta W(t_i)$ will ensure that the proxy constraint holds at time t_{i+1}, even if the diffusion coefficient has changed. However, to obtain a simple formula for the transition probability and thus the Monte Carlo weight $w(t_i)$, it is helpful to take an alternative view to the problem: The idea is similar to what is done in the case of a full proxy scheme (see [70]): We modify the diffusion of the proxy scheme to match the diffusion of the reference scheme and calculate the corresponding change of measure. In other words, we use the unperturbed diffusion coefficient for the (partial) proxy scheme. This adjustment is made prior to the calculation of the mean shift $v(t_i)$ for the corresponding proxy constraint, which will correct additional differences in the drift, if any.

From the previous section it is clear that this is equivalent to specifying an $\mathcal{F}_{t_{i+1}}$-measurable mean shift, being linear in the Brownian increment $\Delta W(t_i)$.

18.2.6 Example: LIBOR Target Redemption Note

We are going to calculate delta and gamma for a TARN[5] swap. The coupon for the period $[T_i, T_{i+1}]$ is an inverse floater $\max(K - 2\,L(T_i, T_{i+1}), 0)$ and it is swapped against

---

[5] TARN: target redemption note; see Section 12.2.5.1.

floating rate $L(T_i, T_{i+1})$ until the accumulated coupon reaches a given target coupon. If the accumulated coupon does not reach the target coupon, then the difference to the target coupon is paid at maturity.

Thus the coupon of the TARN is linked to a trigger feature, similar to the digital caplet. However, here, the trigger depends on more than one rate, so it is not sufficient to set up a proxy constraint for a single forward rate, unlike for the digital caplet.

Our unperturbed scheme is the LIBOR market model with the initial yield curve, evolving the log-LIBOR with an Euler scheme. The natural perturbed scheme is then the same, except for a different initial condition. We will use the following proxy constraint:

$$L^1(T_j, T_{j+1}; t) = L^0(T_j, T_{j+1}; t) \qquad \forall\, t \in (T_{j-1}, T_j],$$

for all periods of the model to obtain the preferred proxy scheme. The constraint is realized by a mean shift of the diffusion of the first factor, and since the forward rate follows a lognormal process, we have $v = (v_1, 0, \ldots, 0)$ with

$$v_1(t) = \frac{\log(L^0(T_j, T_{j+1}; t)) - \log(L^1(T_j, T_{j+1}; t))}{f_{1,j}},$$

where $f_{1,j}$ denotes the j-th component of the first factor. We assume here that $f_{1,j} \neq 0$. A nonzero factor loading exists as long as the forward rate $L(T_j, T_{j+1})$ has a nonzero volatility. The results can be improved if the factor having the largest absolute factor loading is chosen (factor pivoting).

Figure 18.3 shows the delta and gamma of a TARN swap for different shift sizes of finite differences applied to standard resimulation and partial proxy scheme simulation. For this example the interest rate curve was upward sloping from 2% to 10% and for the TARN we took $K = 10\%$ and a target coupon of 10%.

With small shifts the variance of the delta and gamma calculated under full reevaluation increases and the mean becomes unstable, while the mean for delta and gamma calculated under partial proxy scheme remains stable and the variance small. For increasing shift size full re-evaluation stabilizes, but higher order effects give a significant bias. Very high shift increases the Monte Carlo variance of the likelihood ratio and thus increases the variance of the delta and gamma calculated under the partial proxy scheme simulation.

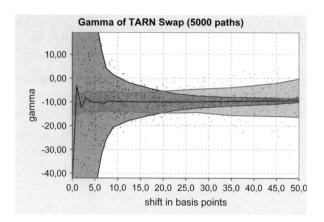

Figure 18.3. *Dependence of the TARN gamma on the shift size of the finite difference approximation. Finite difference is applied to a direct simulation (dark gray) and to a (partial) proxy scheme simulation (gray). Each dot corresponds to one Monte-Carlo simulation with the stated number of paths. The red and green corridors represent the corresponding standard deviation.*
The proxy scheme simulation shows no variance increase for small shift sizes while giving stable expected values for the sensitivity.

18.2.7 Example: CMS Target Redemption Note

Next we will kook at a target redemption note with a coupon $\max(K - 2\,I(T_i), 0)$, where the index $I(T_i)$ is a constant maturity swaprate, i.e., $I(T_i) = S_{i,i+k}(T_i)$ with

$$
\begin{aligned}
S_{i,i+k} &= \frac{P(T_i) - P(T_{i+k})}{\sum_{j=i}^{k-1}(T_{j+1} - T_j)P(T_{j+1})} = \frac{\frac{P(T_i)}{P(T_{i+k})} - 1}{\sum_{j=i}^{k-1}(T_{j+1} - T_j)\frac{P(T_{j+1})}{P(T_{i+k})}} \\
&= \frac{\prod_{l=i}^{i+k-1}(1 + L(T_l)(T_{l+1} - T_l)) - 1}{\sum_{j=i}^{k-1}(T_{j+1} - T_j)\prod_{l=j+1}^{i+k-1}(1 + L(T_l)(T_{l+1} - T_l))}.
\end{aligned}
$$

The swap rate $S_{i,i+k}(t)$ is a nonlinear function of the forward rate curve $L_j(t)$, $j = i, \dots, i + k - 1$ which we denote by S:

$$
S_{i,i+k}(t) = S(L_i(t), \dots, L_{i+k-1}(t)).
$$

From the proxy simulation scheme we require S under L^1 to match S under the reference scheme L^0. Our proxy constraint is therefore

$$
S(L_i^1(t), \dots, L_{i+k-1}^1(t)) = S(L_i^0(t), \dots, L_{i+k-1}^0(t)).
$$

We solve this equation by modifying the first factor, i.e., in each time step t_j we determine a single scalar $v_1(t_j)$ such that

$$S(L_i^*(t_{j+1}) + v_1(t_j) f_{1,i}, \ldots, L_{i+k-1}^*(t_{j+1}) + v_1(t_j) f_{1,i+k-1})$$
$$= S(L_i^0(t_{j+1}), \ldots, L_{i+k-1}^0(t_{j+1})) \quad (18.16)$$

and define $L_i^1(t_{j+1}) := L_i^*(t_{j+1}) + v_1(t_j) f_{1,i}$.

To simplify and speed up the calculation, we (numerically) linearize Equation (18.16) and get an explicit (first-order) formula for v_1; see Equation (18.10).

18.2.7.1 Delta and Gamma of a CMS TARN

The result of the calculation of delta and gamma is depicted in Figure 18.4. Using the simple linearized proxy constraint we see a small increase in Monte Carlo variance for the gamma with very small shifts.

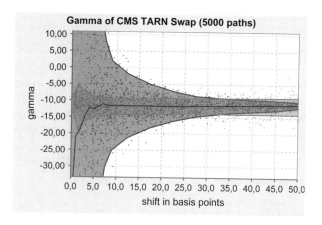

Figure 18.4. *Dependence of the CMS TARN gamma on the shift size of the finite difference approximation. Finite difference is applied to a direct simulation (dark gray) and to a (partial) proxy scheme simulation (gray). The proxy constraint used was a simple (numerical) linearization of (18.16).*

The linearized constraint remains stable for small shifts. However, using a few Newton iterations on the linearization solves the nonlinear constraint and further improves the result for the gamma; see Figure 18.5.

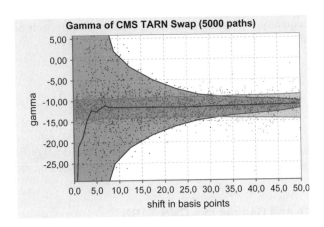

Figure 18.5. *Dependence of the CMS TARN gamma on the shift size of the finite difference approximation. Finite difference is applied to a direct simulation (dark gray) and to a (partial) proxy scheme simulation (gray). The proxy constraint is given by applying a few Newton iterations to the (numerical) linearization of (18.16).*

18.2.7.2 Vega of a CMS TARN

We will calculate the vega of a CMS TARN, i.e., the sensitivity of the CMS TARN with respect to a parallel shift of all instantaneous volatilities. The result is depicted in Figure 18.6. For medium and large shift size the vega calculated from finite differences applied to a partial proxy is similar to the vega calculated from finite differences applied to direct simulation. However, note that for very small shift sizes (around 1 bp), the vega calculated from finite differences applied to direct simulation converges to an incorrect value and that this result occurs with a very small Monte Carlo variance.

The reason for this effect is that the shifts are too small to trigger a change in the exercise strategy. Hence, the vega calculated is the sensitivity conditional on no change in exercise strategy, which is of course a different thing; see Section 17.2.4.

This effect is also present for delta and gamma and for all trigger products, but it has not been visible in the figures so far due to the scale of the shift sizes and the number of paths used there.

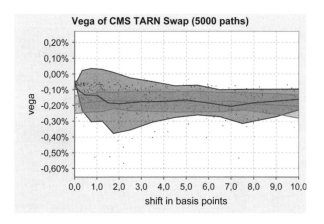

Figure 18.6. *Dependence of the CMS TARN vega on the shift size of the finite difference approximation. Finite difference is applied to a direct simulation (dark gray) and to a (partial) proxy scheme simulation (gray). The proxy constraint was given by applying a few Newton iterations to the (numerical) linearization of (18.16).*

18.3 Localized Proxy Simulation Schemes

18.3.1 Problem Description

Let us consider an asset or nothing option on some underlying S. The asset or nothing pays

$$V(T) = \begin{cases} S(T) & \text{if } S(T) \geq K \\ 0 & \text{else.} \end{cases}$$

in time T, where T is the maturity and K is the strike. Let us assume that our model implies $S(T) > 0$.

Due to the discontinuous payout it seems best to calculate sensitivities using a likelihood ratio method, or - speaking of proxy simulation - to apply a (partial) proxy simulation scheme with a proxy constraint keeping $S(T)$ rigid.

However, for $K \to 0$ the payout of V is $V(T) = S(T)$ and thus smooth. In this case a likelihood ratio method would give extremely noisy results and it is best to calculate sensitivities using the pathwise method.

In Figures 18.7, 18.8 we look at the delta and gamma calculated using direct simulation (pathwise method) or proxy simulation (likelihood ratio method) for a digital caplet with strikes at the forward and away from the forward.

279

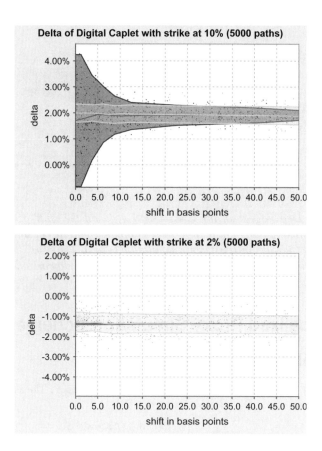

Figure 18.7. *Delta of a digital caplet calculated by finite difference applied to direct simulation (dark gray) and to a partial proxy scheme simulation, internally using the likelihood ratio method (light gray). The forward of the model is at $L(0) = 10\%$. If the the strike K is close to the forward (left figure, $K = L(0) = 10\%$) then the partial proxy scheme (likelihood ratio method) remains stable for small shifts, while the direct simulation (pathwise method) becomes unstable. If the strike K is far from the forward (right figure, $K = 2\%, L(0) = 10\%$) then the partial proxy scheme falls short of the direct simulation due to the huge Monte-Carlo variance introduced by the likelihood ratio.*

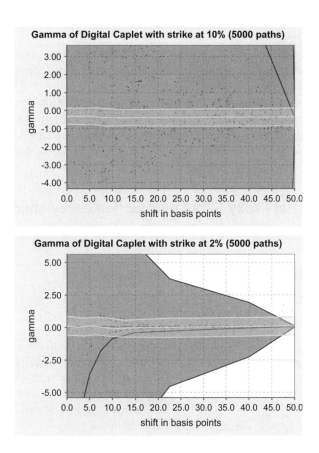

Figure 18.8. *Gamma of a digital caplet calculated by finite difference applied to direct simulation (dark gray) and to a partial proxy scheme simulation, internally using the likelihood ratio (light gray). For gamma the proxy simulation scheme is the method of choice in both cases, $K = L(0) = 10\%$ and $K = 2\%$.*

281

18.3.2 Solution

The idea we present here is to use the likelihood ratio method for those paths ω for which the underlying is close to the discontinuity, while using the pathwise method elsewhere. In other words: we mix the pathwise and likelihood ratio method on a per-path and time-step basis.[6]

Surprisingly, this may be achieved by a simple modification of the partial proxy simulation scheme method, namely through the introduction of a (product dependent) localization function.

Since the location of the discontinuities of a payout is, naturally, known a-priori, it is straightforward to define the localization function as part of the pricing code.

We also suggest an object oriented design that allows the retention of much of the separation of model and product. The model provides a method such that the product can set the localizer before the pricing starts.

18.3.3 Partial Proxy Simulation Scheme (revisited)

We repeat the definition of the partial proxy simulation scheme method.

18.3.3.1 Reference Scheme and Target Scheme

Let a model be given in the form of a stochastic process K_θ. For example an Itô-Process

$$dK_\theta = \mu(t, \theta)dt + \sigma(t, \theta) \cdot dW \tag{18.17}$$

with initial data $K_\theta(0)$, defined over a filtered probability space $(\Omega, \mathcal{F}, \{\mathcal{F}_t \mid t \in [0, T]\}, \mathbb{Q})$ where $\mathbb{Q}$ denotes the pricing measure associated with some numéraire N. Here θ is any model parameter for which we would calculate a sensitivity, i.e. $\frac{\partial}{\partial \theta} E^{\mathbb{Q}}(f(K_\theta)|\mathcal{F}_0)\big|_{\theta=0}$, where f denotes a numéraire relative payout.

Let $0 = t_0 < t_1 < \dots$ denote a time discretization and

$$\{K^\circ(t_i) \mid i = 0, 1 \dots\}$$

a given time discretization scheme of the model K_0. We call K° the primary scheme. Furthermore let

$$\{K^*(t_i) \mid i = 0, 1 \dots\}$$

denote a time discretization scheme for the model K_θ. We call K^* the target scheme.

See Section 13.1 on the time discretizaton of SDEs and Monte-Carlo simulation.

---

[6] The results obtained from using a localized proxy simulation scheme for the test cases in Figures 18.7, 18.8 are shown in Figures 18.9, 18.10 of Section 18.3.7.

18.3.3.2 Transition Probabilities

We assume that the discretized prozess obtained from the discretization scheme is Markovian, such that we may define the transition probility density for the increment $\Delta K^*(t_i)$ as a function of $K^*(t_i)$, $K^*(t_{i+1})$. We will denote the transition probability density of $\Delta K^*(t_i)$ by

$$\phi^{K^*}(t_{i+1}, y, t_i, x) \qquad x = K^*(t_i), \quad y = K^*(t_{i+1})$$

(and correspondingly for $\Delta K^\circ(t_i)$ and all other schemes considered).

18.3.3.3 Proxy Constraint and Proxy Scheme

Let $f : I \times \mathbb{R}^n \to \mathbb{R}^k$ denote a given function, the proxy constraint, fulfilling the following assumption:
  For any $t_i \in I, i > 0$

$$f(t_i, K^\circ(t_i)) = f(t_i, K^P(t_i)) \tag{18.18}$$

has a solution $K^P(t_i)$ such that the transition probability densities of $\Delta K^*(t_i) = K^*(t_{i+1}) - K^*(t_i)$, $\Delta K^P(t_i) = K^P(t_{i+1}) - K^P(t_i)$ fulfill

$$\phi^{K^P}(t_{i+1}, y, t_i, x) = 0 \quad \Longrightarrow \quad \phi^{K^*}(t_{i+1}, y, t_i, x) = 0 \qquad \forall\, i, x, y. \tag{18.19}$$

where

$$K^P(t_0) := K^*(t_0) = K_\theta(0).$$

In other words: Equation (18.18) implicitly defines a scheme $K^P(t_i)$ which coincides with $K^\circ(t_i)$ on the manifold defined through the proxy constraint, but allows a measure transformation to the scheme $K^*(t_{i+1})$.

In the special case where $v(t_i) := \Delta K^*(t_i) - \Delta K^P(t_i)$ is $\mathcal{F}_{T_i}$-measurable, the transition probability of K^P may be given as a modification of the transition probability of K^*, easing calculation. In this case it is

$$\phi^{K^P}(t_{i+1}, y, t_i, x) = \phi^{K^*}(t_{i+1}, y, t_i, x - v) \quad | \quad x = K^P(t_i),\ y = K^P(t_{i+1}).$$

For the general case where $v(t_i)$ depends on $K^*(t_{i+1})$, $K^\circ(t_{i+1})$ we may also derive a simple formula for ϕ^{K^P}.

18.3.3.4 Calculating Expectations using a Proxy Simulation Scheme

For the calculation of expectations we use the simulation scheme K^P in place of K^* and perform a change of measure, i.e. a weighting by $\frac{\phi^{K^*}}{\phi^{K^P}}$. For the expectation operator we have

$$
E^Q(f(K^*(t_0), K^*(t_1), \ldots, K^*(t_n)) | \mathcal{F}_{t_k})
$$

$$
= E^Q \left(f(K^p(t_0), K^p(t_1), \ldots, K^p(t_n)) \cdot \prod_{i=k}^{n-1} \frac{\phi^{K^*}(t_i, K^p(t_i); t_{i+1}, K^p(t_{i+1}))}{\phi^{K^p}(t_i, K^p(t_i); t_{i+1}, K^p(t_{i+1}))} | \mathcal{F}_{t_k} \right).
$$

(18.20)

This is immediately clear using the integral representation of E^Q with the above densities.

18.3.3.5 Example: Euler Schemes

For illustrative purposes, we will assume that K° and K^* are Euler schemes for Itô processes, differing only in the model parameters (initial value, drift and diffusion coefficients), i.e.

$$
\begin{aligned}
K^\circ(t_{i+1}) &= K^\circ(t_i) + \mu^\circ(t_i)\Delta t_i + \Gamma^\circ(t_i)\Delta W(t_i), \quad K^\circ(0) = K_0(0) \\
K^*(t_{i+1}) &= K^*(t_i) + \mu^*(t_i)\Delta t_i + \Gamma^*(t_i)\Delta W(t_i), \quad K^*(0) = K_\theta(0).
\end{aligned}
$$

Let

$$
K^p(t_0) := K^*(t_0).
$$

Let $u(t_i)$ denote the solution of

$$
f(t_{i+1}, K^\circ(t_{i+1})) = f(t_{i+1}, K^p(t_i) + \Delta K^*(t_i) - \Gamma(t_i) \cdot u(t_i)), \quad (18.21)
$$

- implicitly assuming it exists. Then we define

$$
K^p(t_{i+1}) := K^p(t_i) + \Delta K^*(t_i) - \Gamma(t_i) \cdot u(t_i), \quad (18.22)
$$

i.e. with $v(t_i) = \Delta K^*(t_i) - \Delta K^p(t_i)$, $u(t_i)$ solves $\Gamma(t_i) \cdot u(t_i) = v(t_i)$. The scheme K^p has the following properties:

- It coincides with K° on the k-dimensional sub-manifold defined by $f(t_{i+1}, \cdot)$, i.e. $f(t_{i+1}, K^p(t_{i+1})) = f(t_{i+1}, K^\circ(t_{i+1}))$.

- It is given by a mean shift $u(t_i)$ on the Brownian increment $\Delta W(t_i)$ of the target scheme K^*. The change in transition probability is thus trivial to calculate.

18.3.4 Localized Proxy Simulation Scheme

Let K° and K^* be as above. Let $f : I \times \mathbb{R}^n \to \mathbb{R}^k$ denote a given function, the proxy constraint. Let $g : I \times \mathbb{R}^n \to [0, 1]$ denote a given function, the localization function. We define the localized proxy simulation scheme by induction. Let

$$K^{p,loc}(t_0) := K^*(t_0).$$

For $t_i \in I$, $i > 0$ we assume that

$$f(t_{i+1}, K^\circ(t_{i+1})) = f(t_{i+1}, K^{p,loc}(t_i) + \Delta K^p(t_i))$$

has a solution $\Delta K^p(t_i)$. Then we set

$$
\begin{aligned}
K^{p,loc}(t_{i+1}) &:= K^{p,loc}(t_i) + g(t_{i+1}, K^\circ) \cdot \Delta K^p(t_i) + (1 - g(t_{i+1}, K^\circ)) \cdot \Delta K^*(t_i) \\
&= \Delta K^*(t_i) + g(t_{i+1}, K^\circ) \cdot v(t_i),
\end{aligned}
\tag{18.23}
$$

where $v(t_i) = \Delta K^*(t_i) - \Delta K^p(t_i)$ - as above. We assume that f, g allow a solution $K^{p,loc}(t_{i+1})$ such that the transition probability densities of $\Delta K^*(t_i) = K^*(t_{i+1}) - K^*(t_i)$, $\Delta K^{p,loc}(t_i) = K^{p,loc}(t_{i+1}) - K^{p,loc}(t_i)$ fulfill

$$\phi^{K^{p,loc}}(t_{i+1}, y, t_i, x) = 0 \quad \Longrightarrow \quad \phi^{K^*}(t_{i+1}, y, t_i, x) = 0 \qquad \forall\, i, x, y. \tag{18.24}$$

The function g is called the *localization function*. The localized proxy simulation scheme has the following properties:

- At times t_{i+1} and paths ω where $g(t_{i+1}, K^\circ(\omega)) = 1$, the value of f applied to the realization $K^{p,loc}(t_{i+1}, \omega)$ coincides with the value of f applied to the realization $K^\circ(t_{i+1}, \omega)$ of the primary scheme. In other words, at $g = 1$ the quantity f stays rigid.

- At times t_{i+1} and paths ω where $g(t_{i+1}, K^\circ(\omega)) = 0$, the increments of $\Delta K^{p,loc}(t_i)$ coincide with the increments of $\Delta K^*(t_i)$ (as would be the case for a perturbation of a simulation scheme using the pathwise method).

We assume that the localization function g is such that there is a change of measure allowing us to write an expectation of a function of K^* as an expectation of a function of $K^{p,loc}$.

There is a subtle point in the definition of the localized proxy simulation scheme: In (18.23) the localization function depends on K°, thus it does not depend on the model parameter θ. This makes the localization more robust, e.g. if the localization function is not smooth. Note: In most applications the localization function at time t_{i+1} will depend on $K^\circ(t_{i+1})$, i.e. $g(t_{i+1}, K^\circ) = g(t_{i+1}, K^\circ(t_{i+1}))$ in (18.23), but it is also possible to have a localization function that depends on past realizations $g(t_{i+1}, K^\circ) = g(t_{i+1}, \{K^\circ(t_j) | j = 0, \dots, i + 1\})$ (a target redemption note is such an example).

18.3.5 Example: Euler Schemes

As in 18.3.3.5 let us assume that K° and K^* are Euler schemes for Itô processes, differing only in the model parameters (initial value, drift and diffusion coefficients), i.e.

$$K^\circ(t_{i+1}) = K^\circ(t_i) + \mu^\circ(t_i)\Delta t_i + \Gamma^\circ(t_i)\Delta W(t_i), \quad K^\circ(0) = K_0(0)$$
$$K^*(t_{i+1}) = K^*(t_i) + \mu^*(t_i)\Delta t_i + \Gamma^*(t_i)\Delta W(t_i), \quad K^*(0) = K_\theta(0).$$

Let

$$K^{p,loc}(t_0) := K^*(t_0).$$

Let $u(t_i)$ denote the solution of the proxy constraint

$$f(t_{i+1}, K^\circ(t_{i+1})) = f(t_{i+1}, K^{p,loc}(t_i) + \Delta K^*(t_i) - \Gamma(t_i) \cdot u(t_i)), \tag{18.25}$$

- implicitly assuming it exists. Then we define

$$K^{p,loc}(t_{i+1}) := K^{p,loc}(t_i) + \Delta K^*(t_i) - \Gamma(t_i) \cdot g(t_{i+1}) \cdot u(t_i). \tag{18.26}$$

The scheme $K^{p,loc}$ has the following properties:

- At times t_{i+1} and on paths where $g(t_{i+1}) = 1$, it coincides with K° on the k-dimensional sub-manifold defined by $f(t_{i+1}, \cdot)$, i.e. $f(t_{i+1}, K^{p,loc}(t_{i+1})) = f(t_{i+1}, K^\circ(t_{i+1}))$.

- It is given through a mean shift $g(t_{i+1})u(t_i)$ on the Brownian increment $\Delta W(t_i)$ of the target scheme K^*. The change in transition probability is thus trivial to calculate.

18.3.6 Implementation

It may seem that the implementation of the localized proxy simulation scheme is difficult and resource intensive. First, the partial proxy simulation scheme K^p is defined only implicitly by the proxy constraint. Second, $K^{p,loc}$ is calculated as an interpolation of K^p and K^*. So all in all it appears as if we are required to do four simulations.

 However, for the standard Euler scheme at least, the localized proxy simulation is just a simple modification to a standard Monte-Carlo simulation, where a product calculates the required mean shift $v(t_i)$ and provides it to the model. It may be implemented in an object oriented design using just a small amount of additional code. It will not be required to calculate K^* or K^p explicitly.

286

18.3.7 Examples and Numerical Results

18.3.7.1 Localizers

We investigate two simple localization functions. The first based on a piecewise constant function

$$h_{\text{lin}}(x, \epsilon_1, \epsilon_2) = \begin{cases} 1 & \text{for } |x| < \epsilon_1. \\ \frac{|x| - \epsilon_1}{\epsilon_2 - \epsilon_1} & \text{for } \epsilon_1 \leq |x| \leq \epsilon_2. \\ 0 & \text{for } |x| > \epsilon_2. \end{cases}$$

The second being a smooth variant

$$h_{\text{exp}}(x, \epsilon_1, \epsilon_2) = \begin{cases} 1 & \text{for } |x| < \epsilon_1. \\ \exp\left(-\frac{1}{2}\left(\frac{|x| - \epsilon_1}{\epsilon_2 - \epsilon_1}\right)^2\right) & \text{for } \epsilon_1 \leq |x|. \end{cases}$$

In our numerical experiment we found virtually no difference between the use of h_{lin} versus h_{exp}. However the choice of the localization domain given by ϵ_1, ϵ_2 is relevant.

18.3.7.2 Model

As our model SDE we consider a standard LIBOR market model, see Chapter 19.

18.3.7.3 Example: Digital Caplet

We consider a LIBOR market model $L = exp(K)$ with K as in (18.17). The proxy constraint is

$$L_{i+1}^{\circ}(t_{i+1}) = L_{i+1}^{p}(t_{i+1}).$$

We use the localization function

$$g(t_{i+1}) = g(t_{i+1}, L_{i+1}^{\circ}(t_{i+1})) = h_{\text{lin}}(L_{i+1}^{\circ}(t_{i+1}) - K, \epsilon_1, \epsilon_2) \qquad \text{if } t_{i+1} = t_k,$$

$$g(t_{i+1}) = 0 \qquad \text{else.}$$

where t_k is the exercise date of the option, K its strike and $L_{i+1}^{\circ}(t_{i+1})$ is the LIBOR rate calculated from the reference scheme K°.

Numerical Results

We perform a numerical calculation with the simplified model data $L(0) = 10\%$, $\sigma = 20\%$ and the drift μ being chosen as the risk neutral drift under terminal measure.

Using $\epsilon_1 = 1\%$, $\epsilon_2 = 2\%$ (which is a good, but not the optimal choice) we obtain the results shown in Figures 18.9, 18.10. The localized proxy simulation scheme beats all competing methods (direct simulation (pathwise methods) or partial proxy (likelihood ratio method)) for options with strikes both at the forward and distant from the forward. It also gives much better results than the (non-localized) partial proxy simulation scheme for gamma.

18.3.7.4 Example: Target Redemption Note (TARN)

We consider a more sophisticated example: a target remption note with a structured coupon. The target redemption note matures (and pays back the notional) if the cumulated coupon hits a pre-defined target coupon. In contrast to the digital caplet:

- The trigger criteria is (in general) path dependent, e.g. a cumulated coupon.

- The discontinuity is given by a change in maturity (chosen from a discrete set of observation dates). Thus almost all paths will exhibit a discontinuty.

As a consequence, the definition of a localizer is slightly more complex. The localizer itself will be path-dependent.

We give a short definition of the TARN, see also [91]: Let $0 = T_0 < T_1 < T_2 < \ldots < T_n$ denote a given tenor structure. For $i = 1, \ldots, n - 1$ let C_i denote a (generalized) "interest rate" (the coupon) for the periods $[T_i, T_{i+1}]$, respectively. We assume that C_i is a $\mathcal{F}_{T_i}$-measurable random variable (natural fixing). Furthermore let N_i denote a constant value (notional). A *target redemption note* pays

$$N_i \cdot X_i \qquad \text{at} \qquad T_{i+1},$$

where

$$X_i := \begin{cases} C_1 & \text{for } i = 1, \\ \min(C_i , K - \sum_{k=1}^{i-1} C_k) & \text{for } i > 1 \end{cases} \qquad \text{(structured coupon)}$$

$$+ \begin{cases} 1 & \text{for } \sum_{k=1}^{i-1} C_k < K <= \sum_{k=1}^{i} C_k \text{ or } i = n, \\ 0 & \text{else.} \end{cases} \qquad \text{(redemption)}$$

$$+ \begin{cases} \max(0 , K - \sum_{k=1}^{i} C_k) & \text{for } i = n \\ 0 & \text{else.} \end{cases} \qquad \text{(target coupon guarantee).}$$

The payoff of the target redemption note contains the discontinuous (digital) part

$$\begin{cases} 1 & \text{for } \sum_{k=1}^{i-1} C_k < K <= \sum_{k=1}^{i} C_k, \\ 0 & \text{else.} \end{cases}$$

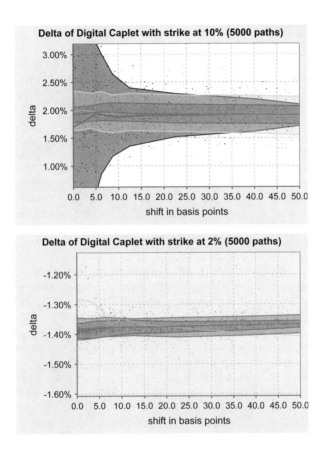

Figure 18.9. *Delta of a digital caplet calculated by finite difference applied to direct simulation (dark gray), to a partial proxy scheme simulation (light gray) and to a localized proxy simulation scheme (gray). The initial forward rate of the model is at 10%. If the the strike K is close to the forward rate (left figure) then the partial proxy scheme (likelihood ratio method) remains stable for small shifts, while the direct simulation (pathwise method) becomes unstable. If the strike K is far away from the forward rate, the partial proxy scheme falls short of the direct simulation due to the huge Monte-Carlo variance introduced by the likelihood ratio.*

289

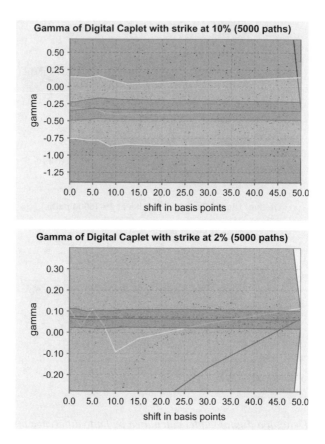

Figure 18.10. *Gamma of a digital caplet calculate by finite difference applied to direct simulation (dark gray), to a partial proxy scheme simulation (light gray) and to a localized proxy simulation scheme (gray). As in Figure 18.7, considering different strikes shows that one or the other methods prevails.*

18.3.7.5 Proxy Constraint and Localizer for the Target Redemption Note

A pathwise payoff of the TARN depends discontinuously on the cumulated coupon $\sum_{k=1}^{i} C_k$. Consequently the proxy constraint is

$$f(t) = \sum_{k=1}^{i-1} C_k(T_k) + C_i(t) \qquad \forall \; T_{i-1} < t \leq T_i.$$

Let

$$K_{i-1} := K - \sum_{k=1}^{i-1} C_k$$

denote the time T_i trigger level. K_{i-1} is T_{i-1}-measurable, so conditional to $\mathcal{F}_{T_{i-1}}$ it is a constant. Then the localizer is given by

$$g(t) = h_{\exp}(K_{i-1} - C_i(t), \epsilon_1, \epsilon_2).$$

Numerical Results

Figure 18.11 shows some numerical results comparing direct simulation, partial proxy simulation and localized proxy simulation for a TARN, calculating Gamma.

 Further Reading: The (full) proxy simulation scheme method was instroduced in [70]. The partial proxy simulation scheme method is found in [69]. The localized partial proxy simulation scheme method is found in [66]. ◁|

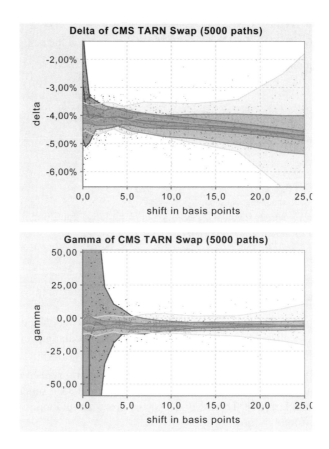

Figure 18.11. *Delta and Gamma of a target redemption note (the coupon is a reverse CMS rate) calculated by finite difference applied to direct simulation (dark gray), to a partial proxy scheme simulation (light gray) and to a localized proxy simulation scheme (gray). Direct simulation produces enormous Monte-Carlo variances for small shift sizes. The method is useless. The partial proxy simulation scheme shows an increase in Monte-Carlo variance if the shift size is large. The localized proxy simulation scheme is an improvement on the partial proxy simulation scheme and shows only small Monte-Carlo variance for large shifts. Note: The localizer used is not the optimal one.*

Part V

Pricing Models for Interest Rate Derivatives

Motivation and Overview - Part III

Up until now we have been considering models of a single scalar stochastic process and options on it: The Black-Scholes model for a stock S, or the Black model for a forward rate L. The true challenge in evaluation of interest rate products lies in the modeling of the whole interest rate curve (instead of a scalar) and in the evaluation of complex derivatives, which depend on the whole curve.

Historically the path to modeling the interest rate curve started with the modeling of the short rate, from which we may calculate the whole interest rate curve; see Remark 104. The initial motivation for considering the short rate derived from the wish to model a scalar quantity, thus to be able to apply familiar numerical methods from stock models, e.g., binomial trees.

For didactic reasons we are not going to present things chronologically. Instead, we consider the LIBOR market model first. It is a high-dimensional model, which discretizes the interest rate curve into a finite number of forward rates. It is highly flexibly due to its huge number of free parameters. It will allow us to study model properties like *mean reversion, number of factors* (Chapter 25), *instantaneous volatility*, and *instantaneous and terminal correlation* (Chapter 21). Despite its presumed complexity, the LIBOR market model is essentially a very simple model: It is nothing more than the simultaneous consideration of multiple Black models under a common measure. So we are carrying on from Chapter 10.

For the *short rate models* the modeled quantity is the *short rate*, a quantity not directly observable. Here we model quantities which are observable as market quotes, like the *LIBOR* or the *swap rate*. The class of models that model quantities which are directly observable on the market are called "market models". We will look at the *LIBOR market model* first.

CHAPTER 19

LIBOR Market Model

> The pure and simple truth
> is rarely pure and never simple.

Oscar Wilde
The Importance of Being Earnest [39].

We assume a time discretization (*tenor structure*)

$$0 = T_0 < T_1 < \cdots < T_n.$$

We model the forward rates $L_i := L(T_i, T_{i+1})$ for $i = 0, \ldots, n - 1$; see Definition 99. This represents a discretization of the interest rate curve, where the continuum of maturities has been discretized.[1]

The *LIBOR market model* assumes a lognormal dynamic for *LIBOR*s $L_i := L(T_i, T_{i+1})$, i.e.,[2]

$$\frac{dL_i(t)}{L_i(t)} = \mu_i^{\mathbb{P}}(t)\, dt + \sigma_i(t)\, dW_i^{\mathbb{P}}(t) \qquad \text{for } i = 0, \ldots, n - 1, \text{ under } \mathbb{P}, \qquad (19.1)$$

with initial conditions

$$L_i(0) = L_{i,0}, \qquad \text{with } L_{i,0} \in [0, \infty), i = 0, \ldots, n - 1,$$

where $W_i^{\mathbb{P}}$ denote (possibly instantaneously correlated) $\mathbb{P}$-Brownian motions with

$$dW_i^{\mathbb{P}}(t)\, dW_j^{\mathbb{P}}(t) = \rho_{i,j}(t)\, dt.$$

Let $\sigma_i : [0, T] \mapsto \mathbb{R}$ and $\rho_{i,j} : [0, T] \mapsto \mathbb{R}$ be deterministic functions and μ_i the drift as $\mathcal{F}_t$-adapted process. By $R(t) := (\rho_{i,j}(t))_{i,j=0,\ldots,n-1}$ we denote the correlation matrix.

---

[1] In practice it is normal to model semiannual or quarterly rates $T_{i+1} - T_i = 0.25$ and to consider these up to a maturity of 20 or 30 years, giving 80 or 120 interest rates to model.
[2] We denote the simulation time parameter of the stochastic process by t.

Motivation: Equation (19.1) is a lognormal model for the *forward rates* L_i. If we consider only a single equation, i.e., fix $i \in \{1, \ldots, n-1\}$, it represents the Black model considered in Chapter 10: Equation (19.1) is identical with Equation (10.1). If we change the measure such that L_i is drift-free (see Chapter 10), we see that the terminal distribution of L_i is lognormal.

Thus, the LIBOR market model is equivalent to the consideration of n Black models under a unified measure.

As was discussed in Chapter 10, to evaluate a caplet under this model it is not relevant that σ_i is time dependent (we have assumed time dependency of σ_i in Chapter 10 for didactical reasons). However, for the value of complex derivatives the time dependency matters. A further degree of freedom introduced in (19.1) is the *instantaneous correlation* $\rho_{i,j}$ of the driving Brownian motions. For the value of a caplet the instantaneous correlation is insignificant (indeed, it does not enter in the Black model). For the evaluation of swaptions the correlation of the forward rates is significant.

For further generalizations of the model, consider nondeterministic σ_i, i.e., stochastic volatility models. In this case the terminal LIBOR distributions no longer correspond to the Black model ones, which is, of course, intended. Equation (19.1) is to be seen as a starting point of a whole model family. The model (19.1) has been chosen as the starting point, because (historically) the lognormal (Black) model is well understood, especially by traders.[3]   ◁|

Remark 212 (Interest Rate Structure): Equation (19.1) models the evolution of the LIBOR $L(T_i, T_{i+1})$. Without further interpolation assumption, these are the shortest forward rates that can be considered in our time discretization (*tenor structure*). The equation system (19.1) thus determines the evolution of all bond prices with maturities T_i and all forward rates for the periods $[T_i, T_k]$, since

$$1 + L(T_i, T_k)(T_k - T_i) = \frac{P(T_i)}{P(T_k)} = \prod_{j=i}^{k-1} \frac{P(T_j)}{P(T_{j+1})}$$

$$= \prod_{j=i}^{k-1} (1 + L(T_j, T_{j+1})(T_{j+1} - T_j)).$$

To shorten notation we write $\delta_i := T_{i+1} - T_i$, $i = 0, \ldots, n-1$ for the period length.

---

[3] Caplet prices are quoted by traders by the implied Black volatility. This is, of course, just another unit of the price, since the Black model is a one-to-one map from price to implied volatility.

19.1 Derivation of the Drift Term

As in Chapters 10 and 11, our first step is to choose some numéraire N and derive the drift under a martingale measure $\mathbb{Q}^N$. If the processes have been derived under the martingale measure $\mathbb{Q}^N$, then the (discretized) interest rate curve may be simulated numerically and a derivative V may be priced through $V(0) = N(0)\mathrm{E}^{\mathbb{Q}^N}(\frac{V}{N}|\mathcal{F}_{T_0})$ (see Chapter 13).

We fix a numéraire N. Let the assumptions of Theorem 74 hold such that there exists a corresponding equivalent martingale measure $\mathbb{Q}^N$ such that N-relative prices are martingales. From Theorem 59 under $\mathbb{Q}^N$ the process (19.1) has a changed drift, namely

$$\frac{\mathrm{d}L_i(t)}{L_i(t)} = \mu_i^{\mathbb{Q}^N}(t)\,\mathrm{d}t + \sigma_i(t)\,\mathrm{d}W_i^{\mathbb{Q}^N}(t) \qquad \text{for } i = 0, \ldots, n-1. \tag{19.2}$$

19.1.1 Derivation of the Drift Term under the Terminal Measure

We fix the T_n-bond $N(t) = P(T_n; t)$ as numéraire. From Theorem 59 under $\mathbb{Q}^{P(T_n)}$ the process (19.1) has a changed drift:

$$\frac{\mathrm{d}L_i(t)}{L_i(t)} = \mu_i^{\mathbb{Q}^{P(T_n)}}(t)\,\mathrm{d}t + \sigma_i(t)\,\mathrm{d}W_i^{\mathbb{Q}^{P(T_n)}}(t) \qquad \text{for } i = 0, \ldots, n-1. \tag{19.3}$$

We need to determine $\mu_i^{\mathbb{Q}^{P(T_n)}}$. The martingale measure $\mathbb{Q}^{P(T_n)}$ corresponding to $N(t) = P(T_n; t)$ is also called *terminal measure* (since T_n is the time horizon of our time discretization).

As in Chapter 10, we will construct relative prices with respect to $P(T_n)$ and obtain equations from which we will derive the drifts μ_i. From Definition 99

$$\prod_{k=i}^{n-1} \underbrace{(1 + \delta_k L_k)}_{= \frac{P(T_k)}{P(T_{k+1})}} = \prod_{k=i}^{n-1} \frac{P(T_k)}{P(T_{k+1})} = \frac{P(T_i)}{P(T_n)} \qquad \text{for } i = 0, \ldots, n-1. \tag{19.4}$$

Since we have a $P(T_n)$-relative price of a traded product on the right-hand side in (19.4), we have for the drifts:

$$\operatorname*{Drift}_{\mathbb{Q}^{P(T_n)}}\left[\prod_{k=i}^{n-1}(1 + \delta_k L_k) \right] = 0, \qquad i = 0, \ldots, n-1.$$

We apply Theorem 48 and obtain $\forall\ i = 0, \ldots, n-1$

$$
d\left(\prod_{k=i}^{n-1} (1 + \delta_k L_k) \right) = \sum_{\substack{j=i}}^{n-1} \prod_{\substack{k=i \\ k \neq j}}^{n-1} (1 + \delta_k L_k)\, \delta_j\, dL_j + \sum_{\substack{j,l=i \\ l>j}}^{n-1} \prod_{\substack{k=i \\ k \neq j,l}}^{n-1} (1 + \delta_k L_k)\, \delta_j\, dL_j\, \delta_l\, dL_l
$$

$$
= \prod_{k=i}^{n-1} (1 + \delta_k L_k) \left(\sum_{j=i}^{n-1} \frac{\delta_j\, dL_j}{1 + \delta_j L_j} + \sum_{\substack{j,l=i \\ l>j}}^{n-1} \frac{\delta_j\, dL_j}{1 + \delta_j L_j} \frac{\delta_l\, dL_l}{1 + \delta_l L_l} \right)
$$

$$
= \prod_{k=i}^{n-1} (1 + \delta_k L_k) \sum_{j=i}^{n-1} \left(\frac{\delta_j\, dL_j}{1 + \delta_j L_j} + \sum_{\substack{l \geq j+1 \\ l \leq n-1}} \frac{\delta_j\, dL_j}{1 + \delta_j L_j} \frac{\delta_l\, dL_l}{1 + \delta_l L_l} \right).
$$

Since $\forall\ i = 0, \ldots, n-1$

$$
\operatorname*{Drift}_{\mathbb{Q}^{P(T_n)}} \left[\prod_{k=i}^{n-1} (1 + \delta_k L_k) \right] = 0 \tag{19.5}
$$

it follows that $\forall\ i = 0, \ldots, n-1$

$$
\sum_{j=i}^{n-1} \operatorname*{Drift}_{\mathbb{Q}^{P(T_n)}} \left[\frac{\delta_j\, dL_j}{1 + \delta_j L_j} + \sum_{\substack{l \geq j+1 \\ l \leq n-1}} \frac{\delta_j\, dL_j}{1 + \delta_j L_j} \frac{\delta_l\, dL_l}{1 + \delta_l L_l} \right] = 0 \tag{19.6}
$$

and thus $\forall\ j = 0, \ldots, n-1$

$$
\operatorname*{Drift}_{\mathbb{Q}^{P(T_n)}} \left[\frac{\delta_j\, dL_j}{1 + \delta_j L_j} + \sum_{\substack{l \geq j+1 \\ l \leq n-1}} \frac{\delta_j\, dL_j}{1 + \delta_j L_j} \frac{\delta_l\, dL_l}{1 + \delta_l L_l} \right] = 0. \tag{19.7}
$$

If we now use

$$
dL_j = L_j \mu_j^{\mathbb{Q}^{P(T_n)}}\, dt + L_j \sigma_j\, dW_j^{\mathbb{Q}^{P(T_n)}} \qquad \text{and} \qquad dL_j\, dL_l = L_j L_l \sigma_j \sigma_l \rho_{j,l}\, dt
$$

in (19.7), then we have

$$
\mu_j^{\mathbb{Q}^{P(T_n)}} \frac{\delta_j \cancel{L_j}}{1 \cancel{+} \delta_j L_j} + \sum_{\substack{l \geq j+1 \\ l \leq n-1}} \frac{\delta_j \cancel{L_j}}{1 \cancel{+} \delta_j L_j} \frac{\delta_l L_l}{1 + \delta_l L_l} \sigma_j \sigma_l \rho_{j,l} = 0,
$$

i.e.,

$$\mu_j^{\mathbb{Q}^{P(T_n)}}(t) = -\sum_{\substack{l \geq j+1 \\ l \leq n-1}} \frac{\delta_l L_l(t)}{1 + \delta_l L_l(t)}\, \sigma_j(t)\sigma_l(t)\rho_{j,l}(t). \tag{19.8}$$

The procedure above may be summarized as follows: To derive the n drifts we write down n independent traded assets as a function of the model quantities. By considering the drifts of their relative prices, we obtain n equations for the drifts of the modeled quantities.

19.1.2 Derivation of the Drift Term under the Spot LIBOR Measure

We fix the *rolled over one period bond* as numéraire, i.e., the investment of 1 at time T_0 into the T_1-bond and after its maturity the reinvestment of the proceeds into the bond of the next period, i.e., in T_j the reinvestment in the T_{j+1}-bond. It is

$$N(t) := P(T_{m(t)+1};t) \prod_{j=1}^{m(t)+1} \frac{1}{P(T_j;T_{j-1})} = P(T_{m(t)+1};t) \prod_{j=0}^{m(t)}(1 + L_j(T_j)\,\delta_j), \tag{19.9}$$

$$= \underbrace{\frac{P(T_{j-1};T_{j-1})}{P(T_j;T_{j-1})}} = (1 + L_{j-1}(T_{j-1})\delta_{j-1})$$

where $m(t) := \max\{i \ : \ T_i \leq t\}$ and $\delta_j := T_{j+1} - T_j$. The corresponding equivalent martingale measure $\mathbb{Q}^N$ is called the *spot measure*.

As before, we consider the processes of N-relative prices of traded products (from which we know that they have drift 0 under $\mathbb{Q}^N$). We consider the N-relative prices of the bonds $P(T_i)$. It is

$$\frac{P(T_i;t)}{N(t)} = \frac{P(T_i;t)}{P(T_{m(t)+1};t)} \prod_{j=0}^{m(t)}(1 + L_j(T_j)\,\delta_j)^{-1}$$

$$= \prod_{j=m(t)+1}^{i-1}(1 + L_j(t)\delta_j)^{-1} \prod_{j=0}^{m(t)}(1 + L_j(T_j)\,\delta_j)^{-1}, \tag{19.10}$$

thus

$$\operatorname*{Drift}_{\mathbb{Q}^N}\left[\prod_{k=m(t)+1}^{i-1}(1 + L_k\delta_k)^{-1}\right] = 0. \tag{19.11}$$

Since

$$d\left(\prod_{j=m(t)+1}^{i-1} (1 + L_j(t)\delta_j)^{-1} \prod_{j=0}^{m(t)} (1 + L_j(T_j)\delta_j)^{-1}\right)$$

$$= d\left(\prod_{j=m(t)+1}^{i-1} (1 + L_j(t)\delta_j)^{-1}\right) \prod_{j=0}^{m(t)} (1 + L_j(T_j)\delta_j)^{-1},$$

we consider

$$d\left(\prod_{k=m(t)+1}^{i-1} \frac{1}{1 + \delta_k L_k}\right)$$

$$= \sum_{j=m(t)+1}^{i-1} \prod_{\substack{k=m(t)+1 \\ k \neq j}}^{i-1} \frac{1}{1 + \delta_k L_k} \left(\frac{-\delta_j\, dL_j}{(1 + \delta_j L_j)^2} + \frac{\delta_j^2\, dL_j\, dL_j}{(1 + \delta_j L_j)^3}\right)$$

$$+ \sum_{\substack{j,l=m(t)+1 \\ l<j}}^{i-1} \prod_{\substack{k=m(t)+1 \\ k \neq j,l}}^{i-1} \frac{1}{1 + \delta_k L_k} \left(\frac{\delta_j\, dL_j}{(1 + \delta_j L_j)^2} + \frac{\delta_j^2\, dL_j\, dL_j}{(1 + \delta_j L_j)^3}\right)\left(\frac{\delta_l\, dL_l}{(1 + \delta_l L_l)^2} + \frac{\delta_l^2\, dL_l\, dL_l}{(1 + \delta_l L_l)^3}\right)$$

$$= \prod_{k=m(t)+1}^{i-1} \frac{1}{1 + \delta_k L_k} \left(\sum_{j=m(t)+1}^{i-1} \frac{-\delta_j\, dL_j}{1 + \delta_j L_j} + \frac{\delta_j^2\, dL_j\, dL_j}{(1 + \delta_j L_j)^2} + \sum_{\substack{j,l=m(t)+1 \\ l<j}}^{i-1} \frac{-\delta_j\, dL_j}{1 + \delta_j L_j} \frac{-\delta_l\, dL_l}{1 + \delta_l L_l}\right)$$

$$= \prod_{k=m(t)+1}^{i-1} \frac{1}{1 + \delta_k L_k} \sum_{j=m(t)+1}^{i-1} \left(-\frac{\delta_j\, dL_j}{1 + \delta_j L_j} + \sum_{l=m(t)+1}^{j} \frac{\delta_j\, dL_j}{1 + \delta_j L_j} \frac{\delta_l\, dL_l}{1 + \delta_l L_l}\right).$$

With (19.11) we have $\forall\, i = 0, \ldots, n-1$

$$\sum_{j=m(t)+1}^{i-1} \operatorname*{Drift}_{\mathbb{Q}^N}\left[-\frac{\delta_j\, dL_j}{(1 + \delta_j L_j)} + \sum_{l=m(t)+1}^{j} \frac{\delta_j\, dL_j}{(1 + \delta_j L_j)} \frac{\delta_l\, dL_l}{(1 + \delta_l L_l)}\right] = 0$$

and thus $\forall\, j = 0, \ldots, n-1$

$$\operatorname*{Drift}_{\mathbb{Q}^N}\left[\frac{-\delta_j\, dL_j}{(1 + \delta_j L_j)} + \sum_{l=m(t)+1}^{j} \frac{\delta_j\, dL_j}{(1 + \delta_j L_j)} \frac{\delta_l\, dL_l}{(1 + \delta_l L_l)}\right] = 0. \qquad (19.12)$$

If we now use

$$dL_j = L_j \mu_j^{\mathbb{Q}^N}\, dt + L_j \sigma_j\, dW_j \qquad \text{and} \qquad dL_j\, dL_l = L_j L_l \sigma_j \sigma_l \rho_{j,l}\, dt$$

in (19.7), then we have[4]

$$-\mu_j^{Q^N}\frac{\delta_j L_j}{1+\delta_j L_j}+\sum_{l=m(t)+1}^{j}\frac{\delta_j L_j}{1+\delta_j L_j}\frac{\delta_l L_l}{1+\delta_l L_l}\sigma_j\sigma_l\rho_{j,l}=0,$$

i.e.,

$$\mu_j^{Q^N}(t)=\sum_{l=m(t)+1}^{j}\frac{\delta_l L_l(t)}{1+\delta_l L_l(t)}\sigma_j(t)\sigma_l(t)\rho_{j,l}(t).$$

19.1.3 Derivation of the Drift Term under the T_k-Forward Measure

Exercise: (Drift under the T_k-forward measure) Consider

$$N(t):=\begin{cases}P(T_k;t), & t\le T_k\\ P(T_{m(t)+1};t)\prod_{j=k+1}^{m(t)+1}\frac{1}{P(T_j;T_{j-1})}, & t>T_k,\end{cases}$$

where $m(t):=\max\{i\ :\ T_i\le t\}$.

1. Give an interpretation of $N(t)$ as traded product.

2. Derive the drift of the model (19.1) under the Q^N measure with the numéraire N.

Solution:

$$\mu_j(t)=-\sum_{l=j+1}^{k-1}\frac{\delta_l L_l}{1+\delta_l L_l}\sigma_j\sigma_l\rho_{j,l}\quad \text{for } j<k-1 \quad \text{and } t\le T_k$$

$$\mu_j(t)=0 \quad \text{for } j=k-1 \quad \text{and } t\le T_k$$

$$\mu_j(t)=\sum_{l=k}^{j}\frac{\delta_l L_l}{1+\delta_l L_l}\sigma_j\sigma_l\rho_{j,l}\quad \text{for } j\ge k \quad \text{and } t\le T_k$$

$$\mu_j(t)=\sum_{l=m(t)+1}^{j}\frac{\delta_l L_l}{1+\delta_l L_l}\sigma_j\sigma_l\rho_{j,l}\quad \text{for } t>T_k.$$

---

[4] Since the coefficient of dt equals 0.

19.2 The Short Period Bond $P(T_{m(t)+1}; t)$

For $t \notin \{T_1, \ldots, T_n\}$ neither the numéraire $N(t)$ of the terminal measure nor the numéraire of the spot measure is not fully described by the processes $L_i(t)$. The unspecified bond $P(T_{m(t)+1}; t)$ occurs in both numéraires. We will now discuss the relevance of $P(T_{m(t)+1}; t)$.

19.2.1 Role of the Short Bond in a LIBOR Market Model

For the modeling of the forward rates $L_i(t) := L(T_i, T_{i+1}; t)$ on the tenor periods $[T_i, T_{i+1}]$, $i = 0, \ldots, n$ the specification of $P(m(t)+1; t)$ is irrelevant. For the derivation of the corresponding drift terms it was not relevant to specify the stochastic of $P(T_{m(t)+1}; t)$, since the term canceled for the relative prices considered.

Conversely, the LIBOR market model does not describe the stochastic of the short bond $P(T_{m(t)+1}; t)$, since it is not given as a function of the processes $L_i(t)$.

19.2.2 Link to Continuous Time Tenors

The specification of the short bond $P(T_{m(t)+1}; t)$ becomes relevant if the model has to describe interest rates of interest rate periods which are not part of the tenor structure. The specification of $P(m(t) + 1; t)$ will determine how the fractional forward rates $L(T_s, T_e; t)$ with $T_s \notin \{T_1, \ldots, T_n\}$ and/or $T_e \notin \{T_1, \ldots, T_n\}$ will evolve (see Section 19.5). It is the link from a model with discrete tenors (LIBOR market model) to a model with continuous time tenors (Heath-Jarrow-Morton framework). In the special case where $P(m(t) + 1; t)$ has zero volatility, the LIBOR market model under spot measure coincides with a Heath-Jarrow-Morton framework with a special volatility structure under the risk-neutral measure (see Section 24.2).

19.2.3 Drift of the Short Bond in a LIBOR Market Model

Within the LIBOR market model there is no constraint on the drift of $P(m(t) + 1; t)$, because in $\frac{P(m(t)+1;t)}{N(t)}$ the term cancels out. The relative price $\frac{P(m(t)+1;t)}{N(t)}$ is always a martingale for any choice of $P(m(t) + 1; t)$. This might come as a surprise, but we have already encountered this behavior: In the Black-Scholes model the drift r of $B(t)$ is a free parameter, because it is the drift of the numéraire. The parameter r is determined by calibration to a market interest rate. In a short rate model the drift is a free parameter. It is determined by calibration to the market interest rate curve; see Chapter 23. Here, similarly, $P(m(t) + 1; t)$ determines the interpolation of the initial interest rate curve.

The trivial fact that the numéraire-relative price of the numéraire, i.e., $\frac{N(t)}{N(t)}$, is always a martingale plays a role in Markov functional models. There, the numéraire is postulated to be a functional of some Markov process.

19.3 Discretization and (Monte Carlo) Simulation

In this section we will discuss the discretization and implementation of the model. Let us therefore assume that the free parameters σ_i, $\rho_{i,j}$, and $L_{i,0}$ $(i, j = 1, \ldots, n)$ are given. Together with the drift formula obtained in the previous section the model is fully specified. Section 19.4 will then discuss how the parameters $L_{i,0}$, σ_i, $\rho_{i,j}$ are obtained.

19.3.1 Generation of the (Time-Discrete) Forward Rate Process

As discussed in Chapter 13, we choose the Euler discretization of the Itô process of $\log(L_i)$. From Lemma 50 we have

$$d(\log(L_i(t))) = (\mu_i^{\mathbb{Q}^N}(t) - \frac{1}{2}\sigma_i^2(t)) \, dt + \sigma_i(t) \, dW_i^{\mathbb{Q}^N}(t) \qquad (19.13)$$

and the corresponding Euler scheme of (19.13) is

$$\log(\tilde{L}_i(t + \Delta t)) = \log(\tilde{L}_i(t)) + (\mu_i(t) - \frac{1}{2}\sigma_i^2(t)) \, \Delta t + \sigma_i(t) \, \Delta W_i(t). \qquad (19.14)$$

Applying the exponential gives us the discretization scheme of L_i as

$$\tilde{L}_i(t + \Delta t) = \tilde{L}_i(t) \exp\left((\mu_i(t) - \frac{1}{2}\sigma_i^2(t)) \, \Delta t + \sigma_i(t) \, \Delta W_i(t) \right). \qquad (19.15)$$

In the special case that the process L_i is considered under the measure $\mathbb{Q}^{P(T_{i+1})}$, i.e., $\mu_i^{\mathbb{Q}^{P(T_{i+1})}}(t) = 0$, and that the given $\sigma_i(t)$ is a known deterministic function, we may use the exact solution for a discretization scheme:

$$L_i(t + \Delta t) = L_i(t) \exp\left(-\frac{1}{2}\bar{\sigma}_i^2(t, t + \Delta t) \, \Delta t + \bar{\sigma}_i(t, t + \Delta t) \, \Delta W_i \right),$$

where

$$\bar{\sigma}_i(t, t + \Delta t) := \sqrt{\frac{1}{\Delta t} \int_t^{t+\Delta t} \sigma_i^2(\tau) \, d\tau}.$$

In the case where L_i is not drift-free, we choose instead of (19.15) the discretization scheme

$$L_i(t + \Delta t) = L_i(t) \exp\left(\left(\mu_i(t) - \frac{1}{2}\bar{\sigma}_i(t, t + \Delta t)^2\right) \Delta t + \bar{\sigma}_i(t, t + \Delta t) \Delta W_i(t)\right) \quad (19.16)$$

(we write L in place of $\tilde{L}$, although (19.16) is an approximation of (19.1)). The diffusion dW is discretized by exact solution; the drift dt is discretized by an Euler scheme. The discretization error of this scheme stems from the discretization of the stochastic drift μ_i only. This discretization error results in a violation of the *no-arbitrage* requirement of the model (the discretized model does not have the correct, arbitrage-free drift). Methods which do not exhibit an arbitrage due to a discretization error are called arbitrage-free discretization; see [73]).

The volatility functions σ_i are usually assumed to be piecewise constant functions on $[T_j, T_{j+1})$, such that $\bar{\sigma}_i(t, t + \Delta t)$ may be calculated analytically. It is $\bar{\sigma}_i(t, t + \Delta t) = \sigma_i(t)$.

19.3.2 Generation of the Sample Paths

Equipped with the time discretization (19.16), realizations of the process are calculated for a given number of paths $\omega_1, \omega_2, \omega_3, \ldots$. To do so, normally distributed random numbers $\Delta W_i(t_j)(\omega_k)$, correlated according to $R = (\rho_{i,j})$, are generated (see Appendices B.1 and B.2). These are used in the scheme (19.16). The result is a three-dimensional tensor $L_i(t_j, \omega_k)$ parametrized by

i : Index of the interest rate period (tenor structure),

j : Index of the simulation time,

k : Index of the simulation path.

19.3.3 Generation of the Numéraire

Given a simulated interest rate curve $L_i(t_j, \omega_k)$, we can calculate the numéraire. Of course, we have to use the numéraire that was chosen for the martingale measure under which the process was simulated (form of the drift in (19.13)). For the *terminal measure* we would calculate

$$N(T_i, \omega_k) = \prod_{j=i}^{n-1}(1 + L_j(T_i, \omega_k) (T_{j+1} - T_j))^{-1}.$$

Note: The numéraire is given only at the tenor times $t = T_i$, since for $t \neq T_i$ we did not define the short period bond $P(T_{m(t)+1}; t)$.[5] An interpolation is possible; see Section 19.5 and [96].

19.4 Calibration—Choice of the Free Parameters

We are now going to explain how the free parameters of the model can be chosen. The free parameters are

- the initial conditions $L_{i,0}$, $i = 0, \ldots, n - 1$,

- the volatility functions or volatility processes[6] σ_i, $i = 1, \ldots, n - 1$,

- the (instantaneous) correlation $\rho_{i,j}$, $i, j = 1, \ldots, n - 1$.

The determination of the free parameters is also called *calibration of the model*.

 Motivation (Reproduction of Market Prices versus Historical Estimation): With the LIBOR market model we have a high-dimensional model framework. The main task is the derivation or estimation of the huge amount of free parameters. Two approaches are possible:

- **Reproduction of Market Prices**: The parameters are chosen such that the model reproduces given market prices.

- **Historical Estimation**: The parameters are estimated from historical data, e.g., time series of interest rate fixings.

It may be surprising at first, but the second approach is not meaningful, being in the context of risk-neutral evaluation. The model is considered under the martingale measure $\mathbb{Q}^N$ and its aim is the evaluation and *hedging* (!) of derivatives. An expectation of the relative value under the martingale measure corresponds to the relative value of the replication portfolio. This replication portfolio has to be set up from traded products, traded at current (!) market prices. If the model did not replicate current market prices, then it would not be possible to buy the replication portfolio of a derivative at the model price of the derivative. The model price would inevitably be wrong.

This remark applies to all free model parameters. In practice, however, it may be difficult or impossible to derive all parameters from market prices. This could be

---

[5] See Section 19.2.

[6] The parameters σ_i may well be stochastic processes. In this case σ_i is called a stochastic volatility model.

because for a specific product no reliable price is known (low liquidity). It could also be that a corresponding product does not exist. This is often the case for correlation-sensitive products from which we would like to derive the correlation parameters. If a parameter cannot be derived from a market price, a historical estimate becomes an option. If in such a case complete hedge is not possible, the residual risk has to be considered, e.g., by a conservative estimate of the parameter.

For the LIBOR market model a parameter reduction is usually applied first, based on historical estimates of rough market assessment. An example of such parameter reduction is the assumption of a family of functional forms for the volatility $\sigma_i(t)$ or the correlation $\rho_{i,j}(t)$. The remaining degrees of freedom are then derived from market prices. ◁|

19.4.1 Choice of the Initial Conditions

19.4.1.1 Reproduction of Bond Market Prices

Let $P^{\text{Market}}(T_i) \in (0, 1]$ denote a market observed (i.e., given) price of a T_i-bond. If we set

$$L_{i,0} := \frac{P^{\text{Market}}(T_i) - P^{\text{Market}}(T_{i+1})}{P^{\text{Market}}(T_{i+1})(T_{i+1} - T_i)},$$

then the model reproduces the given market prices of the bonds P^{Market}. This is ensured by the model having the "right" drift and it is independent of the other parameters.

19.4.2 Choice of the Volatilities

19.4.2.1 Reproduction of Caplet Market Prices

We assume here that the σ_i's are deterministic functions (i.e., not random variables or stochastic processes). The forward rate L_i follows the Itô process

$$dL_i(t) = \mu_i^{\mathbb{Q}}(t)L_i(t)\,dt \ + \ \sigma_i(t)L_i(t)\,dW_i^{\mathbb{Q}}(t) \qquad \text{under } \mathbb{Q} := \mathbb{Q}^N.$$

Thus the model corresponds to the Black model discussed in Chapter 10. Under $\mathbb{Q}^{P(T_{i+1})}$ we have $\mu_i^{\mathbb{Q}^{P(T_{i+1})}} = 0$, the distribution of $L_i(T_i)$ is lognormal, and there exists an analytic evaluation formula for caplets. The only model parameters that enter the caplet price are $L_0(T_i)$ and

$$\sigma_i^{\text{Black,Model}} := \left(\frac{1}{T_i} \int_0^{T_i} \sigma_i^2(t)\,dt \right)^{1/2}. \tag{19.17}$$

If the market price $V_{\text{Caplet},i}^{\text{Market}}$ of a caplet on the forward rate $L_i(T_i)$ is given, then the corresponding implied Black volatility $\sigma_i^{\text{Black,Market}}$ may be calculated by inverting[7] Equation (10.2). If then $\sigma_i(t)$ is chosen such that

$$\sigma_i^{\text{Black,Model}} = \sigma_i^{\text{Black,Market}}, \qquad (19.18)$$

then the model reproduces the given caplet price $V_{\text{Caplet},i}^{\text{Market}}$. A possible trivial choice is, e.g., $\sigma_i(t) = \sigma_i^{\text{Black,Market}} \; \forall \, t$.

Remark 213 (Caplet Smile Modeling):   The fact that the LIBOR market model calibrates to the cap market by a simple boundary condition is one reason for its initial popularity. However, since the model restricted to a single LIBOR is a Black model, the implied volatility does not depend on the strike of an option. Thus, in this form, the model may calibrate to a single caplet per maturity only. It cannot render a caplet smile yet.

To remove this restriction one can extend the model by a local or stochastic volatility or jump-diffusion processes [8]. For an overview on smile modelling in the LIBOR market model see [23].

19.4.2.2 Reproduction of Swaption Market Prices

If the correlation $R = (\rho_{i,j})$ is given and fixed, then we influence swaption prices through the time structure of the volatility function $t \mapsto \sigma_i$. We consider swaptions that correspond to our tenor structure, i.e., option on the swap rates:

$$S(T_i, \ldots, T_j; T_i), \qquad 0 < i < j \le n.$$

From the definition of the swap and swap rate it is obvious that the price of a corresponding swaption with *exercise date* on or before T_i and periods $[T_i, T_{i+1}], \ldots, [T_{j-1}, T_j]$ depends only on the behavior of the forward rates $L_i(t), \ldots, L_{j-1}(t)$ until the *fixing* $t \le T_i$; see Figure 19.1.

If we discretize the volatility function corresponding to the tenor structure and define

$$\sigma_{k,l} := \left(\frac{1}{T_{l+1} - T_l} \int_{T_l}^{T_{l+1}} \sigma_k^2(t)\, dt \right)^{1/2},$$

the price of an option on the swap rate $S(T_i, \ldots, T_j; T_i)$ depends only on $\sigma_{k,l}$ for $k = i, \ldots, j-1$ and $l = 0, \ldots, i-1$.

---

[7] For inversion of a pricing formula we may use a simple numerical algorithm. For the Black formula (10.2) the price is increasing strictly monotone in the volatility.

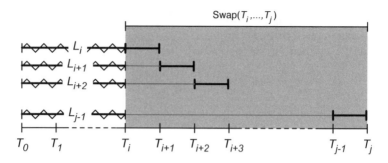

Figure 19.1. *Swaption as a function of the forward rates:* *The swap with periods* $[T_i, T_{i+1}], \ldots, [T_{j-1}, T_j]$ *is a function of the forward rates* $L(T_i, T_{i+1}; T_i), \ldots, L(T_{j-1}, T_j; T_i)$ *(all with fixing in* T_i*). The corresponding swaption depends only on the joint distribution of these forward rates. Under our model, with given initial conditions* $L_{i,0}$ *and correlation* $R = (\rho_{i,j})$, *the swaption price depends on* $\sigma_i(t), \ldots, \sigma_{j-1}(t)$, $t \in [0, T_i]$ *only. The dynamic of these forward rated beyond the* $t > T_i$ *and all other forward rates do not influence the swaption price.*

This allows an iterative calculation of $\sigma_{k,l}$ from given swaption market prices:

For $i = 1, \ldots, n - 1$:

    For $j = i + 1, \ldots, n$:

        Calculate $\sigma_{j-1,i-1}$ from the price of an option on $S(T_i, \ldots, T_j; T_i)$ by considering the already calculated $\sigma_{k,l}$ with $k = i, \ldots, j - 1$ and $l = 0, \ldots, i - 2$ from the previous iterations.

To derive $\sigma_{j-1,i-1}$ from the market price $V_{\text{swaption}}^{\text{market}}(T_i, \ldots, T_j)$ we have to invert the mapping

$$\sigma_{j-1,i-1} \mapsto V_{\text{swaption}}^{\text{model}}(T_i, \ldots, T_j).$$

In principle this mapping may be realized by a Monte Carlo evaluation of the swaption. To allow for faster pricing, and thus faster calibration, analytic approximation formulas for swaption prices within a LIBOR market model have been derived; see also Section 19.4.5.

Remark 214 (Bootstrapping): The above procedure of calculating a piecewise constant instantaneous volatility from swaption prices is called *volatility bootstrapping*.

Remark 215 (Review: *Overfitting*): The calculation of a piecewise constant volatility function $\sigma_{i,j}$ from swaption prices bears the risk of an *overfitting* of the model.

Note that if this procedure is applied, then we accept completely the validity of every single given swaption price, i.e., that the prices are of sufficiently good quality with respect to *topicality* (fixing time of the price) and *liquidity*. If not all prices are of the same quality, then some have to be interpolated, smoothed, or corrected by hand. In this case, the calibration problem has been replaced by an interpolation problem. If the interpolation and maintenance of the market data is not done with care, a calibration that fits to these prices exactly may be useless. See, for example, Chapter 6.

In addition, the bootstrapping of the instantaneous volatility from swaption prices does not allow for a weighting of the swaption prices according to their importance.

A solution to this is the reduction of the free parameters by a reduction of the family of admissible volatility functions with consequent loss of the perfect fit.

19.4.2.3 Functional Forms

To reduce the risk of *overfitting*, the admissible volatility functions may be restricted to a parametrized family of volatility functions. For example, a functional which is empirically motivated by the historical shapes of the volatility and which is common in practice is

$$\sigma_i(t) := (a + b\,(T_i - t))\,\exp(-c\,(T_i - t)) + d.$$

Given a functional form, the calibration of the model consists of a selection of (liquid) market prices of caps and swaptions and an optimization of the remaining parameters (e.g., a, b, c, d above) to fit the model prices to the market prices.

For a detailed discussions of a robust calibration to cap and swaption prices we refer to the literature, especially [7, 30].

19.4.3 Choice of the Correlations

19.4.3.1 Factors

We assumed in (19.1) a model in which (potentially) each forward rate L_i is driven by its own Brownian motion W_i. The model is driven by an n-dimensional[8] Brownian motion

$$W = \begin{pmatrix} W_0 \\ W_1 \\ \vdots \\ W_{n-1} \end{pmatrix}.$$

---

[8] An $n - 1$-dimensional Brownian motion is sufficient here, since we can choose $W_0 = 0$, because the forward rate L_0 is not stochastic. It is fixed in $T_0 = 0$. Formally we achieve this by setting $\sigma_0 \equiv 0$.

The effective number of *factors*, i.e., the number of independent Brownian motions, that are driving the model is determined by the correlation

$$\rho_{i,j}(t)\, dt \;=\; dW_i(t)\, dW_j(t).$$

By an eigenvector decomposition (PCA, *principal component analysis*) of the correlation matrix $R = (\rho_{i,j})_{i,j=1...n}$ we may represent dW as

$$dW(t) = F(t)\, dU(t),$$

where $U := (U_1, \ldots, U_m)^\top$ and $U_1, \ldots, U_m$ denote independent Brownian motions and $F = (f_{i,j})$ denotes a $n \times m$-matrix. In other words, we have

$$dW_i(t) = \sum_{j=1}^{m} f_{i,j}(t) dU_j(t), \qquad \rho_{i,j}(t) = \sum_{k=1}^{m} f_{i,k}(t) \cdot f_{j,k}(t).$$

A proof of this representation is in Appendix B.2. Note that here we can have $m < n$. The columns of the matrix F are called *factors*.

19.4.3.2 Functional Forms

A full rank correlation matrix R is hard to derive from market instruments. As before a common procedure is to reduce the family of admissible correlation matrixes R. One ansatz consists of functional forms, for example

$$\rho_{i,j}(t) := \exp(-\alpha \cdot |T_i - T_j|). \tag{19.19}$$

19.4.3.3 Factor Reduction

The specification of the correlation matrix as a functional form is usually followed by a reduction of the number of factors. This is done in what is known as factor reduction (PCA). There only the eigenvectors corresponding to the largest eigenvalues of R are considered, and a new correlation matrix is formed from these selected factors. For a discussion of the factor reduction see Appendix B.3.

The advantage of the factor reduction is that afterwards only an m-dimensional Brownian motion has to be simulated (and not an n-dimensional Brownian motion). Often $n \geq 40$ is required (e.g., a 20-year interest rate curve with semiannual periods), however, often $m \leq 5$ is sufficient. The choice of the actual number m of factors depends on the application; see Chapter 25.

19.4.3.4 Calibration

The correlation model, e.g., the free parameter a in (19.19), may be chosen such that the fit of model prices to given market prices is improved. Alternatively, it may be chosen to give more realistic interest rate correlations.

It should be stressed that we calibrate the *instantaneous* correlation, i.e., the correlation of the Brownian increments dW, and not the terminal correlation, i.e., the correlations of the distribution of the interest rates at a fixed time. We will consider the relation of the two in Chapter 21.

We have seen that we may calibrate to the swaption matrix solely by the specification of the time-structure of the instantaneous volatility of the LIBORs. The time-structure of the instantaneous correlations of the LIBORs will allow to calibrate correlations of the swaprates with different tenors. In other words: The terminal covariance of swaprates may be calibrated by the time-structure of the instantaneous covariance of the LIBORs. Thus is makes sense to consider the instantaneous LIBOR covariance as the free model parameter. We do so in Section 19.4.4. For an approximation giving the terminal covariance of swaprates in terms of the time-structure of the instantaneous covariance of the LIBORs see Section 19.4.5.

19.4.4 Covariance Structure

In the previous sections we considered volatility and correlation separately. This is not necessary, as both can be considered together in the form of the correlation matrix $(\sigma_i \sigma_j \rho_{i,j})$. Thus the calibration problem consists of the calculation of the (market implied) covariance matrix (or covariance matrix function).

Defining the parametrized functional forms for volatility and correlation, e.g., as

$$\sigma_i(t) := (a + b\,(T_i - t))\,\exp(-c\,(T_i - t)) + d, \quad \rho_{i,j}(t) := \exp(-\alpha\,|T_i - T_j|)$$

reduces the number of degrees of freedom of the covariance model and thus the possible number of products for which an exact fit is possible. This might be a desirable feature, e.g., to avoid an overfitting. A disadvantage is the lack of transparency of the parameters. To derive the parameters numerical optimization has normally to be used, e.g., the minimization of a suitable norm of the error vector of some selected product prices as a function of the model parameters. The optimization of volatility parameters and correlation parameters may occur jointly, i.e., we consider a functional form of the covariance structure.

19.4.5 Analytic Evaluation of Caplets, Swaptions and Swap Rate Covariance

To calculate the calibration error we need to calculate the corresponding model prices. Since a numerical calculation of the model price (e.g., by a full Monte Carlo simulation) is time consuming, and since the optimization requires many calculations of model prices, there is a need for fast analytical pricing formulas for specific calibration products.

19.4.5.1 Analytic Evaluation of a Caplet in the LIBOR Market Model

The analytic evaluation of caplets in the LIBOR market model is provided by the Black formula (10.2) using (19.17) to calculate the Black volatility.

19.4.5.2 Analytic Evaluation of a Swaption in the LIBOR Market Model

The analytic evaluation of swaption in the LIBOR market model is possible only by an approximation formula. An approximation formula can be derived by expressing the volatility of the swap rate as a function of the volatility and correlation of the forward rates. Assuming a lognormal model for the swap rate, which is already an approximation, we can then apply the Black formula for swaptions.[9] Corresponding approximation formulas may be found in [7].

An approximation formula for swaption prices can be derived as follows:

Let $0 = T_0 < T_1 < \ldots$ denote the tenor structure for the forward rates. Consider a swaprate

$$S_{a,b}(t) = S(T_{i_1}, \ldots, T_{i_m}; t) \overset{\text{Def. 122}}{:=} \frac{P(T_a; t) - P(T_b; t)}{\sum_{j=1}^{m-1}(T_{i_{j+1}} - T_{i_j}) \cdot P(T_{i_{j+1}}; t)}$$

associated with the swap tenor $T_a = T_{i_1} < \ldots < T_{i_m} = T_b$. In other words we assume that the swap tenor is a subset of the forward rate tenor, but it is allowed to be coarser.

From the definition of the swap rate it is clear that the swap rate $S_{a,b}$ can be seen as a function of $L_k = L(T_k, T_{k+1})$ with $k = a, \ldots, b - 1$, see also Lemma 123. Using Itô's Lemma we write the swap rate process as

$$dS_{a,b}(t) = \sum_{k=a}^{b-1} \frac{\partial S_{a,b}}{\partial L_k} dL_k + \frac{1}{2} \sum_{k,l=a}^{b-1} \frac{\partial^2 S_{a,b}}{\partial L_k \partial L_l} \underbrace{dL_k dL_l}_{(\ldots)\,dt}$$

---

[9] Assuming lognormal processes for the forward rates, the swap rate is not a lognormal process in general.

314

i.e.,

$$dS_{a,b}(t) = (\ldots)dt + S_{a,b}\sum_{k=a}^{b-1}\frac{\frac{\partial}{\partial L_k}S_{a,b}}{S_{a,b}}dL_k = (\ldots)dt + S_{a,b}\sum_{k=a}^{b-1}w_k(t)dL_k$$

$$= (\ldots)dt + S_{a,b}\sum_{k=a}^{b-1}w_k(t)L_k(t)\sigma_k(t)dW_k^{Q^N},$$

where

$$w_k = \frac{\frac{\partial}{\partial L_k}S_{a,b}}{S_{a,b}} = \frac{\partial}{\partial L_k}\log(S_{a,b}).$$

If the swap rate follows a lognormal model

$$dS_{a,b}(t) = (\ldots)\,dt + S_{a,b}(t)\sigma_{S_{a,b}}(t)\,dW(t).$$

then the price of the corresponding swaption is given by the Black formula using the swaption's Black volatility $\bar{\sigma}_{a,b}$ where

$$\bar{\sigma}_{a,b}^2 = \frac{1}{T_a}\int_0^{T_a}\sigma_{S_{a,b}}^2(t)\,dt.$$

Note that $\sigma_{S_{a,b}}^2(t)\,dt$ is just the relative quadratic variation

$$\left(\frac{dS_{a,b}(t)}{S_{a,b}(t)}\right)\left(\frac{dS_{a,b}(t)}{S_{a,b}(t)}\right) = \sigma_{S_{a,b}}^2(t)\,dt.$$

In other words: If the relative quadratic variation of the swap rate is non-stochastic, we can calculate the swaption price from the Black formula. From the above we see that under a LIBOR Market model

$$\left(\frac{dS_{a,b}(t)}{S_{a,b}(t)}\right)\left(\frac{dS_{a,b}(t)}{S_{a,b}(t)}\right) = \sum_{k,l=a}^{b-1}w_k(t)w_l(t)L_k(t)L_l(t)\sigma_k(t)\sigma_l(t)\rho_{k,l}(t)\,dt.$$

Since $w_k(t)w_l(t)L_k(t)L_l(t)$ is stochastic the swap rate has a stochastic relative quadratic variation and thus stochastic log-volatility. Consequently the Black swaption formula does not hold. In order to apply the Black swaption formula we use the approximation

$$\left(\frac{dS_{a,b}(t)}{S_{a,b}(t)}\right)\left(\frac{dS_{a,b}(t)}{S_{a,b}(t)}\right) \approx \sum_{k,l=a}^{b-1}w_k(0)w_l(0)L_k(0)L_l(0)\sigma_i(t)\sigma_j(t)\rho_{i,j}(t)\,dt,$$

315

i.e., we freeze the random variable $w_i L_i$ to its initial value. Then we can calculate an approximation to the model swaption price using Black's swaption formula and the approximated integrated relative quadratic variation

$$\tilde{\sigma}_{a,b}^2 := \sum_{k,l=a}^{b-1} w_k(0) L_k(0) w_l(0) L_l(0) \frac{1}{T_a} \int_0^{T_a} \sigma_k(t) \sigma_l(t) \rho_{k,l}(t) \, dt. \qquad (19.20)$$

The weights w_k are given as

$$w_k = \frac{\partial}{\partial L_k} \log(S_{a,b})$$

$$= \frac{\partial}{\partial L_k} \log\left(\frac{P(T_a) - P(T_b)}{\sum_{j=1}^{m-1} P(T_{i_{j+1}}) \cdot (T_{i_{j+1}} - T_{i_j})} \right)$$

$$= \frac{\partial}{\partial L_k} \left(\log(P(T_a) - P(T_b)) - \log(\sum_{j=1}^{m-1} P(T_{i_{j+1}}) \cdot (T_{i_{j+1}} - T_{i_j})) \right)$$

$$= \frac{\frac{\partial}{\partial L_k}(P(T_a) - P(T_b))}{P(T_a) - P(T_b)} - \frac{\sum_{j=1}^{m-1} \frac{\partial}{\partial L_k} P(T_{i_{j+1}}) \cdot (T_{i_{j+1}} - T_{i_j})}{\sum_{j=1}^{m-1} P(T_{i_{j+1}}) \cdot (T_{i_{j+1}} - T_{i_j})}.$$

Together with

$$\frac{\partial}{\partial L_k} P(T_{i_{j+1}}) = \begin{cases} -P(T_{i_{j+1}}) \cdot \frac{T_{k+1} - T_k}{1 + L_k(T_{k+1} - T_k)} & \text{if } i_{j+1} > k \\ 0 & \text{otherwise} \end{cases}$$

we find with $l(k) := \min\{l \mid i_l \geq k\}$ (the index $l(k)$ marks the first swap period $[T_{l_k}, T_{l_{k+1}}]$ which contains the forward rate period $[T_k, T_{k+1}]$)

$$w_k = \left(\frac{P(T_b)}{P(T_a) - P(T_b)} + \frac{\sum_{j=l(k)}^{m-1} P(T_{i_{j+1}}) \cdot (T_{i_{j+1}} - T_{i_j})}{\sum_{j=1}^{m-1} P(T_{i_{j+1}}) \cdot (T_{i_{j+1}} - T_{i_j})} \right) \cdot \frac{T_{k+1} - T_k}{1 + L_k(T_{k+1} - T_k)}.$$

With the swap annuities

$$A_{l,m} := \sum_{j=l}^{m-1} P(T_{i_{j+1}}) \cdot (T_{i_{j+1}} - T_{i_j})$$

this can be simplified to

$$w_k = \left(\frac{P(T_b)}{P(T_a) - P(T_b)} + \frac{A_{l(k),m}}{A_{1,m}} \right) \cdot \frac{T_{k+1} - T_k}{1 + L_k(T_{k+1} - T_k)}.$$

The last equation allows the efficient calculation of the weight $w_k(0)$ in our approximation formula.

We summarize the result:

Theorem 216 (Analytic Approximation of Swaptions under a LIBOR Market Model (Hull & White 1999)): Within a LIBOR market model an analytic approximation of the swaption price is given by the Black formula for swaptions using the Black volatility $\tilde{\sigma}_{a,b}^2$ with

$$\tilde{\sigma}_{a,b}^2 T_a := \sum_{k,l=a}^{b-1} v_k(0) \, \bar{\gamma}_{k,l}(T_a) \, v_l(0), \tag{19.21}$$

where

$$\bar{\gamma}_{k,l}(T_a) = \int_0^{T_a} \sigma_k(t)\sigma_l(t)\rho_{k,l}(t) \, dt$$

and

$$v_k = \frac{\partial \log(S_{a,b})}{\partial \log(L_k)} = \left(\frac{P(T_b)}{P(T_a) - P(T_b)} + \frac{A_{l(k),m}}{A_{1,m}} \right) \cdot \frac{L_k(T_{k+1} - T_k)}{1 + L_k(T_{k+1} - T_k)}$$

with $A_{l,m} := \sum_{j=l}^{m-1} P(T_{i_{j+1}}) \cdot (T_{i_{j+1}} - T_{i_j})$ and $l(k) := \min\{l \mid i_l \geq k\}$.

 Interpretation: The term

$$\bar{\gamma}_{i,j}(T_a) = \int_0^{T_a} \sigma_i(t)\sigma_j(t)\rho_{i,j}(t) \, dt$$

is the integrated instantaneous covariance of the log forward rates. Defining the covariance matrix $C = (\bar{\gamma}_{i,j}(T_a))_{i,j=a,\dots,b}$ we can rewrite equation (19.20) as

$$\tilde{\sigma}_{a,b}^2 T_a = v(0)^\top C v(0),$$

where $v = (w_a L_a, \dots, w_b L_b)$. The vector v is the gradient of $\log(S_{a,b})$ in $\log(L_k)$-coordinates:

$$v_k = w_k L_k = \frac{\partial}{\partial L_k} \log(S_{a,b}) L_k = \frac{\frac{\partial}{\partial L_k} \log(S_{a,b})}{\frac{\partial}{\partial L_k} \log(L_k)} = \frac{\partial \log(S_{a,b})}{\partial \log(L_k)}.$$

The approximation we made is

$$\frac{\partial \log(S_{a,b})}{\partial \log(L_k)} \approx \frac{\partial \log(S_{a,b})}{\partial \log(L_k)} \bigg|_{t=0},$$

i.e., we linearize the log-swap rate $\log(S_{a,b})$ as a function of $\log(L_k)$ around $\log(L_k(0))$.

◁|

19.4.5.3 Analytic Calculation of Swap Rate Covariance in the LIBOR Market Model

In the same way we obtain an approximation for the integrated instantaneous swap rate covariance under a LIBOR market model. Considering two swap rates $S_{a,b}$ with tenor $T_a = T_{i_1} < \ldots < T_{i_m} = T_b$ and $S_{c,d}$ with tenor $T_c = T_{j_1} < \ldots < T_{j_m} = T_b$ we have

$$
\int_0^T \left(\frac{\mathrm{d}S_{a,b}(t)}{S_{a,b}(t)}\right)\left(\frac{\mathrm{d}S_{c,d}(t)}{S_{c,d}(t)}\right) \mathrm{d}t = \int_0^T \sum_{k=a}^{b-1}\sum_{l=c}^{d-1} v_k^{S_{a,b}}(t)\, v_l^{S_{c,d}}(t)\, \sigma_k(t)\, \sigma_l(t)\, \rho_{k,l}(t)\, \mathrm{d}t
$$

$$
\approx \sum_{k=a}^{b-1}\sum_{l=c}^{d-1} v_k^{S_{a,b}}(0)\, \bar{\gamma}_{k,l}(T)\, v_l^{S_{c,d}}(0),
$$

where

$$
v_k^{S_{a,b}} = \frac{\partial \log(S_{a,b})}{\partial \log(L_k)}, \quad v_l^{S_{c,d}} = \frac{\partial \log(S_{c,d})}{\partial \log(L_l)}.
$$

The swap rate covariance may then be used in analytic pricing formulas for CMS spread options, i.e., a payout

$$
\max\left(S_{a,b}(T) - S_{c,d}(T) - K, 0\right).
$$

See [7] for the derivation of an analytic pricing formula. This allows to calibrate the LIBOR market model to CMS spread options.

The calibration to caplet, swaptions and swap rate covariance/cms spread options completely determines the instantaneous covariance structure of the LIBOR market model.

 Tip (Efficient Implementation of the Swaption Approximation): To calibrate a LIBOR market model to swaptions we use a multi-dimensional optimizer minimizing the distance of the model swaption prices (given through the swaption approximation) from given market prices as a function of LIBOR covariance structure $(\sigma_i \sigma_j \rho_{i,j})$. This makes it necessary to repeatedly calculate the swaption approximation for different values of $\sigma_i \sigma_j \rho_{i,j}$. The followings two observations are helpful to improve the performance of the calibration:

- For each swaption the weights $v_k(0)$ need to be calculated only once since they do not depend on the LIBOR covariance structure $(\sigma_i \sigma_j \rho_{i,j})$. They only depend on the interest rate curve $L_{i,0}$, which does not change during the calibration (we assume that the model is already calibrated to the interest rate curve by the choice of $L_{i,0}$).

- The integrated instantaneous covariance of the log forward rates (19.21) needs to be calculated only once per iteration since it does not depend on the swaptions. This will improve the performance if there are many swaptions with overlapping tenors, thus sharing a portion of the integral (19.21).

◁|

19.5 Interpolation of Forward Rates in the LIBOR Market Model

 Motivation: An implementation of the LIBOR market model (e.g, as Monte Carlo simulation) allows us to calculate the forward rate $L(T_j, T_k; t_i)$ for interest rate periods $[T_j, T_k]$, $T_j < T_k$ and fixing times t_i.[10] Using these rates, we can evaluate almost all interest rate derivatives that can be represented as a function of these rates.

The discretization of the simulation time $\{t_i\}$ determines at which times we may have interest rates fixings. The discretization of the tenor structure $\{T_j\}$ determines for which periods forward rates are available and, since the numéraire is only defined at $t = T_j$, it determines at which times we may have payment dates. The tenor structure imposes a significant restriction since a change of the tenor structure is essentially a change of the model.

In practice, we desire to calculate as many financial products as possible with the same model. First, the aggregation of risk measures, i.e., of sensitivities[11] of products to the sensitivity of a portfolio, is correct only if the product sensitivities have been calculated using the same model. Second, the setup of a pricing model (calibration, generation of Monte Carlo paths) usually requires much more calculation time than the evaluation of a product, i.e., it is possibly efficient to reuse a model.

Thus, it is desirable to know how to calculate from a given LIBOR market model the quantities $L(T^s, T^e; t)$ for $T^s, T^e \notin \{T_0, T_1, \dots, T_n\}$ (unaligned period) and/or $t \notin \{t_0, t_1, \dots, t_n\}$ (unaligned fixing). ◁|

19.5.1 Interpolation of the Tenor Structure $\{T_i\}$

Let us look at how to interpolate the tenor structure. We will derive an expression for $L(T^s, T^e; t)$ for $T^s \notin \{T_0, T_1, \dots, T_n\}$ and/or $T^e \notin \{T_0, T_1, \dots, T_n\}$. Let $T^s < T^e$.

---

[10] In a Monte Carlo simulation the rates carry, of course, an approximation error of the time discretization scheme.

[11] A sensitivity is a partial derivative of the product price with respect to a model- or product-parameter (e.g., volatility or strike).

319

The forward rates $L(T^s, T^e; t)$ may be derived from corresponding bonds $P(T; t)$. We have

$$1 + L(T^s, T^e; t)\,(T^e - T^s) \;=\; \frac{P(T^s; t)}{P(T^e; t)}$$

For arbitrary $T > t$ the bond $P(T; t)$ is given by

$$P(T; t) = N(t)\,\mathrm{E}^{\mathbb{Q}^N}\left(\frac{1}{N(T)}\,\Big|\,\mathcal{F}_t\right).$$

The definition of the numéraire of the LIBOR market model shows that the specification of the short period bond $P(T_{m(t)+1}; t)$ is sufficient (and necessary) to determine all bonds $P(T; t)$ and thus all forward rates.[12]

19.5.1.1 Assumption 1: No Stochastic Shortly Before Maturity.

We assume that $P(T_{m(t)+1}; t)$ is an $\mathcal{F}_{T_{m(t)}}$-measurable random variable, i.e., the bond has no volatility at time t with $T_{m(t)} \le t \le T_{m(t)+1}$, i.e., shortly before its maturity. In other words, $t \mapsto P(T_{m(t)+1}; t)$ is a deterministic interpolation function of quantities known at the period start $T_{m(t)}$. In this case for the bond with maturity t, seen at time s with $T_{m(t)} \le s \le t \le T_{m(t)+1}$ we have

$$
\begin{aligned}
P(t; s) &= N(s)\,\mathrm{E}^{\mathbb{Q}^N}\left(\frac{1}{N(t)}\,\Big|\,\mathcal{F}_s\right) \\[2mm]
&= N(s)\,\mathrm{E}^{\mathbb{Q}^N}\left(\frac{P(T_{m(t)+1}; t)}{N(t)}\,\frac{1}{P(T_{m(t)+1}; t)}\,\Big|\,\mathcal{F}_s\right) \\[2mm]
&= N(s)\,\mathrm{E}^{\mathbb{Q}^N}\left(\frac{P(T_{m(t)+1}; t)}{N(t)}\,\Big|\,\mathcal{F}_s\right)\frac{1}{P(T_{m(t)+1}; t)} = \frac{P(T_{m(t)+1}; s)}{P(T_{m(t)+1}; t)}.
\end{aligned}
\tag{19.22}
$$

Especially for $s = T_{m(t)}$

$$P(t; T_{m(t)}) = \frac{P(T_{m(t)+1}; T_{m(t)})}{P(T_{m(t)+1}; t)}.\tag{19.23}$$

Thus we see that (under Assumption 1), the interpolation function $t \mapsto P(t; T_{m(t)})$ (interpolation of the maturity t) is derived directly from the interpolation function $t \mapsto P(T_{m(t)+1}; t)$ (interpolation of the evaluation time t) and vice versa. The functions are reciprocal.

---

[12] Note that the considerations on interpolation given in this section do not assume a LIBOR market model. They are valid in general.

19.5.1.2 Assumption 2: Linearity Shortly Before Maturity.

If the chosen interpolation function $T \mapsto P(T; T_{m(t)})$ is linear, then the interpolation of bond prices $P(T; s)$ seen in $s < T_{m(t)}$ is linear too. This follows directly from the linearity of the expectation:

$$P(T; s) = N(s) \, \mathrm{E}^{\mathbb{Q}^N} \left(\frac{P(T; T_{m(t)})}{N(T_{m(t)})} \,\Big|\, \mathcal{F}_s \right). \tag{19.24}$$

In this way the linear interpolation takes a distinct role. With $P(T_{m(t)}; T_{m(t)}) = 1$ and $P(T_{m(t)+1}; T_{m(t)}) = (1 + L_{T_{m(t)}}(T_{m(t)}) \, (T_{m(t)+1} - T_{m(t)}))^{-1}$ the linear interpolation of $t \mapsto P(t; T_{m(t)})$ follows as

$$
\begin{aligned}
P(t; T_{m(t)}) &= \frac{T_{m(t)+1} - t}{T_{m(t)+1} - T_{m(t)}} P(T_{m(t)}; T_{m(t)}) + \frac{t - T_{m(t)}}{T_{m(t)+1} - T_{m(t)}} P(T_{m(t)+1}; T_{m(t)}) \\
&= \frac{1 + L_{T_{m(t)}}(T_{m(t)}) \, (T_{m(t)+1} - t)}{1 + L_{T_{m(t)}}(T_{m(t)}) \, (T_{m(t)+1} - T_{m(t)})}.
\end{aligned}
\tag{19.25}
$$

The corresponding interpolation for the short period bond $P(T_{m(t)+1}; t)$ is thus with (19.23):

$$P(T_{m(t)+1}; t) := \frac{1}{1 + L_{T_{m(t)}}(T_{m(t)}) \, (T_{m(t)+1} - t)}.$$

Applying (19.24) to the (in t) linear interpolation (19.25) we find for all $s \leq T_{m(t)}$

$$\frac{P(t; s)}{P(T_{m(t)}; s)} = \frac{1 + L_{T_{m(t)}}(s) \, (T_{m(t)+1} - t)}{1 + L_{T_{m(t)}}(s) \, (T_{m(t)+1} - T_{m(t)})},$$

and thus

$$\frac{P(t; s)}{P(T_{m(t)+1}; s)} = 1 + L_{T_{m(t)}}(s) \, (T_{m(t)+1} - t).$$

From (19.22) we have for $T_{m(t)} \leq s \leq t$

$$P(t; s) = \frac{1 + L_{T_{m(t)}}(T_{m(t)}) \, (T_{m(t)+1} - t)}{1 + L_{T_{m(t)}}(T_{m(t)}) \, (T_{m(t)+1} - s)}.$$

We summarize this result in a theorem.

Theorem 217 (Interpolation of Forward Rates on Unaligned Periods): Given a tenor structure $T_0 < T_1 < T_2 < \cdots$. For all t let the short bond $P(T_{m(t)+1}; t)$ be given by the interpolation

$$P(T_{m(t)+1}; t) := \frac{1}{1 + L_{T_{m(t)}}(T_{m(t)}) \, (T_{m(t)+1} - t)}. \tag{19.26}$$

321

Then we have for arbitrary $t \le T$ with $k = m(t), l = m(T)$

$$P(T;t) = P(T_{k+1};t) \prod_{j=k+1}^{l} \frac{P(T_{j+1})}{P(T_j)} \frac{P(T;t)}{P(T_{l+1};t)}$$

$$= \frac{1}{1 + L_k(T_k)(T_{k+1} - t)} \prod_{j=k+1}^{l} \frac{1}{1 + L_j(t)(T_{j+1} - T_j)} (1 + L_l(t^*)(T_{l+1} - T)),$$

where $t^* = \min(t, T_l)$. For $T_s \le T_e$ with $T_k < T_s \le T_{k+1}$ and $T_l \le T_e < T_{l+1}$ we have

$$1 + L(T_s, T_e; t)(T_e - T_s) = \frac{P(T_s;t)}{P(T_e;t)} = \frac{P(T_s;t)}{P(T_{k+1};t)} \prod_{i=k+1}^{l} \frac{P(T_i;t)}{P(T_{i+1};t)} \frac{P(T_{l+1};t)}{P(T_e;t)}$$

$$= (1 + L_k(t^*)(T_{k+1} - T_s)) \prod_{i=k+1}^{l} (1 + L_i(t)(T_{i+1} - T_i)) \frac{1}{1 + L_l(t^*)(T_{l+1} - T_e)}.$$

$$(19.27)$$

19.6 Object-Oriented Design

Figures 19.2 and 19.3 show an object-oriented design of a *Monte Carlo LIBOR market model*. The following important aspects are considered in the design:

- **Reuse of implementation**

- **Separation of product and model**

- **Abstraction of model parameters**

- **Abstraction of calibration**

We will describe these aspects in the following.

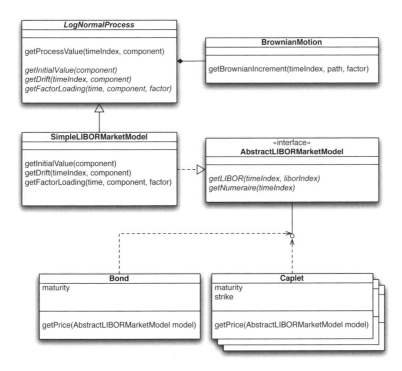

Figure 19.2. *UML Diagram: Evaluation of LIBOR-related products in a LIBOR market model via Monte Carlo simulation.*

19.6.1 Reuse of Implementation

For the Monte Carlo simulation of the lognormal process we use the same classes as in the example of the Black-Scholes model; see Figure 13.4. To do so the classes BROWNIANMOTION and LOGNORMALPROCESS were from the beginning designed for vector-valued, i.e., multifactorial processes, although the Black-Scholes model does not require it. Improvements to the classes BROWNIANMOTION and LOGNORMALPROCESS will result in improvement of both applications.

19.6.2 Separation of Product and Model

The interface ABSTRACTLIBORMARKETMODEL defines how LIBOR-related products communicate with a Monte Carlo LIBOR model. Through this interface the model serves to make the process of the underlyings (the forward rates) and the numéraire available to the product as a Monte Carlo simulation. All corresponding Monte Carlo evaluations of interest rate products expect this interface. All corresponding Monte Carlo LIBOR models implement this interface. This realizes a separation of product and model. The specific LIBOR market model is realized through the class SIMPLELIBORMARKETMODEL. Model extensions may be introduced without the need to change classes that realize LIBOR-related Monte Carlo products.

19.6.3 Abstraction of Model Parameters

The model parameters, i.e., the covariance structure, are encapsulated in their own classes. The model parameter classes implement a simple *interface* LIBORCOVARI-ANCEMODEL. A specific covariance model $(i, j, t) \mapsto \gamma_{i,j}(t) = \sigma_i(t)\sigma_j(t)\rho_{i,j}(t)$ is realized through a class that implements the interface LIBORCOVARIANCEMODEL. This class is then served to the model. The interfaces are designed such that $(i, j, t) \mapsto \gamma_{i,j}(t)$ may be stochastic.[13] See Figure 19.3.

This abstraction of model parameters makes it easy to exchange different modelings of covariance, i.e., volatility and correlation.

 Warning: In cases where the covariance structure is modeled by volatility and correlation, it seems reasonable to define corresponding interfaces LIBORVOLATILITYMODEL and LIBORCORRELATIONMODEL. A simple class LIBORCOVARIANCEMODELFROMVOLATILITYANDCORRLEATION calculates the factor loadings and covariances from given volatility and correlation models. See Figure 19.4. However, the separation of volatility and correlation into their own classes will bring some disadvantages for a joint calibration and general covariance modeling. The corresponding code may become overdesigned. The design in Figure 19.4 would

---

[13] A stochastic volatility model would result in a stochastic covariance model.

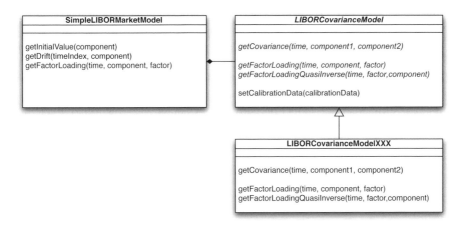

Figure 19.3. *UML Diagram: LIBOR Market Model: Abstraction of model parameters.*

make sense if one wished to explore many combinations of different volatility and correlation models. ◁|

19.6.4 Abstraction of Calibration

The abstraction of model parameters allows for the abstraction of calibration. The algorithm calibrating the covariance model is clearly a part of the covariance model. Thus each covariance model object can carry calibration data (e.g., market data) that, once set, is used to calibrate the model. The calibration data themselves may be anything from given correlation and volatility parameters to a list of products with associated target values. A generic calibration for parametric models may be implemented in an abstract class defining the properties of parametric covariance models; see Figure 19.5.

Experiment: At http://www.christian-fries.de/finmath/applets/LMMPricing.html several interest rate products can be priced using a LIBOR market model. ◁|

Further Reading: The original articles on the LIBOR market model are [50] and [88]; for the calibration of the LIBOR market model see [7, 30]; for the arbitrage-free discretization see [73]; for the interpolation of forward rates see [96]. The evaluation of Bermudan options in Monte Carlo is considered in Chapter 15; see also [44, 45].

We will use the LIBOR market model as foundation for further investigations into general interest rate model properties. In Chapter 21 we will investigate the instantaneous correlation $\rho_{i,j}$ and volatility σ_i and their effect on terminal correlation. In Chapter 25 we will investigate the influence of mean reversion and multifactoriality on the shape of interest rate curve. ◁|

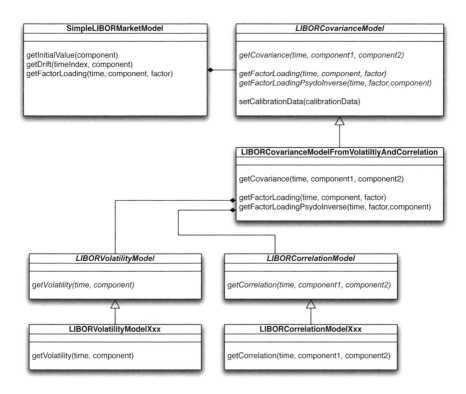

Figure 19.4. *UML Diagram: LIBOR market model: Abstraction of model parameters as volatility and correlation. Introducing separate classes for volatility and correlation has some disadvantages for joint calibration and general covariance modeling. The design above might be considered overdesigned.*

327

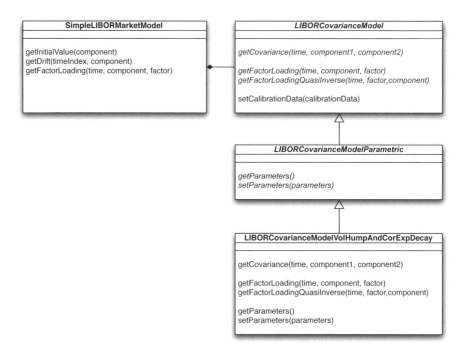

Figure 19.5. *UML Diagram: LIBOR market model: Abstraction of model parameters: Parametric covariance models.*

CHAPTER 20

Swap Rate Market Models

 Motivation: The LIBOR market model postulates as lognormal dynamic for the forward rate $L_i := L(T_i, T_{i+1})$. In other words, each single forward rate follows a Black model. This allows an easy calibration of the LIBOR market model to caplet prices. We only have to fulfill condition (19.18).

If, however, swap options (i.e., swaption or swaption-related products like Bermudan swaptions) are in the focus, then a model that simulates the swap rate directly might be a better choice.[1] If, for example, the swap rate follows a lognormal process, then the corresponding swaptions may be calibrated by a simple condition involving the implied Black-volatility of the swap rate. ◁|

Instead of a lognormal dynamic for the forward rate $L(T_i, T_{i+1})$, which is the starting point of the LIBOR market model, we postulate here a lognormal dynamic of the swap rate:

$$S_{i,k}(t) := S(T_i, T_k; t) := \frac{P(T_i; t) - P(T_k; t)}{\sum_{j=i}^{k-1}(T_{j+1} - T_j) P(T_{j+1}; t)}, \quad k > i. \qquad (20.1)$$

Since the set of swap rates defined for a given tenor structure $T_0 < T_1 < \cdots < T_n$ is a two parametric family of $((n-1)\,n)/2$ rates which are related by functional dependencies, a meaningful dynamic can be given only for a subset of swap rates.[2]

When choosing a system of swap rates $S_{i,k}$, for which we wish to specify the dynamics, we have to take care that the system is neither overdetermined nor, with

---

[1] Later, we will explain why a forward-rate-based model might be the choice even for swap-rate-related products; see Remark 219.

[2] For example, the swap rate $S_{i,i+4}$ is a function of the swap rates $S_{i,i+2}$ and $S_{i+2,i+4}$, which in turn are functions of the swap rates $S_{i,i+1}, \ldots S_{i+3,i+4}$. The swap rates with one period are forward rates $L_i = S_{i,i+1}$.

Co-sliding:	$S_{i,\min(i+k,n)}$	$i = 0, 1, \ldots, n - 1$
Co-terminal:	$S_{i,n}$	$i = 0, 2, \ldots, n - 1$

Table 20.1. *Co-sliding and co-terminal swap rates*

respect to the given tenor structure, underdetermined. The system of rates has to consist of n swap rates since on the tenor structure $0 = T_0 < T_1 < \cdots < T_n$ we have n degrees of freedom in terms of bond prices.

Two common variants are given by the set of *co-sliding* swap rates and *co-terminal* swap rates; see Table 20.1. When specifying co-sliding swap rates, it is necessary to close the system. Our definition achieves this by first considering the swap rates $S_{i,i+k}$ with k periods (co-sliding, $i < n - 1 - k$) and starting with $i = n - k$ we consider co-terminal swap rates.

If the selection of swap rates is made, we model each $S_{i,k}$ from the selection as a lognormal process:

$$dS_{i,k}(t) = \mu_{i,k}^{\mathbb{P}}(t) S_{i,k}(t)\, dt + \sigma_{i,k}(t) S_{i,k}(t)\, dW_{i,k}^{\mathbb{P}}(t) \qquad \text{under } \mathbb{P}, \qquad (20.2)$$

with initial conditions

$$S_{i,k}(0) = S_{i,k,0}.$$

 Interpretation: The modeling of co-terminal swap rates is a suitable choice if, e.g., we have to price Bermudan swaptions, which have these swap rates as underlying. The modeling of co-sliding swap rates is a suitable choice if we have to price products relying on swap rates with constant time to maturity (CMS rate[3]). ◁|

20.1 The Swap Measure

If we consider the definition of the swap rate in (20.1), it is apparent that $S_{i,k}$ is a martingale under the martingale measure $\mathbb{Q}^N$ corresponding to the numéraire

$$N(t) := A(T_i, \ldots, T_k; t) := \sum_{j=i}^{k-1} (T_{j+1} - T_j)\, P(T_{j+1}; t), \quad k > i,\, t \le T_{i+1}. \qquad (20.3)$$

[3] See Definition 160.

The right-hand side in (20.3) is a portfolio of zero bonds and thus a traded product and the swap rate is the N-relative price of a traded product.

Definition 218 (Swap Measure, Swap Annuity):
The equivalent martingale measure $\mathbb{Q}^N$ corresponding to the numéraire N in (20.3) is called *swap measure* corresponding to the swap rate $S(T_1, \ldots, T_k)$. The expression on the right-hand side in (20.3) is also called *swap annuity*.

The numéraire is, so far, defined for $t \leq T_{i+1}$ only, since at $t = T_{i+1}$ the first bond $P(T_{i+1})$ is at its maturity and we have to specify how its payment has to be reinvested.[4] A continuation of the numéraire definition to $t > T_{i+1}$ can be given by a reinvestment into the next swap annuity. This is the analog to the numéraire (19.9) of the spot measures. For $i = 1, \ldots, k-1$ we have

$$N(t) = A(T_i, \ldots, T_k; t) \prod_{j=1}^{i-1} \frac{A(T_j, \ldots, T_k; T_{j+1})}{A(T_{j+1}, \ldots, T_k; T_{j+1})}, \quad T_{i-1} \leq t < T_i$$

where $T_0 := 0$. The swap rates we are considering here are co-terminal. Of course, we may consider co-sliding swaps in a similar way, using the swap annuities $A(T_j, \ldots, T_{j+k}; t)$. The corresponding numéraire of reinvestment in co-sliding swap annuities, i.e., a *rolling co-sliding swap annuity* then is

$$N(t) = A(T_i, \ldots, T_{i+k}; t) \prod_{j=1}^{i-1} \frac{A(T_j, \ldots, T_{j+k}; T_{j+1})}{A(T_{j+1}, \ldots, T_{j+1+k}; T_{j+1})}, \quad T_{i-1} \leq t < T_i.$$

For $k = i + 1$ this corresponds to (19.9).

20.2 Derivation of the Drift Term

For the swap rate market model we have multiple sets of swap rates, which may be modeled and (as in the LIBOR market model) multiple possible choices of numéraires. This section does not give a detailed derivation of the drift terms. The derivation is done similarly to the derivation of the drift in the LIBOR market model by expressing a martingale through the elementary swap rate processes $S_{i,j}$. If for example $A_{k,l}$ is the numéraire, we consider the $\mathbb{Q}^{A_{k,l}}$-martingale $\left(S_{i,j} \frac{A_{i,j}}{A_{k,l}} \right)$.

---

[4] The reinvestment determines the evolution of the numéraire for $t > T_{i+1}$: For example, if we compare the investment of the paid 1 in $\frac{1}{P(T_k; T_{i+1})}$ parts of a T_k-bond with the investment in $\frac{1}{P(T_{k+1}; T_{i+1})}$ parts of a T_{k+1}-bonds, then the evolution of the numéraire will differ by the evolution of the T_k forward rate, i.e., by the factor $\frac{P(T_k; t)}{P(T_k; T_{i+1})} / \frac{P(T_{k+1}; t)}{P(T_{k+1}; T_{i+1})} = \frac{1 + L(T_k, T_{k+1}; t) \cdot (T_{k+1} - T_k)}{1 + L(T_k, T_{k+1}; T_{i+1}) \cdot (T_{k+1} - T_k)}$.

20.3 Calibration—Choice of the Free Parameters

20.3.1 Choice of the Initial Conditions

20.3.1.1 Reproduction of Bond Market Prices or Swap Market Prices

If we set t to the preset time in the definition of the swap rate (20.1), i.e., $t = 0$ following our convention, then we get an equation relating today's bond prices to today's swap rates $S_{i,k}(0)$, and the latter are just the initial conditions of the chosen swap rate processes. Thus the initial conditions of the processes are given by (20.1) with $t = 0$ and today's bond prices, i.e., today's interest rate curve.

Although we regard the family of zero bonds as the natural description of the interest rate curve and we see swap rates and swap prices as derived quantities, it is in this case natural to calculate today's swap rates directly from today's swap prices (assuming they are given). In this case the initial conditions are given by today's swap prices. With this choice, the model will reproduce these prices.

20.3.2 Choice of the Volatilities

20.3.2.1 Reproduction of Swaption Market Prices

The calibration of the model to swaption prices is analog to the calibration of the LIBOR market model to caplet prices. Let the dynamic of the swap rate $S_{i,k}$ be given by (20.2). Furthermore let $\sigma_{i,k}^{\text{Black,Market}}$ denote the market prices of an option on $S_{i,k}$ given as implied Black-volatility. If we calculate

$$
\sigma_{i,k}^{\text{Black,Model}} := \left(\frac{1}{T_i} \int_0^{T_i} \sigma_{i,k}^2(t)\, dt \right)^{1/2},
$$

then the model reproduces the given swaption market prices if

$$
\sigma_{i,k}^{\text{Black,Model}} = \sigma_{i,k}^{\text{Black,Market}}.
$$

This statement is trivial since, if we consider only a single swap rate $S_{i,k}$, then (20.2) is a Black model for this swap rate, and under this model the implied volatility is defined by inverting the pricing formula. The inversion of the pricing formula is what a calibration should achieve.

Remark 219 (LIBOR Market Model versus Swaprate Market Model): The question of whether one should choose a LIBOR market model or a swap rate market model seems to depend on the application only, to be precise, on whether the model should calibrate to caplets or swaptions—and whether or not one sees a lognormal forward rate or a lognormal swap rate as a realistic model.[5]

Therefore, the criterion that defines the choice of the model thus is the quality of the model calibration to the specific application.

However, the swap rate market model has a disadvantage compared to the LIBOR market model: If we calculate a forward rate L_i in a swap rate market model, then the forward rate tends to suffer from numerical instabilities. Conversely the calculation of a swap rate from forward rates in a LIBOR market model is generally much more stable.

 Interpretation: The reason lies in the representation of the swap rate as a convex combination of the forward rates. From Lemma 123 we have

$$S_{i,j} = \sum_{k=i}^{j-1} \alpha_k^{i,j} L_k, \quad \text{with} \quad \alpha_k^{i,j} \geq 0 \quad \text{and} \quad \sum_{k=i}^{j-1} \alpha_k^{i,j} = 1,$$

with

$$\alpha_k^{i,j} := \frac{P(T_{k+1}) \cdot (T_{k+1} - T_k)}{\sum_{k=i}^{j-1} P(T_{k+1}) \cdot (T_{k+1} - T_k)}.$$

If we calculate a forward rate L_i from (e.g., co-terminal) swap rates $S_{j,n}$, we have

$$L_i = \frac{1}{\alpha_i^{i,n}} S_{i,n} - \frac{1}{\alpha_i^{i+1,n}} S_{i+1,n+1}$$

$$= \frac{1}{\alpha_i^{i,n}} (S_{i,n} - S_{i+1,n}) + \left(\frac{1}{\alpha_i^{i,n}} - \frac{1}{\alpha_i^{i+1,n}} \right) S_{i+1,n}.$$

Assuming for simplicity $\alpha_j^{i,n} = \frac{1}{n-i-1}$, which is with $\sum_{j=i}^{n-1} \alpha_j^{i,n} = 1$ plausible[6], then we have

$$S_{i,n} = \frac{1}{n-i-1} \sum_{k=i}^{n-1} L_k,$$

$$L_i = (n-i-1)(S_{i,n} - S_{i+1,n}) + S_{i+1,n}.$$

This shows:

---

[5] In general both assumptions cannot hold, and it is necessary to modify the models with respect to their distribution assumption. Such a modification of the model is called *smile modeling*.

[6] Indeed we have

$$\frac{1}{\alpha_i^{i,n}} = \frac{P(T_{i+2})(T_{i+2} - T_{i+1})}{P(T_{i+1})(T_{i+1} - T_i)} \left(\frac{1}{\alpha_i^{i+1,n}} + 1 \right).$$

- The calculation of a swap rate $S_{i,n}$ from forward rates L_k corresponds to the calculation of an average (rate)—the swap rate can be interpreted as an integral of the forward rates. Errors in L_k are averaged and thus smoothed. The variance of an unsystematic error is reduced.

- The calculation of a forward rate L_i from swap rates $S_{i,n}$, $S_{i+1,n}$ consists of a finite difference term—this part of the forward rate may be interpreted as a derivative. The calculation of a difference is very sensitive to errors in the swap rates (e.g. small jumps) and the error is scaled up by the factor $(n - i - 1)$ for n large and i small. Thus forward rates for short periods in a model of long period swap rates have a tendency to numerical instability.

Tip: If there is no strong reason for a swap rate market model, a *generic* LIBOR market model with calculation of the corresponding swap rates from forward rates is preferable. This provides a single, thus consistent, model for multiple applications (products), which allows the aggregation of risk parameters (delta, gamma). The difference in the distributional properties is often negligible (see [7]).

Further Reading: The original article on the swap rate market model is [81].

334

CHAPTER 21

Excursus: Instantaneous Correlation and Terminal Correlation

In this chapter we will use the LIBOR market model to discuss the influence of instantaneous volatility and instantaneous correlation on option prices. Although our study is based on the LIBOR market model, the intuition gained from our experiments is universally valid.

We will experiment with different (extreme) parameter configurations, and we will see how a single-factor model in which all interest rates $L(T_i, T_{i+1})$ move (instantaneously) perfectly correlated may, however, exhibit at time $t > 0$ (terminal) perfectly decorrelated random variables $L(T_j, T_{j+1}; t)$, $L(T_k, T_{k+1}; t)$.

We will start by repeating some basic concepts.

21.1 Definitions

Definition 220 (Covariance, Correlation):

Let X, Y denote two (numeric) random variables, $\bar{X} = E(X)$, $\bar{Y} = E(Y)$. Then

$$\text{Cov}(X, Y) := E((X - \bar{X}) \cdot (Y - \bar{Y}))$$

is called the *covariance* of X and Y, $\text{Var}(X) := \text{Cov}(X, X)$ is called the variance of X and

$$\text{Cor}(X, Y) := \frac{E((X - \bar{X}) \cdot (Y - \bar{Y}))}{\sqrt{\text{Var}(X)} \cdot \sqrt{\text{Var}(Y)}}$$

is called the *correlation* of X and Y.

Let $L = (L_1, \ldots, L_n)$ denote an n-dimensional m-factorial Itô process of the form

$$dL_i = \mu_i \, dt + \sigma_i \, dW_i, \quad \text{where} \quad dW_i = \sum_{k=1}^{m} f_{i,k} \, dU_k \qquad (21.1)$$

and U_k denote independent Brownian motions. Furthermore, let $f_{i,k}$ be such that

$$R := \left(\rho_{i,j}(t)\right)_{i,j=1,\ldots,n} = \left(\sum_{k=1}^{m} f_{i,k} f_{j,k}\right)_{i,j=1,\ldots,n}$$

is a correlation matrix (i.e., $\sum_{k=1}^{m} f_{i,k}^2 = 1$). We have

$$< dW(t), dW(t) > = R \, dt.$$

Definition 221 (Instantaneous Covariance, Instantaneous Correlation):
With the notation above we call $\rho_{i,j}$ defined by

$$\rho_{i,j}(t) := \left(\sum_{k=1}^{m} f_{i,k} f_{j,k}\right)_{i,j=1,\ldots,n}$$

the *instantaneous correlation* of the processes L_i and L_j, and we call $\sigma_i \sigma_j \rho_{i,j}$ the *instantaneous covariance* of the processes L_i and L_j.

Definition 222 (Terminal Covariance, Terminal Correlation):
With the notation above we call $\rho_{i,j}^{\text{Term}}$ defined by

$$\rho_{i,j}^{\text{Term}}(t) := \text{Cor}(L_i(t), L_j(t))$$

the *terminal correlation* of the processes L_i and L_j. Correspondingly we call $t \mapsto \text{Cov}(L_i(t), L_j(t))$ the *terminal covariance* of the processes L_i and L_j.

21.2 Terminal Correlation Examined in a LIBOR Market Model Example

We are considering a LIBOR market model with semiannual tenor structure $T_i := 0.5 \, i$ and investigating the behavior of the two rates $L_{10} = L(5.0, 5.5)$ and $L_{11} = L(5.5, 6.0)$. Under the numéraire $N = P(T_{12}) = P(6.0)$ we have for the dynamic of these rates (see (19.3), (19.8))

$$dL_i(t) = \mu_i(t) L_i(t) \, dt + \sigma_i(t) L_i(t) \, dW_i^{Q^N}(t) \qquad (i = 10, 11) \qquad (21.2)$$

$$\mu_{10} = -\frac{\delta_{11} L_{11}(t)}{1 + \delta_{11} L_{11}(t)} \sigma_{10}(t) \, \sigma_{11}(t) \, \rho_{10,11}(t), \qquad \delta_{11} := T_{11} - T_{10}$$

$$\mu_{11} = 0.$$

If we neglect the drift (i.e., set $\mu_{10} = 0$) and assume a constant instantaneous covariance $\sigma_{10}\sigma_{11}\rho_{10,11} = $ const., then it follows from (21.1) that the terminal correlation is

$$\rho_{i,j}^{\text{Term}}(t) = \rho_{10,11} \quad \forall\, t.$$

As one might have expected, the terminal correlation is given by the choice of the instantaneous correlation. In this case, to achieve a terminal correlation different from zero we need at least a two-factor model. Figure 21.1 shows a scatter plot for a one-factor and a five-factor model[1] of the interest rates $L_{10}(t)$, $L_{11}(t)$ at time $t = T_{10} = 5.0$.

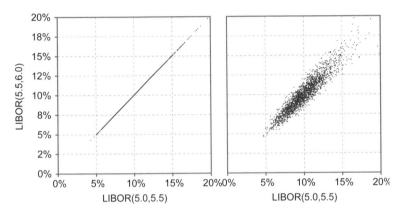

Figure 21.1. *The two (adjacent) rates $L_{10} = L(5.0, 5.5)$ and $L_{11} = L(5.5, 6.0)$ in a one- and a multifactor model for constant instantaneous volatility $\sigma_{10}(t) = \sigma_{11}(t) = $ const. In a one-factor model both random variables are perfectly correlated (left). In a five-factor model both random variables show a correlation different from 1. This is a consequence of the instantaneous correlation $\rho_{10,11}$ being different from 1.*

21.2.1 Decorrelation in a One-Factor Model

It is possible to achieve a terminal de-correlation for processes which have perfect instantaneous correlation. Consider

$$\sigma_{10}(t)\begin{cases} > 0 & \text{for } t < 2.5, \\ = 0 & \text{for } t \geq 2.5, \end{cases} \qquad \sigma_{11}(t)\begin{cases} = 0 & \text{for } t < 2.5, \\ > 0 & \text{for } t \geq 2.5, \end{cases} \tag{21.3}$$

---

[1] The exact model specification is: $L_{i,0} = 0.1$, $\sigma_i = 0.1$, and $\rho_{i,j} = \exp(-0.5|i - j|)$, followed by a factor reduction as given in Section B.3. For the five-factor model we have $\rho_{10,11} = 0.94$.

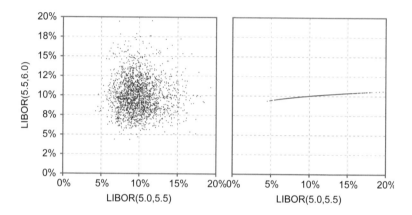

Figure 21.2. *The two (adjacent) rates $L_{10} = L(5.0, 5.5)$ and $L_{11} = L(5.5, 6.0)$ in a one-factor model. Left: The two random variables exhibit a correlation close to 0 (perfect decorrelation). Right: The two random variables exhibit very different variances. The covariance is close to zero since the variance of L_{11} is close to 0. Both scenarios are the consequence of a very special choice for the instantaneous volatility.*

i.e., the processes receive the Brownian increment $dW(t)$ at different times t; thus the increments received are independent. Since in this case we have $\mu_{10} = \mu_{11} = 0$ in (21.2), the two random variables $L_{10}(5.0)$, $L_{11}(5.0)$ are given by

$$\log(L_{10})(5.0) = -\frac{1}{2}\bar{\sigma}_{10}^2\, 2.5 + \bar{\sigma}_{10}\, (W_{10}(2.5) - W_{10}(0))$$

$$\log(L_{11})(5.0) = -\frac{1}{2}\bar{\sigma}_{11}^2\, 2.5 + \bar{\sigma}_{11}\, (W_{11}(5.0) - W_{11}(2.5)),$$

where $\bar{\sigma}_{10}^2 = \frac{1}{2.5}\int_0^{2.5}\sigma_{10}^2(t)\,dt$ and $\bar{\sigma}_{11}^2 = \frac{1}{2.5}\int_{2.5}^{5.0}\sigma_{11}^2(t)\,dt$.

Since, even for a one-factor model, the increments $(W(2.5)-W(0))$, $(W(5)-W(2.5))$ are independent, $L_{10}(5.0)$, $L_{11}(5.0)$ are independent as well; see Figure 21.2, left.

21.2.2 Impact of the Time Structure of the Instantaneous Volatility on Caplet and Swaption Prices

The previous example of the decorrelation of the rates L_{10}, L_{11} in a one-factor model shows the importance of the time structure of the instantaneous volatility for the (terminal) distribution of (L_{10}, L_{11}) at time $t = 5.0$. Now we will look at the corresponding caplets and a swaption with maturity 5.0 and payment dates 5.5, 6.0, which is dependent on L_{10} and L_{11}:

Scenario	$\sigma_i(t)$	$\rho_{10,11}$	Caplet 5.0–5.5	Caplet 5.5–6.0	Swaption 5.0–6.0
1	0.1	1.0	0.26%	0.26%	0.51%
2	0.1	0.94	0.26%	0.26%	0.50%
3	as in (21.3)	1.0	0.26%	0.26%	0.36%
4	$0.7\exp(4.9(T_i - t))$	1.0	0.26%	0.26%	0.27%

Table 21.1. *Caplet and swaption prices for different instantaneous correlations and volatilities.*

In all scenarios we have $\int_0^{T_i}\sigma_i(t)^2\,dt = 0.05$ for $i = 10$, thus all caplet prices are the same.[2]

Figures 21.1 and 21.2 are generated with these parameters.

---

[2] We have $\int\limits_0^{T_i}(b\,\exp(-c\,(T_i - t)))^2\,dt = \frac{b^2}{2c}(1 - \exp(-2\,c\,T_i))$. For $T_i = 5.0$, $b = 0.7$, $c = 4.9$ we thus have $\int_0^{T_i}\sigma_i(t)^2\,dt = 0.05(1 - \exp(-49)) \approx 0.05\,(1 - 5\times 10^{-22}) \approx 0.05$.

21.2.3 Swaption Value as a Function of Forward Rates

To interpret these results we analyze the dependency of the swaption value from the rates L_{10}, L_{11}.

For the value of a swaption $V_{\text{Swaption}}(T_0)$ with fixed swap rate (strike) K we have

$$V_{\text{Swaption}}(T_0) \;=\; N(0)\mathrm{E}^{\mathbb{Q}^N}\left(\frac{\max(S(T_i) - K, 0)\, A(T_i)}{N(T_i)} \mid \mathcal{F}_{T_0}\right)$$

with

$$A(T_i) = \sum_{j=i}^{n-1}(T_{j+1} - T_j)P(T_{j+1}; T_i) \qquad\qquad \text{(swap annuity)}$$

$$S(T_i) = \frac{1 - P(T_n; T_i)}{A(T_i)} \qquad\qquad \text{(par swap rate)}.$$

With the numéraire $N = P(T_n)$ we have

$$\frac{A(T_i)}{N(T_i)} = \frac{A(T_i)}{P(T_n; T_i)} = \sum_{j=i}^{n-1}(T_{j+1} - T_j)\frac{P(T_{j+1}; T_i)}{P(T_n; T_i)}$$

$$= \sum_{j=i}^{n-1}(T_{j+1} - T_j)\prod_{k=j+1}^{n-1}(1 + L_k(T_i)(T_{k+1} - T_k))$$

and

$$S(T_i)\frac{A(T_i)}{N(T_i)} = \frac{1 - P(T_n; T_i)}{P(T_n; T_i)} = \prod_{j=i}^{n-1}(1 + L_j(T_i)(T_{j+1} - T_j)) - 1,$$

i.e.,

$$V_{\text{Swaption}}(T_0) \;=\; P(T_n; T_0)\mathrm{E}^{\mathbb{Q}^{P(T_n)}}\left(\max(\frac{(S(T_i) - K)\, A(T_i)}{P(T_n; T_i)}, 0) \mid \mathcal{F}_{T_0}\right)$$

with

$$\frac{(S(T_i) - K)\, A(T_i)}{P(T_n; T_i)} = \prod_{j=i}^{n-1}(1 + L_j(T_i)(T_{j+1} - T_j)) - 1$$

$$- K\sum_{j=i}^{n-1}(T_{j+1} - T_j)\prod_{k=j+1}^{n-1}(1 + L_k(T_i)(T_{k+1} - T_k)).$$

340

If we apply this to the special case of a swaption with a two-period tenor $\{T_i, \ldots, T_n\} = \{T_{10}, T_{11}, T_{12}\} = \{5.0, 5.5, 6.0\}$, we get

$$\max\left(\frac{S(T_i) - K\,A(T_i)}{P(T_n; T_i)}, 0\right)$$

$$= \max((1 + L_{10}\,\Delta T)(1 + L_{11}\,\Delta T) - K(\Delta T(1 + L_{11}\,\Delta T) + \Delta T), 0)$$

$$= \max((L_{10} - K)\,\Delta T + (L_{11} - K)\,\Delta T + L_{11}(L_{10} - K)(\Delta T)^2), 0). \qquad (21.4)$$

From (21.4) we can derive the following observations for the value of the swaption:

- If $L_{11}(T_{10}) = K$, then the value of the swaption corresponds to the value of a caplet paying $\max(L_{10} - K, 0)$. If L_{11} has at time T_{10} no or small variance and if $L_{11}(T_{10})$ is close to K, then the value of the swaption is close to the value of a caplet with payoff $\max(L_{10} - K, 0)$.

- Neglecting the term $L_{11}(T_{10})(L_{10}(T_{10}) - K)(\Delta T)^2$, which is justified for small rates and short periods ΔT, and considering thus only

$$(L_{10}(T_{10}) - K)\,\Delta T + (L_{11}(T_{10}) - K)\,\Delta T = (L_{10}(T_{10}) + L_{11}(T_{10}) - 2K)\,\Delta T,$$

we see that the option price is determined by the variance of $L_{10}(T_{10}) + L_{11}(T_{10})$. For this we have

$$\text{Var}(L_{10}(T_{10}) + L_{11}(T_{10}))$$
$$= \text{Var}(L_{10}(T_{10})) + \text{Var}(L_{11}(T_{10})) + 2\,\text{Cov}(L_{10}(T_{10}), L_{11}(T_{10})).$$

- From the previous we know that the option value is maximal for $\rho_{10,11}^{\text{Term}}(T_{10}) = 1$ and minimal (even 0) for $\rho_{10,11}^{\text{Term}}(T_{10}) = -1$ (still neglecting the term $L_{11}(T_{10})(L_{10}(T_{10}) - K)\Delta T^2$).

From these remarks the results in Table 21.1 become plausible. In scenario 4 the rate $L_{11}(T_{10})$ has a negligible small variance (compare Figure 21.2, right). The swaption value is close to the caplet value. The caplet on the period $[T_{11}, T_{12}]$, however, has the same price as in the other scenarios, since the high instantaneous volatility for $t \in [T_{10}, T_{11}]$ will give the rate $L_{11}(T_{11})$ the required (terminal) variance.

While for the swaption the rate $L_{11}(T_{10})$ is relevant, for the caplet it is the rate $L_{11}(T_{10})$.

Experiment: The influence of the instantaneous volatility and instantaneous correlation on terminal correlation, caplet and swaption prices may be investigated at http://www.christian-fries.de/finmath/applets/LMMCorrelation.html. ◁|

21.3 Terminal Correlation Is Dependent on the Equivalent Martingale Measure

The terminal correlation is dependent on the martingale measure and thus the numéraire used. The whole (terminal) probability density is, of course, measure dependent; see also Lemma 81 in Chapter 5. Thus an interpretation of terminal correlation and other terminal quantities should be made with caution.

How the chosen martingale measure influences the terminal distribution, especially the terminal correlation, may easily be seen in a LIBOR market model. Consider the processes $L_i = L(T_i, T_{i+1})$ and $L_{i+1} = L(T_{i+1}, T_{i+2})$, i.e., two adjacent forward rates, under the martingale measure $\mathbb{Q}^{P(T_n)}$ corresponding to the numéraire $P(T_n)$ (*terminal measure*). It is

$$d \log(L_i) = -\sum_{i<j<n} \frac{\delta_j L_j(t)}{1 + \delta_j L_j(t)} \sigma_i(t) \sigma_j(t) \rho_{i,j}(t) \, dt + \frac{1}{2}\sigma_i(t)^2 \, dt + \sigma_i(t) \, dW_i^{\mathbb{Q}^{P(T_n)}}.$$

With $dK_j(t) := \dfrac{\delta_j L_j(t)}{1 + \delta_j L_j(t)} \, dt$ we thus have

$$d \log(L_i) = \sigma_i \left(-\sigma_{i+1}\rho_{i,i+1} \, dK_{i+1} - \sum_{i+1<j<n} \sigma_j \rho_{i+1,j} \, dK_j + \tfrac{1}{2}\sigma_i(t) \, dt + dW_i^{\mathbb{Q}^{P(T_n)}} \right)$$

$$d \log(L_{i+1}) = \sigma_{i+1} \left(-\sum_{i+1<j<n} \sigma_j \rho_{i+1,j} \, dK_j + \tfrac{1}{2}\sigma_{i+1} \, dt + dW_{i+1}^{\mathbb{Q}^{P(T_n)}} \right).$$

The terminal correlation is influenced by the common drift term $\sum_{i+1<j<n} dK_j$ and this influence can be increased arbitrarily through the factor σ_j in front of K_j. If and how this term is present depends on the chosen martingale measure: For $n = i + 2$ the sum is empty and the term is $= 0$, for $n > i + 2$ the term is > 0. In theory it might be possible that L_i and L_{i+1} appear almost perfectly correlated under $\mathbb{Q}^{P(T_{i+3})}$ and perfectly uncorrelated under $\mathbb{Q}^{P(T_{i+2})}$.

21.3.1 Dependence of the Terminal Density on the Martingale Measure

How the chosen martingale measure influences the terminal distribution function is shown in the following examples. In Figure 21.3 we look at the density of a forward rate under a one-factor LIBOR market model with constant instantaneous volatility, equal for all rates. Under different martingale measures (spot measure, terminal measure) the distribution is slightly different. If, however, the volatility of the *other* rates is increased, then, depending on the chosen martingale measure, the distribution will change, see Figure 21.4. As before, the change in the distribution function stems from the drift of the LIBOR market model.

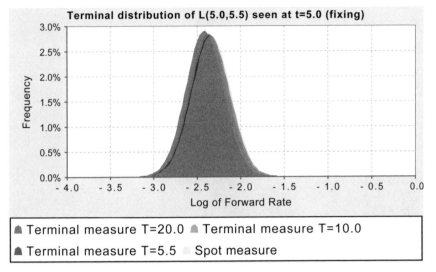

Figure 21.3. *The terminal distribution function of a forward rate under different martingale measures. Shown is the rate $L(5.0, 5.5)$ upon its fixing at $t = 5.0$. All rates are simulated in a one-factor LIBOR market model with constant instantaneous volatility $\sigma = 10\%$.*

 Tip (Terminal Quantities Independent of the Martingale Measure): In place of martingale measure-dependent quantities, like the terminal distribution of the terminal correlation, we can define meaningful alternatives. The implied (Black) volatility is an example of a martingale measure-independent quantity. Apart from the scaling with the square root of the maturity $\sqrt{T_k}$, it corresponds to the terminal standard deviation under the T_{k+1}-forward measure. If, for example,

$$d \log(L_i(t)) = (\ldots) \, dt + \sigma_i(t) \, dW_i(t)$$
$$d \log(L_j(t)) = (\ldots) \, dt + \sigma_j(t) \, dW_j(t)$$

are given processes, then the *integrated instantaneous covariance*

$$\int_0^T d \log(L_i(t)) \, d \log(L_j(t)) = \int_0^T \sigma_i(t) \sigma_j(t) \rho_{i,j}(t) \, dt$$

is independent of the chosen martingale measure.[3] It would correspond to the covariance of $\log(L_i(t))$ and $\log(L_j(t))$, if both were martingales. ◁|

---

[3] This is clear because a change of martingale measure is a change of drift only.

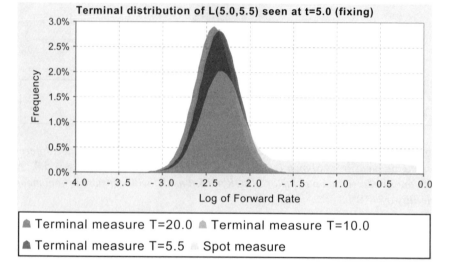

Figure 21.4. *The terminal distribution function of a forward rate under different martingale measures. Shown is the rate $L(5.0, 5.5)$ upon its fixing at $t = 5.0$. In contrast to Figure 21.3 the rates $L(T_i, T_{i+1})$ for $T_i < 5.0$ are simulated differently. They are simulated with a high volatility of 150%. All other rates are simulated as in Figure 21.3 with volatility $\sigma = 10\%$. The change of the simulation of the other rates has an significant impact on the distribution of $L(5.0, 5.5)$ under the spot measure.*

CHAPTER 22

Heath-Jarrow-Morton Framework: Foundations

The Heath-Jarrow-Morton (HJM) Framework [78] postulates an Itô process as a model for the instantaneous forward rate[1]:

$$
\begin{aligned}
\mathrm{d}f(t, T) &= \alpha^{\mathbb{P}}(t, T)\, \mathrm{d}t + \sigma(t, T) \cdot \mathrm{d}W^{\mathbb{P}}(t) \\
f(0, T) &= f_0(T)
\end{aligned}
\tag{22.1}
$$

for $0 \le t < T$, where $W^{\mathbb{P}} = (W_1^{\mathbb{P}}, \ldots, W_m^{\mathbb{P}})$ is an m-dimensional $\mathbb{P}$-Brownian motion with instantaneously uncorrelated components.[2] Furthermore we assume that $\sigma(t, T) = (\sigma_1(t, T), \ldots, \sigma_m(t, T))$ and $\alpha^{\mathbb{P}}(t, T)$ are adapted processes.

In case of its existence, let $\mathbb{Q}$ denote the *risk-neutral measure*, i.e., the martingale measure $\mathbb{Q} = \mathbb{Q}^B$ corresponding to the numéraire B with

$$
B(t) := \exp\left(\int_0^t f(\tau, \tau)\, \mathrm{d}\tau \right) = \exp\left(\int_0^t r(\tau)\, \mathrm{d}\tau \right),
\tag{22.2}
$$

where r denotes the short rate—see Definition 103.

Girsanov's theorem (Theorem 59), gives the process (22.1) under $\mathbb{Q}$ as

$$
\begin{aligned}
\mathrm{d}f(t, T) &= \alpha^{\mathbb{Q}}(t, T)\, \mathrm{d}t + \sigma(t, T) \cdot \mathrm{d}W^{\mathbb{Q}}(t), \\
f(0, T) &= f_0(T).
\end{aligned}
\tag{22.3}
$$

Equation (22.3) represents a family of stochastic processes parametrized by T, which give a complete description of the interest rate curve: From Definition 101 we have

$$
f(t, T) = -\frac{\partial \log(P(T; t))}{\partial T},
$$

---

[1] Definition 101 on Page 127.
[2] I.e., that $\mathrm{d}W^\top \cdot \mathrm{d}W = I\, \mathrm{d}t$. See Section 2.7.

i.e. (using $P(t; t) = 1$)

$$P(T; t) = \exp\left(-\int_t^T f(t, \tau) d\tau\right).$$

Apart from the requirement that the processes are Itô processes, we do not consider a specific model or its implementation. A *specific model* would be given if we had specified the form of $(t, T) \mapsto \sigma(t, T)$. With a specific choice of $\sigma(t, T)$ (22.3) may become a known short-rate model or the LIBOR market model; see Chapter 24.

In this chapter we will discuss the no-arbitrage conditions of (22.3) and discuss how other models fit into this *framework*.

22.1 Short-Rate Process in the HJM Framework

The specification of the families of processes $f(\cdot, T)$ implies a process for the short-rate r. We write Equation (22.3) in integral form:[3]

$$f(t, T) = f_0(T) + \int_0^t \alpha(s, T) \, ds + \int_0^t \sigma(s, T) \cdot dW(s). \tag{22.4}$$

With $T \to t$ we find for the short-rate $r(t) := \lim_{T \searrow t} f(t, T)$ that

$$r(t) = f(t, t) = f_0(t) + \int_0^t \alpha(s, t) \, ds + \int_0^t \sigma(s, t) \cdot dW(s), \tag{22.5}$$

and thus the short-rate process is in differential notation given as

$$dr(t) = \left(\frac{\partial f_0}{\partial T}(t) + \alpha(t, t) + \int_0^t \frac{\partial \alpha}{\partial T}(s, t) \, ds + \int_0^t \frac{\partial \sigma}{\partial T}(s, t) \cdot dW(s)\right) dt$$
$$+ \sigma(t, t) \cdot dW(t). \tag{22.6}$$

Remark 223 (Notation): Equation (22.6) follows from (22.5) by differentiating with respect to t. Since t enters into the second argument of α and σ, we have to calculate the partial derivative of α and σ with respect to their second argument. In accordance with the notation in (22.1) we denote the partial derivative of α with respect to its second argument by $\frac{\partial \alpha}{\partial T}$ and the partial derivative of σ with respect to its second argument by $\frac{\partial \sigma}{\partial T}$. Likewise we denote the (partial) derivative of f_0 with respect to its argument by $\frac{\partial f_0}{\partial T}$.

---

[3] We are dropping the superscript Q^B on the drift α and the diffusion W for a while.

22.2 The HJM Drift Condition

Theorem 224 (Heath-Jarrow-Morton—HJM Drift Condition): For the family of bond price processes $P(T)$ the following holds: The B-relative price $\frac{P(T)}{B}$ is a $\mathbb{Q}^B$-martingale, if and only if

$$\int_s^T \alpha(s, S)\, dS \; = \; \frac{1}{2} \int_s^T \sigma(s, S)\, dS \, \cdot \, \int_s^T \sigma(s, S)^\top \, dS.$$

From this we have: All bond price processes of the bond curve $T \mapsto P(T)$ are $\mathbb{Q}^B$-martingales, i.e., the model is arbitrage-free, if and only if

$$\alpha^{\mathbb{Q}^B}(t, T) \; = \; \sigma(t, T) \cdot \int_t^T \sigma(t, \tau)^\top \, d\tau \quad \forall \, T. \tag{22.7}$$

Equation (22.7) is called the HJM drift condition.[4]

Proof (of the HJM Drift Condition): Let T denote a fixed maturity. With $B(t) = \exp\left(\int_0^t r(s)\, ds\right)$ and $P(T; t) = \exp\left(-\int_t^T f(t, S)\, dS\right)$ it follows for the B-relative price of the bond $P(T)$ that:

$$\frac{P(T; t)}{B(t)} = \exp(X(t)) \qquad \text{with} \qquad X(t) = -\int_t^T f(t, S)\, dS - \int_0^t r(s)\, ds.$$

From (22.4) and (22.5) follows

$$
\begin{aligned}
X(t) = & -\int_t^T f(t, S)\, dS - \int_0^t r(s)\, ds \\
= & -\int_t^T f_0(S)\, dS - \int_t^T \int_0^t \alpha(s, S)\, ds\, dS - \int_t^T \int_0^t \sigma(s, S) \cdot dW(s)\, dS \\
& - \int_0^t f_0(S)\, dS - \int_0^t \int_0^u \alpha(s, u)\, ds\, du - \int_0^t \int_0^u \sigma(s, u) \cdot dW(s)\, du.
\end{aligned}
$$

---

[4] Note that σ is a row-vector, i.e., (22.7) involves a scalar product.

With $\int_0^t \int_0^u dW(s) \, du = \int_0^t \int_s^t du \, dW(s)$ and the interchange of the integrals this is

$$
\begin{aligned}
= & -\int_t^T f_0(S) \, dS - \int_0^t \int_t^T \alpha(s,S) \, dS \, ds - \int_0^t \int_t^T \sigma(s,S) \, dS \cdot dW(s) \\
& -\int_0^t f_0(S) \, dS - \int_0^t \int_s^t \alpha(s,u) \, du \, ds - \int_0^t \int_s^t \sigma(s,u) \, du \cdot dW(s) \\
= & -\int_0^T f_0(S) \, dS - \int_0^t \int_s^T \alpha(s,S) \, dS \, ds - \int_0^t \int_s^T \sigma(s,S) \, dS \cdot dW(s) \\
= & X(0) + \int_0^t A(s) \, ds + \int_0^t \Sigma(s) \cdot dW(s),
\end{aligned}
$$

thus

$$
dX(t) = A(t) \, dt + \Sigma(t) \cdot dW(t),
$$

where

$$
X(0) = -\int_0^T f_0(S) \, dS,
$$

$$
A(s) = -\int_s^T \alpha(s,S) \, dS,
$$

$$
\Sigma(s) = -\int_s^T \sigma(s,S) \, dS.
$$

Let the B-relative price of $P(T)$ be a martingale under $\mathbb{Q}^B$, i.e., the process $\exp(X(t))$ is drift-free. From Itô's lemma we have $d\exp(X(t)) = \exp(X(t)) \, dX(t) + \frac{1}{2} \exp(X(t)) \, dX(t) \, dX(t)$, i.e.,

$$
d\exp(X(t)) = \exp(X(t)) \cdot \left(\left(A(t) + \frac{1}{2}\Sigma(t)\Sigma(t)^\top \right) dt + \Sigma(t) \cdot dW(t) \right).
$$

That $\exp(X(t))$ is drift-free thus implies $A(t) + \frac{1}{2}\Sigma(t)\Sigma(t)^\top = 0$, i.e.,

$$
\int_t^T \alpha(t,S) \, dS = \frac{1}{2} \int_t^T \sigma(t,S) \, dS \cdot \int_t^T \sigma(t,S)^\top \, dS.
$$

If this equation is valid for all T, we get by differentiation $\frac{\partial}{\partial T}$ the HJM drift condition

$$
\alpha(t,T) = \sigma(t,T) \cdot \int_t^T \sigma(t,S)^\top \, dS.
$$

□

Interpretation (Bond Volatility): The expression $\Sigma(t) = -\int_t^T \sigma(t, S) \, dS$ corresponds to the volatility of the bond price process $P(T)$ at time t (bond volatility), since we have

$$
\begin{aligned}
dP(T;t) &= d(B(t) \exp(X(t))) = B(t) \, d \exp(X(t)) \\
&= B(t) \exp(X(t)) \left(dX(t) + \frac{1}{2} dX(t) \, dX(t) \right) \\
&= P(T;t) \left((\ldots) \, dt + \Sigma(t) \cdot dW(t) \right) \\
&= P(T;t) (\ldots) \, dt + P(T;t) \Sigma(t) \cdot dW(t).
\end{aligned}
$$

 Motivation (Embedding other Models): If an interest rate model is arbitrage-free and if the processes of the instantaneous forward rates $f(\cdot, T)$ are Itô processes, then the model has to fulfill the HJM drift condition (22.7). Thus, these interest rate models may be derived as a special case of the HJM framework. Since the volatility structure $(t, T) \mapsto \sigma(t, T)$ and the initial conditions $f(0, T)$ are the only free parameters of the HJM framework, this embedding of arbitrage-free interest rate models can be achieved by choosing the HJM volatility structure and the initial interest rate curve. We will show in Chapter 24 how short-rate models and the LIBOR market model can be interpreted as special HJM volatility structures.

CHAPTER 23

Short-Rate Models

23.1 Introduction

At a fixed point t in time the *short rate* is given by

$$r(t) := -\left.\frac{\partial P(T;t)}{\partial T}\right|_{T=t}.$$

See Definition 103. Thus $r : t \mapsto r(t)$ is a real-valued stochastic process. We make the following assumptions:

1. Given is a model for r (*short-rate model*), e.g., in the form of an Itô process

   $$dr = \mu^{\mathbb{P}}(t,r)\,dt + \sigma(t,r)\,dW^{\mathbb{P}}(t), \quad r(0) = r_0, \tag{23.1}$$

   where $\mathbb{P}$ denotes the real probability measure.

2. The *continuously compounding money market account* $B(t)$,

   $$dB(t) = r(t)B(t)\,dt, \quad B(0) = 1, \quad \text{i.e.,} \quad B(t) = \exp\left(\int_0^t r(\tau)\,d\tau\right),$$

   is a traded asset.[1]

3. Corresponding to the numéraire $N(t) = B(t)$, there exists a martingale measure $\mathbb{Q} = \mathbb{Q}^B$ equivalent to $\mathbb{P}$.

---

[1] The short-rate r is, as an interest rate for an infinitesimal period dt, an idealized quantity. Correspondingly the product B is an idealized quantity: The continuous reinvestment of an initial value of 1 over infinitesimal periods $[t, t+dt]$ with rate $r(t)$.

351

From Girsanov's Theorem[2] the process of r under $\mathbb{Q}$ is

$$dr = \mu^{\mathbb{Q}}(t, r)\, dt + \sigma(t, r)\, dW^{\mathbb{Q}}(t), \quad r(0) = r_0, \tag{23.2}$$

with $\mu^{\mathbb{Q}}(t, r) = \mu^{\mathbb{P}}(t, r) + C(t)$. Since under $\mathbb{Q}$ all B-relative prices of traded assets are martingales, all bond prices are given by

$$P(T; t) = B(t) \mathbb{E}^{\mathbb{Q}}\left(\frac{1}{B(T)} \mid \mathcal{F}_t\right) = \mathbb{E}^{\mathbb{Q}}\left(\exp\left(-\int_t^T r(\tau)\, d\tau\right) \mid \mathcal{F}_t\right).$$

From the bond prices $P(T; t)$ we can calculate all derived quantities such as forward rates or swap rates; see Section 8.2. Thus, the *short-rate model* (23.2) gives a complete description of the interest rate curve dynamic.

Short-rate models were and are popular, since the underlying stochastic process r is one-dimensional (i.e., scalar valued). Thus many techniques that are known from the modeling of (also one-dimensional) stock price processes can be used (e.g., finite difference implementations). Depending on the specific model (i.e., the form of $\mu^{\mathbb{Q}}$ and σ), analytic formulas for bond prices or simple European interest rate options may be derived, similar to the Black-Scholes formula for European stock options under a Black-Scholes model.

Instead of specifying the model (23.1) of the short-rate process under the real measure $\mathbb{P}$ and applying the measure transformation to $\mathbb{Q}$, it is usual to specify the model (23.2) directly under $\mathbb{Q}$ and calibrate given model parameters.

23.2 The Market Price of Risk

Consider a bond with maturity T. Under a short-rate model its price process $P(T)$: $t \mapsto P(T; t)$ is a function of $(t, r(t))$ and if Itô's lemma is applicable, we have[3]

$$dP(T) = \alpha_T^{\mathbb{P}}(t, r)P(T)\, dt + \sigma_T(t, r)P(T)\, dW^{\mathbb{P}}(t), \tag{23.3}$$

where the price process is considered under the real measure $\mathbb{P}$.

Let $P(T_1)$ and $P(T_2)$ denote two bonds with different maturities $T_1 \neq T_2$. We construct a portfolio process (ϕ_0, ϕ_1) for a self-financing portfolio of B and $P(T_1)$, which replicates $P(T_2)$. The portfolio process (ϕ_0, ϕ_1) has to satisfy the following equations:

$$\phi_0 B + \phi_1 P(T_1) = P(T_2) \qquad \text{("replicating")} \tag{23.4}$$

$$d(\phi_0 B + \phi_1 P(T_1)) = \phi_0\, dB + \phi_1\, dP(T_1) \qquad \text{("self-financing")}. \tag{23.5}$$

---

[2] Theorem 59 on page 39.

[3] At this point, it is not obvious that Itô's lemma is applicable, especially if the functional dependence of $P(T; t)$ from $r(t)$ is sufficiently smooth. However, for the short rate models presented this is the case.

From Itô's lemma we then have $\alpha_T = \dfrac{\frac{\partial}{\partial t}P(T) + \mu\frac{\partial}{\partial r}P(T) + \frac{1}{2}\sigma^2\frac{\partial^2}{\partial r^2}P(T)}{P(T)}$ and $\sigma_T = \dfrac{\sigma\frac{\partial}{\partial r}P(T)}{P(T)} = \sigma\frac{\partial}{\partial r}\log(P(T))$.

From (23.4) we find $dP(T_2) = d(\phi_0 B + \phi_1 P(T_1))$ and with

$$d(\phi_0 B + \phi_1 P(T_1)) \overset{(23.5)}{=} (\phi_0 r B + \phi_1 \alpha^{\mathbb{P}}_{T_1} P(T_1))\, dt + \phi_1 \sigma_{T_1} P(T_1)\, dW^{\mathbb{P}}(t)$$

$$dP(T_2) \overset{(23.3)}{=} \alpha^{\mathbb{P}}_{T_2} P(T_2)\, dt + \sigma_{T_2} P(T_2)\, dW^{\mathbb{P}}(t),$$

we have, by comparing coefficients,

$$\alpha^{\mathbb{P}}_{T_2} P(T_2) = \phi_0 r B + \phi_1 \alpha^{\mathbb{P}}_{T_1} P(T_1) \tag{23.6}$$

$$\sigma_{T_2} P(T_2) = \phi_1 \sigma_{T_1} P(T_1). \tag{23.7}$$

While (23.7) and (23.4) uniquely determine the portfolio process (ϕ_0, ϕ_1), (23.6) is a consistency condition for r, α_{T_1} and α_{T_2}. If (23.6) were violated, then the model would not be arbitrage-free. We rewrite the consistency condition (23.6) as:

$$\Leftrightarrow \qquad \alpha^{\mathbb{P}}_{T_2} P(T_2) = \phi_0 r B + \phi_1 \alpha^{\mathbb{P}}_{T_1} P(T_1)$$

$$\Leftrightarrow \qquad \alpha^{\mathbb{P}}_{T_2} P(T_2) = \phi_0 r B + \phi_1 r P(T_1) + \phi_1 (\alpha^{\mathbb{P}}_{T_1} - r) P(T_1)$$

$$\overset{(23.4)}{\Leftrightarrow} \qquad \alpha^{\mathbb{P}}_{T_2} P(T_2) = r P(T_2) + \phi_1 (\alpha^{\mathbb{P}}_{T_1} - r) P(T_1)$$

$$\Leftrightarrow \qquad (\alpha^{\mathbb{P}}_{T_2} - r) P(T_2) = \phi_1 (\alpha^{\mathbb{P}}_{T_1} - r) P(T_1)$$

$$\overset{(23.7)}{\Leftrightarrow} \qquad \frac{\alpha^{\mathbb{P}}_{T_2} - r}{\sigma_{T_2}} = \frac{\alpha^{\mathbb{P}}_{T_1} - r}{\sigma_{T_1}}.$$

It follows that there exists a $\lambda^{\mathbb{P}}$ such that for *all* bond price processes

$$dP(T) = \alpha^{\mathbb{P}}_T(t,r) P(T)\, dt + \sigma_T(t,r) P(T)\, dW^{\mathbb{P}}(t)$$

we have

$$\frac{\alpha^{\mathbb{P}}_T - r}{\sigma_T} =: \lambda^{\mathbb{P}}.$$

Since $\alpha^{\mathbb{P}}$ is the local rate of return of the bond, we may interpret $\lambda^{\mathbb{P}}$ as the local excess return rate over r per risk unit σ_T.

Definition 225 (Market Price of Risk):
The quantity $\lambda^{\mathbb{P}} := \frac{\alpha^{\mathbb{P}}_T - r}{\sigma_T}$, which is independent of T, is called the *market price of risk*.

If we consider the bond price process

$$dP(T) = \alpha^{\mathbb{Q}}_T(t)P(T)\, dt + \sigma_T(t)P(T)\, dW^{\mathbb{Q}}(t)$$

under the measure $\mathbb{Q}$, it is obvious that $\alpha_T^{\mathbb{Q}} = r$ for all T, since all B-relative prices are $\mathbb{Q}$-martingales. Thus, under $\mathbb{Q}$ the *price of risk* $\lambda^{\mathbb{Q}} = 0$. It follows that

$$\mu_T^{\mathbb{Q}} = \lambda^{\mathbb{Q}}\, \sigma_T + r = 0 + r = \lambda^{\mathbb{P}}\, \sigma_T + r - \lambda^{\mathbb{P}}\, \sigma_T = \mu_T^{\mathbb{P}} - \lambda^{\mathbb{P}}\, \sigma_T,$$

and we find that market price of risk $\lambda^{\mathbb{P}}$ appears in the change of drift to the measure $\mathbb{Q}$, i.e., we have $C(t) = -\lambda^{\mathbb{P}}\, \sigma_T$ in Theorem 59.

Definition 226 (Risk Neutral Measure):
Let $r(t)$ denote the short rate. The martingale measure $\mathbb{Q}^B$ corresponding to the numéraire $B(t) = \exp\left(\int_0^t r(\tau)\, d\tau\right)$ is called the *risk-neutral measure*.

Remark 227 (Risk-Neutral Measure):   The *continuously compounding money market account* B is locally risk-free, since the process $dB(t) = r(t)B(t)\, dt$ does not exhibit a $dW(t)$ term. However, $r(t)$ may be stochastic. If r were not stochastic, then B would be globally risk-free.

23.3 Overview: Some Common Models

Table 23.3 gives a selection of the most common short-rate models.

Name	Model
Vasicek Model	$dr = (b - ar)\, dt + \sigma\, dW^{\mathbb{Q}}$
Hull-White Model	$dr = (\phi(t) - ar)\, dt + \sigma(t)\, dW^{\mathbb{Q}}$
Ho-Lee Model	$dr = a(t)\, dt + \sigma(t)\, dW^{\mathbb{Q}}$
Dothan Model	$dr = ar\, dt + \sigma r\, dW^{\mathbb{Q}}$
Black-Derman-Toy Model	$d\log(r) = \phi(t)\, dt + \sigma(t)\, dW^{\mathbb{Q}}$
Black-Karasinski Model	$d\log(r) = (\phi(t) - a\log(r))\, dt + \sigma(t)\, dW^{\mathbb{Q}}$
Cox-Ingersoll-Ross Model	$dr = (b - ar)\, dt + \sigma(t)\, \sqrt{r}\, dW^{\mathbb{Q}}$

Table 23.1. *Selection of Short-Rate Models*

The Hull-White model is sometimes called *extended Vasicek model*. The Vasicek, Hull-White, and Ho-Lee models allow for negative short rates. The Black-Derman-Toy (BDT) and Black-Karansinski (BK) models use a lognormal process, and the Cox-Ingersoll-Ross model uses a square-root process. Neither of these two processes allow for negative rates.[4]

---

[4] This result holds for the time-continuous process. A time discretization of the process may allow for negative rates. See, for example, Section 13.1.2.

23.4 Implementations

23.4.1 Monte Carlo Implementation of Short-Rate Models

A short-rate model gives a description of the dynamics of the short rate. To obtain a complete interest rate curve at a given simulation time t, we have to calculate the bond prices from (23.1) as conditional expectation. To calculate a conditional expectation in a Monte Carlo simulation numerically requires additional, numerically expensive methods; see Chapter 15. To obtain a Monte Carlo simulation of the full interest rate curve from a Monte Carlo simulation of the short rate, analytic formulas for bond prices are indispensable. The popularity of short-rate models is thus partly due to the need for a simple and efficient implementation.

For a fast calibration to a given interest rate curve it is also required to calculate bond prices analytically.

23.4.2 Lattice Implementation of Short-Rate Models

If the short-rate model is Markovian in low dimensions, then it is best to implement the short-rate model on a lattice, allowing for a backward algorithm.[5] Depending on the model, implementations using binomial or trinomial trees or general finite differences for PDE's are used. See [35] for a detailed discussion.

 Further Reading: Björg's book [6] contains a discussion of short-rate models with affine term structure. Tavella and Randal's book [35] gives an introduction to finite difference methods, as well as applications to interest rate models. ◁|

---

[5] See Section 13.3.2.

CHAPTER 24

Heath-Jarrow-Morton Framework: Immersion of Short-Rate Models and LIBOR Market Model

> You're going to find that many of the truths we cling to depend greatly on our own point of view.
>
> *Obi-Wan Kenobi / George Lucas*
> *Star Wars: Episode VI (Wikiquote).*

24.1 Short-Rate Models in the HJM Framework

The Heath-Jarrow-Morton (HJM) framework

$$\begin{aligned} df(t,T) &= \alpha(t,T)\,dt + \sigma(t,T) \cdot dW(t) \\ f(0,T) &= f_0(T) \end{aligned}$$

(was 22.3)

implies the short-rate process

$$dr(t) = \left(\frac{\partial f_0}{\partial T}(t) + \alpha(t,t) + \int_0^t \frac{\partial \alpha}{\partial T}(s,t)\,ds + \int_0^t \frac{\partial \sigma}{\partial T}(s,t)\,dW(s) \right) dt$$

(was 22.6)

$$+ \sigma(t,t) \cdot dW(t),$$

both under the measure $\mathbb{Q}^B$—see Equations (22.3), (22.4), (22.5), and (22.6). The short-rate model is thus given by the specific choice of the HJM volatility structure $\sigma(t,t)$ ($\to$ short-rate volatility) and initial conditions f_0 ($\to$ short-rate drift).

24.1.1 Example: The Ho-Lee Model in the HJM Framework

Consider the simple case of a constant volatility function

$$\sigma(t, T) = \sigma = \text{const.}$$

From (22.7) we have $\alpha(t, T) = \sigma \int_t^T \sigma \, d\tau = \sigma^2 (T - t)$, i.e.,

$$df(t, T) = \sigma^2 (T - t) \, dt + \sigma \, dW(t), \qquad f(0, T) = f_0(T).$$

For the short rate it follows that

$$r(t) = f(t, t) = f_0(t) + \left. \int_0^t \sigma^2 (T - s) \, ds \right|_{T=t} + \int_0^t \sigma \, dW(s)$$

$$= f_0(t) + \frac{1}{2}\sigma^2 t^2 + \sigma W(t),$$

i.e.,

$$dr(t) = \left(\frac{df_0}{dT}(t) + \sigma^2 t \right) dt + \sigma \, dW(t).$$

Using the notation from the Ho-Lee model, $dr(t) = \phi(t) \, dt + \sigma \, dW(t)$, it is

$$\phi(t) = \frac{df_0}{dT}(t) + \sigma^2 t. \qquad (24.1)$$

 Interpretation: Equation (24.1) allows a calibration of the Ho-Lee model to a given curve of bond prices $P(T)$ by setting

$$\phi(t) = -\frac{d^2 \log(P(T))}{dT^2}(t) + \sigma^2 t.$$

With this choice the model reproduces the given bond prices.

If we consider the interest rate curve $f_{T_1}(T) := f(T_1, T_1 + T), T \geq 0$ at a later time $T_1 > 0$, then from

$$\frac{df_{T_1}}{dT}(t) + \sigma^2 t = \phi(T_1 + t) = \frac{df_0}{dT}(T_1 + t) + \sigma^2 (T_1 + t),$$

we find that $f_{T_1}(t) = f_{T_1}(0) + f_0(T_1 + t) - f_0(t) + \sigma^2 (T_1 t + \frac{1}{2}t^2)$.

So to summarize, the model reproduces all bond prices, but in the evolution the interest rate curve gets steeper and steeper—a rather unrealistic behavior. ◁|

24.1.2 Example: The Hull-White Model in the HJM Framework

Consider the case of an exponential volatility function

$$\sigma(t, T) = \sigma \, e^{-a \, (T-t)}, \qquad (a > 0).$$

Then we have $\frac{\partial \sigma}{\partial T}(t, T) = -a \, \sigma \, e^{-a \, (T-t)} = -a \, \sigma(t, T)$. For the drift $\mu(t)$ of the short-rate process $dr(t) = \mu(t) \, dt + \sigma(t, t) \, dW(t)$ we get

$$
\begin{aligned}
\mu(t) \quad \overset{(22.6)}{=} \quad & \frac{\partial f_0}{\partial T}(t) + \alpha(t, t) + \int_0^t \frac{\partial \alpha}{\partial T}(s, t) \, ds + \int_0^t \frac{\partial \sigma}{\partial T}(s, t) \, dW(s) \\
= \quad & \frac{\partial f_0}{\partial T}(t) + \alpha(t, t) + \int_0^t \frac{\partial \alpha}{\partial T}(s, t) \, ds - \int_0^t a \, \sigma(s, t) \, dW(s) \\
\overset{(22.5)}{=} \quad & \frac{\partial f_0}{\partial T}(t) + \alpha(t, t) + \int_0^t \frac{\partial \alpha}{\partial T}(s, t) \, ds - a \, r(t) + a \, f_0(t) + \int_0^t a \, \alpha(s, t) \, ds,
\end{aligned}
$$

i.e.,

$$dr(t) = (\phi(t) - a r(t)) \, dt + \sigma \, dW(t)$$

with

$$\phi(t) = \frac{\partial f_0}{\partial T}(t) + a \, f_0(t) + \alpha(t, t) + \int_0^t \frac{\partial \alpha}{\partial T}(s, t) \, ds + \int_0^t a \cdot \alpha(s, t) \, ds.$$

With the HJM drift condition (22.7) it follows that $\alpha(t, T) = \sigma^2 \, e^{-a \, (T-t)} \frac{1}{a}(1 - e^{-a \, (T-t)}) = \sigma^2 \frac{1}{a}(e^{-a \, (T-t)} - e^{-2a \, (T-t)})$ and thus

$$
\begin{aligned}
\phi(t) &= \frac{\partial f_0}{\partial T}(t) + a \, f_0(t) + \alpha(t, t) + \int_0^t \frac{\partial \alpha}{\partial T}(s, t) \, ds + \int_0^t a \, \alpha(s, t) \, ds \\
&= \frac{\partial f_0}{\partial T}(t) + a \, f_0(t) + \int_0^t \sigma^2 e^{-2a \, (t-s)} \, ds \\
&= \frac{\partial f_0}{\partial T}(t) + a \, f_0(t) + \frac{\sigma^2}{2a}(1 - e^{-2a \, t}).
\end{aligned}
$$

Altogether we have

$$dr(t) = \left(\frac{\partial f_0}{\partial T}(t) + a \, f_0(t) + \frac{\sigma^2}{2a}(1 - e^{-2a \, t}) - a r(t) \right) dt + \sigma \, dW(t).$$

Note that this equation allows a calibration of the Hull-White model to a given curve of bond prices. From the bond price curve we can calculate $\frac{\partial f_0}{\partial T}(t) + a \, f_0(t)$.

Interpretation (Mean Reversion): The derivation of a Hull-White model from a Heath-Jarrow-Morton model gives an important insight to the relevance of the time structure of the volatility function:

A volatility function of the instantaneous forward rate $f(t, T)$, which is exponentially decaying in $(T - t)$ (*time to maturity*), i.e., $\sigma(t, T) = \exp(-a(T - t))$, corresponds to a *mean reversion* term for the short-rate process $r(t)$ with mean reversion speed a.

Correspondingly, this effect is visible in the LIBOR market model; see Chapter 25.

◁|

24.2 LIBOR Market Model in the HJM Framework

24.2.1 HJM Volatility Structure of the LIBOR Market Model

In the specification (19.1) of the LIBOR market model dW denoted the increment of a n-dimensional Brownian motion with instantaneous correlation R. In the specification (22.3) of the HJM framework dW denoted the increment of an m-dimensional Brownian motion with instantaneous uncorrelated components. To resolve this conflict we employ the notation of Section 2.7: Let U denote an m-dimensional Brownian motion with instantaneous uncorrelated components and W denote an n-dimensional Brownian motion with $dW(t) = F(t) \cdot dU(t)$, i.e., the instantaneous correlation of W is $R := FF^{\top}$. Consider the HJM model

$$
\begin{aligned}
df(t, T) &= \alpha^{Q}(t, T)\, dt + \sigma(t, T) \cdot dU^{Q}(t) \\
f(0, T) &= f_{0}(T)
\end{aligned}
\tag{24.2}
$$

with $dU = (dU_1, \ldots, dU_m)$. From

$$
P(T; t) = \exp\left(-\int_{t}^{T} f(t, \tau)\, d\tau\right)
$$

(see Remark 102) it follows that the forward rate $L_i(t) := L(T_i, T_{i+1}; t)$ is given by

$$
1 + L_i(t)\, \Delta T_i = \frac{P(T_i; t)}{P(T_{i+1}; t)} = \exp\left(\int_{T_i}^{T_{i+1}} f(t, \tau)\, d\tau\right).
$$

Note that for $X(t) := \int_{T_i}^{T_{i+1}} f(t, \tau)\, d\tau$ we have by the linearity of the integral that $dX = \int_{T_i}^{T_{i+1}} df(t, \tau) d\tau$, thus we find from Itô's lemma that within the HJM framework the process of the forward rate $L_i(t)$ is

$$dL_i(t)\,\Delta T_i = d\exp(X) = \exp(X)\,(dX + \frac{1}{2}dX\,dX)$$

$$= \exp\left(\int_{T_i}^{T_{i+1}} f(t,\tau)d\tau\right)\left[\int_{T_i}^{T_{i+1}} (df(t,\tau))\,d\tau\right.$$

$$\left. + \frac{1}{2}\int_{T_i}^{T_{i+1}} (df(t,\tau))\,d\tau\int_{T_i}^{T_{i+1}} (df(t,\tau))\,d\tau\right]$$

$$= (1 + L_i(t)\,\Delta T_i)\left[(A(t) + \frac{1}{2}\Sigma(t)\cdot\Sigma(t)^\top)\,dt + \Sigma\cdot dU^Q\right]$$

where $A(t) = \int_{T_i}^{T_{i+1}} \alpha^Q(t,\tau)\,d\tau$ and $\Sigma(t) = \int_{T_i}^{T_{i+1}} \sigma(t,\tau)\,d\tau$.

$$dL_i(t) = \frac{1 + L_i(t)\,\Delta T_i}{\Delta T_i}\left((A(t) + \frac{1}{2}\Sigma(t)\cdot\Sigma(t)^\top)\,dt + \int_{T_i}^{T_{i+1}} \sigma(t,\tau)\,d\tau\,dU^Q(t)\right).$$
(24.3)

We will now choose the volatility structure such that (24.3) corresponds to the process of a LIBOR market model: Let $W = (W_1,\ldots,W_n)^\top$ denote an n-dimensional Brownian motion as given in Section 2.7.

$$dW(t) = F(t)\cdot dU(t), \qquad \text{with correlation matrix } R := FF^\top,$$

i.e.

$$dW_i(t) = F_i(t)\cdot dU(t), \qquad \text{with } F = \begin{pmatrix} F_1 \\ \vdots \\ F_n \end{pmatrix}.$$

Let the volatility structure be chosen as

$$\sigma(t,\tau) = \begin{cases} \frac{L_i(t)}{1+L_i(t)\,\Delta T_i}\sigma_i(t)\,F_i(t) & \text{for } t \leq T_i \\ (0,\ldots,0) & \text{for } t > T_i, \end{cases}$$
(24.4)

where i is such that $\tau \in [T_i, T_{i+1})$. Then we have

$$\frac{1 + L_i(t)\,\Delta T_i}{\Delta T_i}\int_{T_i}^{T_{i+1}} \sigma(t,\tau)\,d\tau\cdot dU = \begin{cases} L_i(t)\,\sigma_i(t)\,dW_i(t) & \text{for } t \leq T_i, \\ 0 & \text{for } t > T_i, \end{cases}$$

The forward rate then follows the process

$$dL_i = \mu_i^{Q^B}(t)L_i(t)\,dt + \sigma_i(t)L_i(t)\,dW_i(t),$$

where

$$
\mu_i^{\mathbb{Q}^B} = \frac{1 + L_i(t)\,\Delta T_i}{L_i(t)\,\Delta T_i} \left(\int_{T_i}^{T_{i+1}} \alpha^{\mathbb{Q}}(t,\tau)\,d\tau + \frac{1}{2}\int_{T_i}^{T_{i+1}} \sigma(t,\tau)\,d\tau \cdot \int_{T_i}^{T_{i+1}} \sigma(t,\tau)^{\top}\,d\tau \right).
$$

 Interpretation (LIBOR Market Model as HJM Framework with Discrete Tenor Structure): Apart from the factor $\frac{L_i(t)}{1+L_i(t)\,\Delta T_i}$, (24.4) gives the volatility structure $\sigma(t,T)$ of $f(t,T)$ as piecewise constant in T. The factor $L_i(t)$ results from the requirement to have a lognormal process for L_i. The factor $\frac{1}{1+L_i(t)\,\Delta T_i}$ results from the discretization of the tenor structure. This shows that the LIBOR market model can be interpreted as *HJM framework with discrete tenor structure*. In the limit $\Delta T_i \to 0$ the factor $\frac{1}{1+L_i(t)\,\Delta T_i}$ vanishes and we obtain (apart from the restriction to a lognormal model) the HJM framework. ◁|

24.2.2 LIBOR Market Model Drift under the $\mathbb{Q}^B$ Measure

The HJM drift condition states that

$$
\alpha^{\mathbb{Q}^B}(t,T) = \sigma(t,T) \cdot \int_t^T \sigma(t,\tau)^{\top}\,d\tau.
$$

Since for fixed t, $\sigma(t,T)$ is a piecewise constant function in T—namely constant on $[T_i, T_{i+1})...$, we have for $T \in [T_i, T_{i+1})$

$$
\alpha^{\mathbb{Q}^B}(t,T) = \sigma(t,T_i) \cdot \left(\underbrace{\sigma(t,t)^{\top}(T_{m(t)+1} - t)}_{=0} + \sum_{j=m(t)+1}^{i-1} \sigma(t,T_j)^{\top}\Delta T_j + \sigma(t,T_i)^{\top}(T - T_i) \right)
$$

where $m(t) := \max\{i\ :\ T_i \le t\}$. Thus we have

$$
\int_{T_i}^{T_{i+1}} \alpha^{\mathbb{Q}^B}(t,\tau)\,d\tau = \sigma(t,T_i)\,\Delta T_i \cdot \left(\sum_{j=m(t)+1}^{i-1} \sigma(t,T_j)^{\top}\Delta T_j + \frac{1}{2}\sigma(t,T_i)^{\top}\Delta T_i \right).
$$

With

$$
\sigma(t,T_i) \cdot \sigma(t,T_j)^{\top} = \frac{\sigma_i L_i}{1 + L_i\,\Delta T_i}\frac{\sigma_j L_j}{1 + L_j\,\Delta T_j}\rho_{i,j}
$$

we find

$$
\begin{aligned}
\mu_i^{\mathbb{Q}^B} &= \frac{1 + L_i(t)\,\Delta T_i}{L_i(t)\,\Delta T_i}\left(\int_{T_i}^{T_{i+1}} \alpha^{\mathbb{Q}^B}(t,\tau)\,\mathrm{d}\tau + \frac{1}{2}\int_{T_i}^{T_{i+1}} \sigma(t,\tau)\,\mathrm{d}\tau \cdot \int_{T_i}^{T_{i+1}} \sigma(t,\tau)^{\top}\,\mathrm{d}\tau\right) \\[2mm]
&= \frac{1 + L_i(t)\,\Delta T_i}{L_i(t)\,\Delta T_i}\,\frac{\sigma_i L_i\,\Delta T_i}{1 + L_i\,\Delta T_i}\left(\sum_{j=m(t)+1}^{i-1} \frac{\sigma_j L_j\,\Delta T_j}{1 + L_j\,\Delta T_j}\rho_{i,j} + \frac{\sigma_i L_i\,\Delta T_i}{1 + L_i\,\Delta T_i}\right) \\[2mm]
&= \frac{1 + L_i(t)\,\Delta T_i}{L_i(t)\,\Delta T_i}\,\frac{\sigma_i L_i\,\Delta T_i}{1 + L_i\,\Delta T_i}\left(\sum_{j=m(t)+1}^{i} \frac{\sigma_j L_j\,\Delta T_j}{1 + L_j\,\Delta T_j}\rho_{i,j}\right) \\[2mm]
&= \sigma_i\left(\sum_{j=m(t)+1}^{i} \frac{\sigma_j L_j\,\Delta T_j}{1 + L_j\,\Delta T_j}\,\rho_{i,j}\right).
\end{aligned}
$$

Interpretation: Surprisingly, we find that the drift under $\mathbb{Q}^B$ is identical to the drift under the spot LIBOR measure (see Section 19.1.2)

$$
N(t) = P(T_{m(t)+1};t)\prod_{j=0}^{m(t)}(1 + L_j(T_j)\,\Delta T_j). \tag{was 19.9}
$$

The reason is simple: Under the assumed volatility structure the numéraires $B(t)$ and $N(t)$ are identical. To be precise, it is the assumption

$$
\sigma(t,T) = 0 \quad \text{for } T_{m(t)} \le t \le T < T_{m(t)+1} \tag{24.5}
$$

which implies that the two numéraires coincide. By this the HJM drift implies

$$
\alpha^{\mathbb{Q}^B}(t,T) = 0 \quad \text{for } T_{m(t)} \le t \le T < T_{m(t)+1}
$$

and thus for $T_{m(t)} \le T < T_{m(t)+1}$:

$$
f(t,T) = f(T_{m(t)},T) + \underbrace{\int_{T_i}^{t} \alpha^{\mathbb{Q}^B}(\tau,T)\,\mathrm{d}\tau}_{=0} + \underbrace{\int_{T_i}^{t} \sigma(\tau,T)\,\mathrm{d}U^{\mathbb{Q}^B}(\tau)}_{=0}.
$$

From $f(t,T) = f(T_{m(t)},T)$ we have

$$
\begin{aligned}
\frac{B(t)}{B(T_{m(t)})} &= \exp\left(\int_{T_{m(t)}}^{t} f(\tau,\tau)\mathrm{d}\tau\right) = \exp\left(\int_{T_{m(t)}}^{t} f(T_i,\tau)\mathrm{d}\tau\right) \\[2mm]
&= \frac{P(T_{m(t)+1};t)}{P(T_{m(t)+1};T_{m(t)})} = \frac{N(t)}{N(T_{m(t)})},
\end{aligned}
$$

with $B(0) = N(0) = 1$, i.e., $B(t) = N(t)$. ◁|

We will summarize this result as a theorem:

Theorem 228 (Equivalence of the Risk-Neutral Measure and the Spot LIBOR Measure): Given a tenor structure $0 = T_0 < T_1 < \cdots < T_n$. Under the assumption that the T_{i+1}-bond $P(T_{i+1}; t)$ has volatility 0 on $t \in [T_i, T_{i+1}]$ for all $i = 0, 1, 2, \ldots$, we have

$$B(t) = N(t)$$

for $B(t)$ as in (22.2) and $N(t)$ as in (19.9).

Proof: The claim follows from the considerations above, since the assumption in the theorem is equivalent to (24.5). □|

24.2.3 LIBOR Market Model as a Short Rate Model

In Section 24.2.1 we have given the volatility structure for $(t, T) \mapsto f(t, T)$ under which the forward rates L_i evolve as in a LIBOR market model. Since the short rate is given as $r(t) := \lim_{T \searrow t} f(t, T)$, the volatility structure also implies a short-rate model. Furthermore, the numéraire $B(t) = \exp\left(\int_0^t r(\tau)\, d\tau\right)$ is fully determined by the short rate, thus the short-rate process under $\mathbb{Q}^B$ gives a complete description of all bond prices (and all derivatives):

$$P(T; t) = B(t)\mathrm{E}^{\mathbb{Q}^B}\left(\frac{1}{B(T)} \,\Big|\, \mathcal{F}_t\right).$$

The short-rate process r implied by the volatility structure (24.4) generates a LIBOR market model. The short-rate process under $\mathbb{Q}^B$ is given by (22.6):

$$dr(t) = \left(\frac{\partial f_0}{\partial T}(t) + \alpha^{\mathbb{Q}^B}(t, t) + \int_0^t \frac{\partial \alpha^{\mathbb{Q}^B}}{\partial T}(s, t)\, ds + \int_0^t \frac{\partial \sigma}{\partial T}(s, t)\, dW^{\mathbb{Q}^B}(s)\right) dt$$

$$+ \sigma(t, t) \cdot dW^{\mathbb{Q}^B}(t),$$

$$\text{(was 22.6)}$$

The drift of this short-rate model is, as a function of $\{r(s)|0 \le s \le t\}$, path-dependent. Only in high dimensions, namely as a function of $\{L_i(t)|i = 0, \ldots, n\}$, will the model be Markovian (i.e., the drift is no longer path-dependent).

CHAPTER 25

Excursus: Shape of the Interest Rate Curve under Mean Reversion and a Multifactor Model

In this chapter we are considering the influence of model properties like *mean reversion*, *number of factors*, *instantaneous correlation*, and *instantaneous volatility* on the possible future shapes of the interest rate curve.

As in Chapter 21, which discussed the relation of instantaneous correlation and instantaneous volatility to the terminal correlation, our goal is to develop an understanding of the significance of the model properties rather than looking at them rigorously in abstract mathematical terms. We thus pose the question of how the interest rate curve differs *qualitatively* under different model configurations.

25.1 Model

As a model framework we will use the LIBOR market model. Due to its many parameters it gives us enough freedom to play with. We will restrict the set of parameters and concentrate on three (important) parameters that are sufficient to create the phenomena we are interested in.

Let us restrict the model to a simple volatility structure, namely

$$\sigma_i(t) = \sigma^* \, \exp\left(-a \, (T_i - t) \right) \tag{25.1}$$

with $\sigma^* = 0.1$ and $a = 0.05$. We will choose an equally simple correlation model, namely $dW_i \, dW_j = \rho_{i,j}dt$ with

$$\rho_{i,j} = \exp(-r|T_i - T_j|). \tag{25.2}$$

365

To this correlation model we apply a factor reduction (principal component analysis); see Appendix B.3. The number of factors is the number of independent Brownian motions (effectively) entering the model; see Definition 51. Upon a factor reduction the m largest eigenvalues of the correlation matrix are determined. Together with the corresponding eigenvectors a new correlation matrix is constructed, having at most m nonzero eigenvalues. This process guarantees that the resulting correlation model defines a valid correlation matrix.

We simulate under the terminal measure and start with an initially flat interest rate curve $L_i(0) = 0.1$, $i = 0, 1, 2, \ldots$.

To summarize, our model framework consists of three degrees of freedom which will be varied in our analysis (see Table 25.1).

Parameter	Effect
a	Damping of the exponentially decaying, time-homogenous volatility
r	Damping of the exponentially decaying instantaneous correlation
m	Number of factors extracted from the correlation matrix

Table 25.1. *Free parameters of the LIBOR market model considered.*

25.2 Interpretation of the Figures

Figures 25.1, 25.2, 25.3, and 25.4 show 100 paths of a Monte Carlo simulation of the interest rate curve. The simulation was frozen at a fixed point in time ($t = 7.5$ in Figures 25.2, 25.3, and 25.4 and $t = 17.5$ in Figure 25.1). To the left of this point the forward rates $L_i(T_i)$ are shown, each upon their individual maturity—this is a discrete analog of the short rate. To the right of this point the future forward rate curve $L_j(t)$ is drawn.

The figures differ only in the parameters used to generate the paths. The same random numbers are used, thus the simulated paths depend smoothly on a and r.

To improve the visibility of the individual paths, each path is given a different color, where the hue of the color depends smoothly on the level of the last rate $L_n(t)$.[1] This makes it very easy to check if the interest rate curves are parallel or exhibit some regular structure; see Figure 25.2.

---

[1] The choice of the last rate is arbitrary.

25.3 Mean Reversion

We will consider the example of a simple one-factor Brownian motion ($\rho_{i,j} = 1$, i.e., $r = 0$). Figure 25.1 shows the simulated forward rates for different parameters a in Equation (25.1).

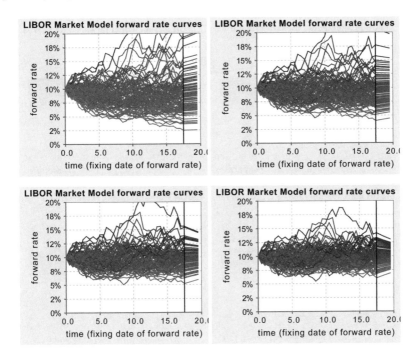

Figure 25.1. *Shape of the fixed rates $L_i(T_i)$ and the interest rate curve for different instantaneous volatilities (corresponds to different mean reversion) frozen at time $t = 17.5$ using a one-factor mode. We used $a = 0$ (upper left), $a = 0.05$ (upper right), $a = 0.10$ (lower left) and $a = 0.15$ (lower right). For interpretation see Section 25.3.*

From the derivation of the Hull-White model from the HJM framework it became obvious that an exponentially decreasing volatility structure of the forward rate corresponds to a *mean reversion* of the *short rate*; see Section 24.1.2. We rediscover this property qualitatively here. Figure 25.1 shows 100 paths of a Monte Carlo simulation of a LIBOR market model with different values for the parameter a: $a = 0$, $a = 0.05$ in the upper, and $a = 0.1$, $a = 0.15$ in the lower row (left to right). Observe the fixed rates $L_i(T_i)$ left from the simulation time. They may be interpreted as a direct analog of the short rate. In Figure 25.1 it becomes obvious that with an increasing parameter *a* the paths develop a tendency to revert to the mean (*mean reversion*).

25.4 Factors

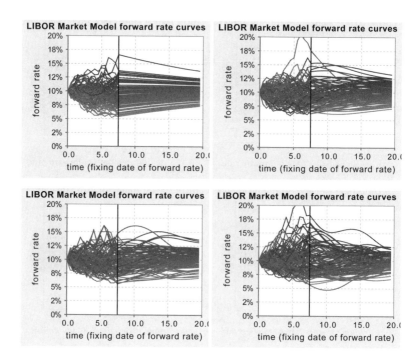

Figure 25.2. *Shape of the interest rate curve with different factor configurations, seen at time t = 7.5: One, two, three, and five factors (from upper left to lower right). For interpretation see Section 25.4.*

Figure 25.2 shows a Monte Carlo simulation with the parameters above and varying numbers of factors *m*. The possible shapes of the interest rate curve are given by combinations of the factors *parallel shift*, *tilt*, *bend*, and oscillations with increasing frequencies; see also Figure B.1.

25.5 Exponential Volatility Function

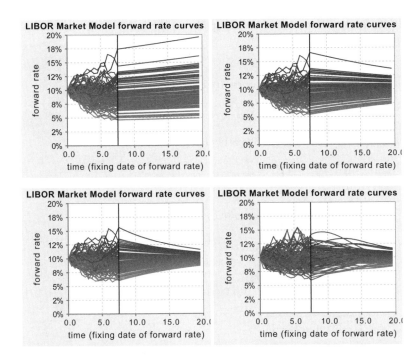

Figure 25.3. *Shape of the fixed rates $L_i(T_i)$ and the interest rate curve with different instantaneous volatilities (corresponds to mean reversion) at time $t = 7.5$ in a one-factor model (upper row and lower left) with $a = 0.0$, $a = 0.05$ and $a = 0.1$ and a three-factor model (lower right) with $a = 0.1$. For interpretation see Section 25.5.*

As in Figure 25.1 we consider the Monte Carlo simulation under different parameters a. First $a = 0$, $a = 0.05$, and $a = 0.1$ in a one-factor model ($r = 0, m = 1$), and last $a = 0.1$ in a three-factor model. We are observing this at simulation time $t = 7.5$ and concentrate here on the section right of the simulation time, i.e., the interest rate curve $L_j(t)$ for $j > m(t)$.

It is apparent that the curve $\{L_j(t) \mid j > m(t)\}$ shows a shape similar to an exponential in j, depending on the parameter a; see Figure 25.3, lower left ($a = 0.1$) to the right of the simulation time. If we consider a one-factor model (as used in the figure), we have

$$L_j(t) = L_j(0) \exp\left(\int_0^t \mu_j(\tau) \, d\tau + \sqrt{\int_0^t \sigma_j^2(\tau) \, d\tau} \, W(t) \right).$$

For a fixed point in time t (and a state (path) ω) the interest rate curve shows the following dependence on j:

$$j \mapsto L_j(0) \exp\left(\int_0^t \mu_j(\tau, \omega) \, d\tau \right) \exp\left(k \sqrt{\int_0^t \sigma_j^2(\tau) \, d\tau} \right)$$

where $k := W(t, \omega)$.

For the volatility structure (25.1) particularly, we find

$$\int_0^t \sigma_j^2(\tau) = \int_0^t \exp(-2a(T_j - \tau)) \, d\tau$$

$$= \frac{1}{2a}\left(\exp(-2a(T_j - t)) - \exp(-2a(T_j - 0)) \right)$$

$$= \frac{1}{2a} \left(\exp(2a\,t) - 1 \right) \exp(-2aT_j)$$

$$i \mapsto L_j(0) \exp\left(\int_0^t \mu_j(\tau, \omega) \, d\tau \right) \exp\left(\tilde{k} \exp(-aT_j) \right), \qquad (25.3)$$

where $\tilde{k} = k \sqrt{\frac{1}{2a}(\exp(2a\,t) - 1)}$.

The drift $\int_0^t \mu_j(\tau, \omega) \, d\tau$ is monotone increasing in j; see Equation (19.8). This explains the shape of the interest rate curve in Figure 25.3, upper left. With increasing parameter a the interest rate curve is multiplied by the double exponential (25.3) with increasing steepness. This explains the shape of the interest rate curve in Figure 25.3, upper right and lower left. Only the addition of more driving factors allows for a richer family of possible curves. If the parallel movement (level) remains the dominant factor, then the shape (25.3) still dominates the interest rate curve, Figure 25.3, lower right.

25.6 Instantaneous Correlation

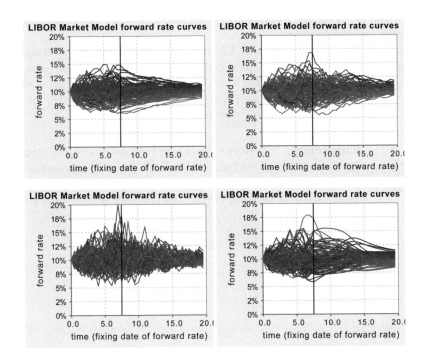

Figure 25.4. *Shape of the fixed rates $L_i(T_i)$ and the interest rate curve with different instantaneous correlations, seen at time $t = 7.5$. We used a correlation matrix with (all) 40 factors and $r = 0.01$ (upper left, high correlation), $r = 0.1$ (upper right) and $r = 1.0$ (lower left, high de-correlation). In the lower right we used a correlation matrix with $r = 1.0$ (the same as in lower left), but reduced the number of factors to three. For interpretation see Section 25.6.*

We fix a slightly decreasing volatility structure (25.1) with $a = 0.1$ and vary the parameter r of the correlation function (25.2). We do not apply a factor reduction, thus keep all 40 factors. The parameter $r = 0.01$ corresponds to an almost perfect correlation of the processes. Thus the possible shapes of the curve are almost parallel; the curve is very smooth since we started from a smooth (namely flat) curve. If the correlation parameter is increased to $r = 1.0$, then the distribution of rates within the curve is almost independent. See Figure 25.4, upper left, upper right, and lower left.

It should be noted that this (terminal) decorrelation is also achievable under $r = 0.01$ by an appropriate choice of the volatility structure; see Chapter 21. The instantaneous decorrelation introduces an *additional* terminal decorrelation. The statement that a model with perfect instantaneous correlation exhibits perfect terminal decorrelation of the forward rates is wrong.

Finally we have chosen in Figure 25.4, lower right, the parameter $r = 1.0$ again (as for the lower left with strong decorrelation) but have applied a reduction to the three largest factors. It is obvious that this strongly reduces the possibility of decorrelation. The three factors only allow that the beginning, the middle, and the end of the curve attain different values. Adjacent rates are still on similar levels.

 Experiment: At http://www.christian-fries.de/finmath/ applets/LMMSimulation.html a simulation of an interest rate curve with the model framework above is to be found. The parameters may be chosen at will to study the different shapes of the interest rate curve. ◁|

CHAPTER 26

Ritchken-Sakarasubramanian Framework: HJM with Low Markov Dimension

26.1 Introduction

 Motivation: The LIBOR market model is, with respect to the flexibility of the modeling, much more advanced than the short-rate models
discussed in Chapter 23. Under the LIBOR market model all forward
rates are modeled directly. Their volatility and correlation structure may
be specified directly.

Like all models which derive from the HJM framework, the LIBOR market model
may be interpreted as a short rate model; see Section 24.2.3. In this formulation the
price that has to be paid for its modeling flexibility becomes apparent: The model is
non-Markovian in the short rate. The drift is path-dependent. Only by the addition of
all forward rates does the model become Markovian. Since a Markovian representation thus requires a high-dimensional state space, a numerical implementation on a
lattice cannot be achieved.[1]

On the other hand, all the short rate models that were discussed in Chapter 23 were
one-dimensional Markov processes.

If we now reconsider the derivation of the short rate models and the LIBOR
market model from the Heath-Jarrow-Morton framework, then the question arises:
*What is the HJM volatility structure that results in a model, i.e. short rate, being a
low-dimensional Markov process?*

---

[1] For an implementation using a lattice the complexity, i.e., the requirements on memory and CPU time,
grows exponentially in the Markov dimension.

One answer to this question was given by Cheyette [61], Ritchken and Sakarasub-ramanian [92], and others. ◁|

26.2 Cheyette Model

Let $\mathbb{Q}$ denote the risk-neutral measure, i.e., the equivalent martingale measure corre-sponding to the numéraire

$$N(t) \;=\; \exp\!\left(\int_0^t r(\tau)\, d\tau \right).$$

Consider an HJM framework

$$
\begin{aligned}
df(t,T) &= \alpha^{\mathbb{Q}}(t,T)\, dt + \sigma(t,T)\, dW^{\mathbb{Q}}(t) \\
f(0,T) &= f_0(T)
\end{aligned}
\qquad \text{(was 22.3)}
$$

with a special volatility structure

$$\sigma(t,T) := g(t)\, h(T), \qquad (26.1)$$

where $g : [0, T^*] \mapsto \mathbb{R} \setminus \{0\}$ denotes a deterministic function and $h : [0, T^*] \times \Omega \mapsto \mathbb{R}^m$ an m-dimensional Markov process.

Remark 229 (Separability of Volatility):   The property (26.1) is called *separability of volatility*.

Theorem 230 (Cheyette Model (Single Factor)):   Given an HJM dynamic with the special volatility structure (26.1). Then the short-rate process is given by

$$r(t) \;=\; f(0,t) + X(t), \qquad (26.2)$$

where

$$
\begin{aligned}
dX(t) &= (Y(t) - \kappa(t)X(t))\, dt + \eta(t)\, dW^{\mathbb{Q}}(t), \qquad X(0) = 0, \\
dY(t) &= (\eta^2(t) - 2\kappa(t)Y(t))\, dt, \qquad Y(0) = 0,
\end{aligned}
\qquad (26.3)
$$

and

$$\eta(t) = \sigma(t,t) = g(t)\, h(t), \qquad \kappa(t) = -\frac{g'(t)}{g(t)}.$$

Remark 231 (Cheyette Model):   The representation of the short rate by Equa-tions (26.2) and (26.3) gives a complete model of the interest rate curve, since the numéraire depends on r alone.

374

The interest rate model (26.2), (26.3) is called the *Cheyette model*.

Remark 232 (Markov Dimension):   Within the Cheyette model the short rate $r(t)$ is a function of $X(t)$. The increment dX of $X(t)$ depends on $X(t)$, $Y(t)$, and $h(t)$. If h is deterministic, then Y is deterministic too, and the Markov dimension is 1; the time t state of the model is represented by state variable $X(t)$. If $h(t)$ is a function of $X(t)$ (local volatility), then Y is stochastic via the link to X through η, and the Markov dimension is 2; the time t state of the model is represented by the state variable $(X(t), Y(t))$. If h is a stochastic process such that $h(t)$ is not a function of $(X(t), Y(t))$ (stochastic volatility), then the Markov dimension of h has to be added; the time t state of the model is represented by the state variable $(X(t), Y(t))$ and the state variables of $h(t)$.

26.3 Implementation: PDE

If the Markov dimension is low (say ≤ 2), the model is an ideal candidate to be implemented using PDE methods. See [84] for an in depth discussion of the PDE implementation of the Cheyette model.

CHAPTER 27

Markov Functional Models

27.1 Introduction

 Motivation: From Chapter 5 we have a relation between the prices of European options and the probability distribution function (or probability density) of the underlying (under the martingale measure). If we consider a European option on some underlying, say the forward rate $L_i :=$ $L(T_i, T_{i+1}; T_i)$ (i.e., a caplet), then Lemma 81 allows us to calculate the probability density of L_i from the given market prices. It seems as if this allows perfect calibration of a "model" to a continuum of given market prices. However, the terminal distribution alone does not determine a pricing model. What is missing is the specification of the dynamics, i.e., the specification of transition probabilities, and, of course, the specification of the numéraire. This is the motivation for the *Markov functional modeling*. There we postulate a simple Markov process, e.g., $dx = \sigma(t) \, dW(t)$ for which the distribution function $\xi \mapsto P(x(T) \le \xi)$ is known analytically. Then we require the underlying L_i to be a function of $x(T_i)$. Let us denote this function by g_i, i.e., let $L_i(\omega) = g_i(x(T_i, \omega))$ for all paths ω. If the functional g_i is strictly monotone, then with $K = g_i(\xi)$:

$$F_{L_i}(K) := P(L_i \le K) = P(g_i(x(T)) \le K)$$
$$= P(x(T) \le \xi) =: F_{x(T_i)}(\xi) = F_{x(T_i)}(g_i^{-1}(K)).$$

With a given distribution function F_{L_i} of L_i (e.g., extracted from market prices through Lemma 81), the choice of the functional g_i allows the calibration to the distribution of L_i, while the process x (and the sequence of functionals $\{g_i\}$) describe the dynamics. To achieve a fully specified pricing model we further require the specification of the numéraire as function of the Markov process x. To achieve this we may use Theorem 79, if

---

- x is given under the equivalent martingale measure $\mathbb{Q}$ and

- x generates the filtration.

◁|

Given a filtered probability space $(\mathbb{R}, \mathcal{B}(\mathbb{R}), \mathbb{Q}, \{\mathcal{F}_t\})$. Consider a time discretization $0 = T_0 < T_1 < \cdots < T_n$. Financial products beyond T_n are not considered.

Let $t \mapsto N(t)$ denote the price process of a traded asset, which we choose as numéraire and let $\mathbb{Q}$ denote the corresponding equivalent martingale measure. Then for any replicable asset price process $V(t)$ (see Definition 73 and Theorem 79)

$$\frac{V(T_i)}{N(T_i)} = \mathrm{E}^{\mathbb{Q}^N}\left(\frac{V(T_k)}{N(T_k)} \,\big|\, \mathcal{F}_{T_i}\right).$$

In particular for every zero-coupon bond $P(T_k)$, paying 1 in T_k

$$\frac{P(T_k; T_i)}{N(T_i)} = \mathrm{E}^{\mathbb{Q}^N}\left(\frac{1}{N(T_k)} \,\big|\, \mathcal{F}_{T_i}\right), \text{ i.e.,}$$

$$P(T_k; T_i) = N(T_i)\, \mathrm{E}^{\mathbb{Q}^N}\left(\frac{1}{N(T_k)} \,\big|\, \mathcal{F}_{T_i}\right).$$

Let x denote an $\mathcal{F}_t$-adapted stochastic process with

$$dx(t) = \sigma(t)\, dW(t), \qquad x(0) = x_0.$$

The filtration should be generated by x. On this space we consider time-discrete stochastic processes, namely those for which their T_i realization is a function of $(x(T_0), \ldots, x(T_i))$, for all i. We particularly consider processes for which their time T_i realization is a function of $x(T_i)$ alone (i.e., independent of the process's history).

27.1.1 The Markov Functional Assumption (Independent of the Model Considered)

We assume that the time T_i realization of the numéraire process is a function of $x(T_i)$, i.e.,

$$N(T_i, \omega) = N(T_i, x(T_i, \omega)), \tag{27.1}$$

where we use the same letter N for the (deterministic) functional $\xi \mapsto N(T_i, \xi)$.

Then, for any payoff $V(T_k)$ that is itself a function of $x(T_k)$ for some k, the value process $V(T_i)$ for $i \leq k$ is

$$V(T_i) = N(T_i)\, \mathrm{E}\left(\frac{V(T_k)}{N(T_k)} \,\big|\, \mathcal{F}_{T_i}\right) = N(T_i, x(T_i))\, \mathrm{E}\left(\frac{V(T_k, x(T_k))}{N(T_k, x(T_k))} \,\big|\, \sigma(x(T_i))\right).$$

Thus, the time T_i realization of the value process $V(T_i)$ is also a functional of $x(T_i)$, which we denote by the same letter V. The functional $\xi \mapsto V(T_i, \xi)$ of the value process is

$$\xi \mapsto N(T_i, \xi)\, E\left(\frac{V(T_k, x(T_k))}{N(T_k, x(T_k))} \,\Big|\, \{x(T_i) = \xi\}\right).$$

Note: The Markov functional assumption (27.1) may be relaxed such that the numéraire is allowed to depend on $x(T_0), \dots, x(T_i))$. This relaxation is used in the LIBOR Markov functional model in spot measure.

27.1.2 Outline of This Chapter

In Section 27.2 we consider a Markov functional model for a stock (or any other non-interest-rate-related (single) asset). In Section 27.3 we will then consider a Markov functional model for the forward rate $L(T_i, T_{i+1}; T_i)$, which may be viewed as a time-discrete analog of the short rate. Both sections are essentially independent of each other. In Section 27.4 we will discuss how to implement a Markov functional model using a lattice in the state space.

27.2 Equity Markov Functional Model

27.2.1 Markov Functional Assumption

Consider a simple one-dimensional Markov process, e.g.,

$$dx(t) = \sigma(t)\, dW^{\mathbb{Q}}(t), \quad x(0) = x_0, \tag{27.2}$$

where σ is a deterministic function and $W^{\mathbb{Q}}$ denotes a $\mathbb{Q}$-Brownian motion. Without loss of generality we may assume $x_0 = 0$. Equation (27.2) is the most simple choice of a Markovian driver process. We will consider the addition of a drift term to (27.2) in our discussion of model dynamics in Section 27.2.5.

Let $S(t)$ denote the time t value of some asset for which we assume that we have a continuum of European option prices. Let x and S be adapted stochastic processes defined on $(\Omega, \mathbb{Q}, \mathcal{F}_t)$, where $\{\mathcal{F}_t\}$ denotes the filtration generated by $W^{\mathbb{Q}}$.

We assume that the time t value of the asset S is a function of $x(t)$, i.e., we assume the existence of a functional $(t, \xi) \mapsto S(t, \xi)$ such that

$$S(t, \omega) = S(t, x(t, \omega)),$$

where the left-hand side denotes our asset value at time t on path ω, and the right-hand side denotes some functional of our Markovian driver x, which we ambiguously name

S. We allow some ambiguity in notation here. From here on S will also denote a deterministic mapping (the functional)

$$(t, \xi) \mapsto S(t, \xi).$$

It will be clear from the arguments of S if we speak of the functional $(t, \xi) \mapsto S(t, \xi)$ or of the process $t \mapsto S(t)$.

For $t_1 < t_2$ we trivially have that

$$\frac{S(t_1)}{S(t_1)} = \mathrm{E}^{\mathbb{Q}}\left(\frac{S(t_2)}{S(t_2)} \mid \mathcal{F}_{t_1}\right).$$

We now postulate that $\mathbb{Q}$ is the equivalent martingale measure with respect to the numéraire S and that a universal pricing theorem holds for all other traded products, i.e., that their S relative price is a $\mathbb{Q}$-martingale.

This implies that the zero-coupon bond $P(T; t)$ having maturity T and being observed in $t < T$ fulfills

$$\frac{P(T; t)}{S(t)} = \mathrm{E}^{\mathbb{Q}}\left(\frac{1}{S(T)} \mid \mathcal{F}_t\right).$$

Using the functional representation of S we find that $P(T; t)$ is represented as a functional of $x(t)$ too, namely

$$(t, \xi) \mapsto P(T; t)$$

with

$$\frac{P(T; t, \xi)}{S(t, \xi)} = \mathrm{E}^{\mathbb{Q}}\left(\frac{1}{S(T, x(T))} \mid \{x(t) = \xi\}\right). \tag{27.3}$$

27.2.2 Example: The Black-Scholes Model

Let us assume a Markovian driver with constant instantaneous volatility $\sigma(t) = \sigma$. For the Black-Scholes model we have

$$S(t, \xi) = S(0) \exp\left(r\,t + \frac{1}{2}\sigma_{\mathrm{BS}}^2 t + \frac{\sigma_{\mathrm{BS}}}{\sigma}\,\xi\right), \tag{27.4}$$

where σ_{BS} denotes the (constant) Black-Scholes volatility. Plugging this into (27.3) we find

$$P(T; t, \xi) = \exp(-r(T - t)),$$

so that interest rates are indeed deterministic here.

This is the Black-Scholes model: From the definition of the Markovian driver we have $\frac{1}{\sigma} x(t) = W(t)$ and thus

$$S(t, x(t)) = S(0) \exp\left(r t + \frac{1}{2}\sigma_{BS}^2 t + \sigma_{BS} W(t)\right).$$

In other words, the $\mathbb{Q}$ dynamics of S is[1]

$$dS(t) = rS(t)\, dt + \sigma_{BS}^2 S(t)\, dt + \sigma_{BS} S(t)\, dW^{\mathbb{Q}}(t).$$

Introducing a new numéraire

$$dB(t) = rB(t)\, dt, \qquad B(0) = 1$$

we find for the change of numéraire process $\frac{S}{B}$ that

$$d\frac{S}{B} = \sigma_{BS}^2 S(t)\, dt + \sigma_{BS} S(t)\, dW^{\mathbb{Q}}(t).$$

For $\frac{S}{B}$ to be a martingale under $\mathbb{Q}^B$ it has to be $dW^{\mathbb{Q}}(t) = dW^{\mathbb{Q}^B} - \sigma_{BS}^2\, dt$ and thus

$$dS(t) = r S(t)\, dt + \sigma_{BS} S(t)\, dW^{\mathbb{Q}^B}(t),$$
$$dB(t) = r B(t)\, dt.$$

Note: $dW^{\mathbb{Q}}(t)$ is a $\mathbb{Q}$-Brownian motion, where $\mathbb{Q}$ is the equivalent martingale measure with respect to the numéraire S, while $dW^{\mathbb{Q}^B}(t)$ is a $\mathbb{Q}^B$-Brownian motion, where $\mathbb{Q}^B$ is the equivalent martingale measure with respect to the numéraire B.

27.2.3 Numerical Calibration to a Full Two-Dimensional European Option Smile Surface

As for the interest rate Markov functional model we are able to calculate the functionals numerically from a given two-dimensional smile surface. Our approach here is similar to the approach for the one-dimensional LIBOR Markov functional model under spot measure [71]. Consider the following time T payout:

$$V(T, K; T) := \begin{cases} S(T) & \text{if } S(T) > K \\ 0 & \text{otherwise.} \end{cases} \tag{27.5}$$

---

[1] Note that $\mathbb{Q}$ is the equivalent martingale measure with respect to the numéraire S.

Obviously

$$V(T, K; T) = \max(S(T) - K, 0) + K \begin{cases} 1 & \text{if } S(T) > K \\ 0 & \text{else,} \end{cases}$$

i.e. the value of V is given by the value of a portfolio of one call option and K digital options, all having strike K. This is our calibration product.

27.2.3.1 Market Price

Let $\bar{\sigma}_{BS}(T, K)$ denote the Black-Scholes implied volatility surface given by market prices. Then the market price of V is

$$V^{\text{market}}(T, K; 0) = \underbrace{S(0)\Phi(d_+) - \exp(-rT)K\Phi(d_-)}_{\text{call option part}}$$

$$+ \underbrace{K \exp(-rT)\left(\Phi(d_-) + S(0)\sqrt{T}\Phi'(d_+)\frac{\partial\bar{\sigma}_{BS}(T, K)}{\partial K}\right)}_{\text{digital part}}$$

$$= S(0)\Phi(d_+) + KS(0)\sqrt{T}\Phi'(d_+)\frac{\partial\bar{\sigma}_{BS}(T, K)}{\partial K},$$

where

$$\Phi(x) := \frac{1}{\sqrt{2\pi}} \int_{-\infty}^{x} \exp\left(-\frac{y^2}{2}\right) dy \quad \text{and} \quad d_\pm = \frac{\log(\frac{\exp(rT)S(0)}{K}) \pm \frac{1}{2}\bar{\sigma}_{BS}^2(T, K)T}{\bar{\sigma}_{BS}(T, K)\sqrt{T}}.$$

27.2.3.2 Model Price

Within our model the price of the product (27.5) is

$$V^{\text{model}}(T, K; 0) = S(0) E^Q\left(\frac{S(T, x(T))\, \mathbf{1}_{\{S(T,x(T))>K\}}}{S(T)} \,\Big|\, \{x(0) = x_0\}\right)$$

$$= S(0) E^Q\left(\mathbf{1}_{\{S(T,x(T))>K\}} \,\big|\, \{x(0) = x_0\}\right).$$

Assuming that our functional $(T, \xi) \mapsto S(T, \xi)$ is monotonely increasing in ξ, we may write

$$V^{\text{model}}(T, K; 0) = S(0) E^Q(\mathbf{1}_{\{x(T)>x^*\}} \mid \{x(0) = x_0\}), \tag{27.6}$$

where x^* is the (unique) solution of $S(T, x^*) = K$. Note that (27.6) depends on x^* and the probability distribution of $x(T)$ only and that $x(T)$ is known due to the simple form of our Markovian driver. It does not depend on the functional S! Thus for given x^* we can calculate

$$V^{\text{model}}(T, x^*; 0) := S(0) \, E^Q(\mathbf{1}_{\{x(T)>x^*\}} \mid \{x(0) = x_0\}).$$

27.2.3.3 Solving for the Functional

For given x^* we now solve the equation

$$V^{\text{market}}(T, K^*; 0) = V^{\text{model}}(T, x^*; 0)$$

to find $S(T, x^*) = K^*$ and thus the functional form $(T, \xi) \mapsto S(T, \xi)$. This can be done very efficiently using fast one-dimensional root finders, e.g., bisection or Newton's method; see Section 30.3 and Appendix B.4.

27.2.4 Interest Rates

27.2.4.1 A Note on Interest Rates and the No-Arbitrage Requirement

Functional models for equity option pricing have been investigated before; see, e.g., [57] and references therein. However, the approach considered there chooses deterministic interest rates and the bank account as numéraire. As suggested in Section 27.2.2, this will impose a very strong self-similarity requirement on the functionals (which is fulfilled by the Black-Scholes model). Such models may calibrate only to a one-dimensional submanifold of a given implied volatility surface; see [58]. For the Markov functional model this follows directly from (27.3). Assuming that the Markovian driver x is given and that the interest rate dynamic $P(T; t, \xi)$ is given, we find from (27.3) that

$$S(t, \xi) = \frac{P(T; t, \xi)}{E^Q\left(\frac{1}{S(T, x(T))} \mid \{x(t) = \xi\}\right)}.$$

So once a terminal time T functional $\xi \mapsto S(T, \xi)$ has been defined, all other functionals are implied by the interest rate dynamics P and the dynamics of the Markovian driver.

Sticking to prescribed interest rates, the only way to allow for more general functional is to violate the no-arbitrage requirement (27.3) or change the Markovian driver. The latter will be considered in Section 27.2.5.

27.2.4.2 Where Are the Interest Rates?

Our model calibrates to a continuum of options on S. We do not even specify interest rates. This is not necessary, since the specification of the interest rates is already contained in the specification of a continuum of options on S. Consider options on $S(T)$, i.e., options with maturity T. First note that from a continuum $K \mapsto V_{\text{call}}^{\text{market}}(T, K; 0)$ of market prices for call option payouts:

$$V_{\text{call}}^{\text{market}}(T, K; T) = \max(S(T) - K, 0)$$

we obtain prices for the corresponding digital payouts

$$V_{\text{digital}}^{\text{market}}(T, K; T) = \begin{cases} 1 & S(T) > K \\ 0 & \text{else} \end{cases}$$

by

$$V_{\text{digital}}^{\text{market}}(T, K; 0) = -\frac{\partial}{\partial K} V_{\text{call}}^{\text{market}}(T, K; 0).$$

Thus the value of the zero-coupon bond with maturity T is

$$P(T; 0) = \lim_{K \searrow 0} V_{\text{digital}}^{\text{market}}(T, K; 0) = -\lim_{K \searrow 0} \frac{\partial}{\partial K} V_{\text{call}}^{\text{market}}(T, K; 0). \qquad (27.7)$$

Note that this argument is model-independent.

Within the functional model, Equation (27.7) holds locally in each state. Given that we are at time t in state $x(t) = \xi$, we have for the corresponding bond

$$P(T; t, \xi) = \lim_{K \searrow 0} V_{\text{digital}}^{\text{model}}(T, K; t, \xi).$$

From this it becomes clear why specifying interest rates would represent a violation of the no-arbitrage requirement.[2]

In the next section we show that the model-implied interest rate dynamics are likely to be undesirable. However, as is known from interest rate hybrid Markov functional models [71], it is possible to calibrate to different model dynamics by changing the Markovian driver x.

27.2.5 Model Dynamics

27.2.5.1 Introduction

Markov functional models calibrate perfectly to a continuum of option prices, i.e., to the market-implied probability density of the underlying; see Chapter 5 and [52]. Indeed, the functional $(t, \xi) \mapsto S(t, \xi)$ is nothing more than the transformation of the measure from the probability density of $x(t)$ to the market-implied probability density of the underlying $S(t)$.

---

[2] This is precisely the reason why the model in [57] allows for arbitrage.

While calibration to terminal probability densities is a desirable feature, it is not the only requirement imposed on a model, specifically if the model is used to price complex derivatives like Bermudan options. Here the transition probabilities play a role, i.e., the model dynamics. The most prominent aspects of model dynamics are:

- **Interest Rate Dynamics**: For an equity Markov functional model the joint movement of the interest rate and the asset has to be analyzed. It is possible to calibrate to given interest rate dynamics by adding a drift to the Markovian driver; see Section 27.2.5.2.

- **Forward Volatility**: This is the implied volatility of an option with maturity T and strike K, given we are in state (t, ξ), i.e.,

$$S(t, \xi) \, \mathrm{E}^{\mathbb{Q}} \left(\frac{\max(S(T, x(T)) - K, 0)}{S(T)} \, \middle| \, \{x(t) = \xi\} \right).$$

Obviously it will play an important role for compound options and Bermudan options. The forward volatility may be calibrated by changing the instantaneous volatility of the Markovian driver; see Section 27.2.5.3.

- **Auto Correlation/Forward Spread Volatility**: The autocorrelation of the process S impacts the forward spread volatility. This is the implied volatility of an option on $S(T_2) - S(T_1)$ with maturity T_2, given we are in state (t, ξ), i.e.,

$$S(t, \xi) \, \mathrm{E}^{\mathbb{Q}} \left(\frac{\max(S(T_2, x(T_2)) - S(T_1, x(T_1)), 0)}{S(T_2, x(T_2))} \, \middle| \, \{x(t) = \xi\} \right).$$

Markov functional models allow limited calibration to different model dynamics by changing the dynamics of the Markovian driver x. For our choice

$$\mathrm{d}x = \sigma(t) \, \mathrm{d}W(t)$$

we can change the autocorrelation of x by choosing different instantaneous volatility functions σ. Since the calibration of the functionals is scale invariant with respect to the terminal standard deviation $\bar{\sigma}(t)$ of $x(t)$, the calibration to the terminal probability densities is independent of the choice of σ. See the Black-Scholes example in Section 27.2.2 for an example of this invariance.

Time Copula

The specification of the autocorrelation of x (through σ) is sometimes called *time copula* [57], since it may be specified through the joint distribution of $(x(t_1), x(t_2))$. For this reason similar functional models are sometimes called *copula models*, a term that is more associated with credit models, where joint default distribution is constructed from marginal default distributions.

In addition to a specification of the instantaneous volatility, the Markovian driver may be endowed with a drift.

Time-Discrete Markovian Driver

We assume a given time discretization $\{0 = t_0 < t_1 < t_2 < \ldots\}$ and consider the realizations $x(t_i)$ of the Markovian driver x given by increments $\Delta x(t_i) = x(t_{i+1}) - x(t_i)$. It is natural that a practical implementation of the model will feature a certain time discretization. Thus, speaking of calibration of a specific time-discretized implementation, it is best to consider the Markovian driver given by an Euler scheme (as in Equations (27.8) and (27.9))

$$
x(t_{i+1}) = x(t_i) + \mu(t_i, x(t_i))\, \Delta t_i + \sigma(t_i, x(t_i))\, \Delta W(t_i).
$$

27.2.5.2 Interest Rate Dynamics

Example: Black-Scholes Model with a Term Structure of Volatility

Let us first assume that the Markovian driver is given by

$$
x(t_{i+1}) = x(t_i) + \sigma_i\, \Delta W(t_i). \tag{27.8}
$$

Consider a term structure of Black-Scholes implied volatilities, i.e., let $\bar{\sigma}_{\mathrm{BS}}(t_i)$ denote the implied volatility of an option with maturity t_i. With the simple Markovian driver (27.8), the corresponding functionals that calibrate to these options are

$$
S(t_i, \xi) = S(0)\, \exp\!\left(r\, t_i + \frac{1}{2}\bar{\sigma}_{\mathrm{BS}}(t_i)^2 t + \frac{\bar{\sigma}_{\mathrm{BS}}(t_i)}{\bar{\sigma}_i}\, \xi \right),
$$

where $\bar{\sigma}_i^2 := \frac{1}{t_i} \sum_{j=0}^{i-1} \sigma_j^2\, \Delta t_j$.

Within this model a stochastic interest rate dynamic is already implied. From (27.3) we find

$$
\begin{aligned}
P(t_{i+1}; t_i, \xi) &= S(t_i, \xi)\, \mathrm{E}^{Q}\!\left(\frac{1}{S(t_{i+1}, x(t_{i+1}))} \,\Big|\, \{x(t_i) = \xi\} \right) \\
&= \exp\!\left[-r\,(t_{i+1} - t_i) - \frac{1}{2}\left(\bar{\sigma}_{\mathrm{BS}}(t_{i+1}) \right)^2 \left(t_{i+1} - \frac{\sigma_i^2}{\bar{\sigma}_{i+1}^2}\, \Delta t_i \right) - \bar{\sigma}_{\mathrm{BS}}(t_i)^2\, t_i \right. \\
&\qquad\left. - \left(\frac{\bar{\sigma}_{\mathrm{BS}}(t_{i+1})}{\bar{\sigma}_{i+1}} - \frac{\bar{\sigma}_{\mathrm{BS}}(t_i)}{\bar{\sigma}_i} \right) \xi \right],
\end{aligned}
$$

If the volatility of the Markovian driver x decays faster than the implied Black-Scholes volatility, then the interest rate will move positively correlated with the stock. If the volatility of the Markovian driver x decays slower than the implied Black-Scholes volatility, then the interest rate will move in a negatively correlated way.[3] If we choose the instantaneous volatility of x such that $\bar{\sigma}_i = \bar{\sigma}_{BS}(t_i)$,

$$P(t_{i+1}; t_i, \xi) = \exp(-r\,(t_{i+1} - t_i)),$$

i.e., we have recovered a model with deterministic interest rates. Reconsidering the case of a continuous driver $dx = \sigma(t)\,dW(t)$, we see that for $\frac{1}{t}\int_0^t \sigma(\tau)^2\,d\tau = \bar{\sigma}_{BS}(t)^2$ the functionals above define a Black-Scholes model with instantaneous volatility σ.

However, we do not need to sacrifice the instantaneous volatility of x to match the interest rate dynamics. A much more natural choice is to add a suitable drift to the Markovian driver x. Consider a Markovian driver x such that

$$x(t_{i+1}) = x(t_i) + \alpha_i x(t_i)\,\Delta t_i + \sigma_i\,\Delta W(t_i), \qquad x(t_0) = x_0 = 0. \tag{27.9}$$

Note that the $x(t_i)$'s are normally distributed with mean 0 (assuming $x_0 = 0$) and standard deviations $\gamma_i \sqrt{t_i}$ where

$$\gamma_{i+1}^2 t_{i+1} = \gamma_i^2 t_i\,(1 + \alpha_i\,\Delta t_i) + \sigma_i^2\,\Delta t_i.$$

Together with the functionals

$$S(t_i, \xi) = S(0)\exp\left(r\,t_i + \frac{1}{2}\bar{\sigma}_{BS}(t_i)^2 + \frac{\bar{\sigma}_{BS}(t_i)}{\gamma_i}\,\xi\right),$$

$$S(t_{i+1}, \xi) = S(0)\exp\left(r\,t_{i+1} + \frac{1}{2}\bar{\sigma}_{BS}(t_{i+1})^2 + \frac{\bar{\sigma}_{BS}(t_{i+1})}{\gamma_{i+1}}\,\xi\right)$$

we have

$$P(t_{i+1}; t_i, \xi) = S(t_i, \xi)\, E^{\mathbb{Q}}\left(\frac{1}{S(t_{i+1}, x(t_{i+1}))}\,\Big|\,\{x(t_i) = \xi\}\right)$$

$$= \exp\left[-r\,\Delta t_i - \frac{1}{2}(\bar{\sigma}_{BS}(t_{i+1})^2\left(t_{i+1} - \frac{\sigma_i^2}{\gamma_{i+1}^2}\,\Delta t_i\right) - \bar{\sigma}_{BS}(t_i)^2\,t_i\right.$$
$$\left. -\left(\frac{\bar{\sigma}_{BS}(t_{i+1})}{\gamma_{i+1}}(1 + \alpha_i\,\Delta t_i) - \frac{\bar{\sigma}_{BS}(t_i)}{\gamma_i}\right)\xi\right]$$

$$= \exp\left[-r\,\Delta t_i - \frac{1}{2}\left(\bar{\sigma}_{BS}(t_{i+1})^2\,\frac{\gamma_i^2}{\gamma_{i+1}^2}(1 + \alpha_i\,\Delta t_i)^2 - \bar{\sigma}_{BS}(t_i)^2\,t_i\right)\right.$$
$$\left. -\left(\frac{\bar{\sigma}_{BS}(t_{i+1})}{\gamma_{i+1}}(1 + \alpha_i\,\Delta t_i) - \frac{\bar{\sigma}_{BS}(t_i)}{\gamma_i}\right)\xi\right].$$

---

[3] Note that on average the interest rate is still r.

Choosing α_i such that

$$\frac{\bar{\sigma}_{BS}(t_{i+1})}{\gamma_{i+1}}(1 + \alpha_i \, \Delta t_i) - \frac{\bar{\sigma}_{BS}(t_i)}{\gamma_i} = 0 \tag{27.10}$$

we have

$$P(t_{i+1}; t_i, \xi) = \exp(-r \, (t_{i+1} - t_i)).$$

Interestingly, the Markov functional model (27.9)–(27.10) does not necessarily need to be a Black-Scholes model having the $\mathbb{Q}^B$-dynamics:

$$\begin{aligned} dS(t) &= rS(t) \, dt + \sigma_{BS}(t) \, S(t) \, dW^{\mathbb{Q}^B}(t), \\ dB(t) &= rB(t) \, dt, \end{aligned} \tag{27.11}$$

where $\bar{\sigma}_{BS}(t_i)^2 = \frac{1}{t_i} \int_0^{t_i} \sigma_{BS}(t)^2 \, dt$. The two models are not the same, although their terminal probability densities (European option prices) and their interest rate dynamics agree. The difference lies in the forward volatility, which may be changed for the Markov functional model by the instantaneous volatility of x. Only for $\sigma(t) = \sigma_{BS}(t)$ we have the dynamics (27.11). In this case the α_i in (27.10) will be zero, i.e., we are in the situation of the previous example.

Calibration to Arbitrary Interest Rate Dynamics

Within the no-arbitrage constraints, it is possible to calibrate the model to a given arbitrary interest rate dynamics by choosing the appropriate drift. To do so, we have to find $\mu(t_i, \xi)$ such that

$$P(t_{i+1}; t_i, \xi) = \mathrm{E}^{\mathbb{Q}}\left(\frac{S(t_i, \xi)}{S(t_{i+1}, x(t_{i+1}))} \; \middle| \; \{x(t_i) = \xi - \mu(t_i, \xi) \, \Delta t_i\} \right).$$

This can be done numerically by means of a one-dimensional root finder. The functional $S(t_{i+1})$ has to be recalibrated in every iteration.[4]

27.2.5.3 Forward Volatility

The calibration to European option prices (Section 27.2.3) and joint movements of asset and interest rates (Section 27.2.5.2) still leaves the instantaneous volatility σ of the Markovian driver x a free parameter. It may be used to calibrate the forward volatility, i.e., the volatility of an option; conditionally we are at time $t > 0$ in state ξ.

---

[4] The procedure is the same as in the calibration of the FX forward within the cross currency Markov functional model, [71].

Example: Black-Scholes Model with a Term Structure of Volatility

Consider the simple Black-Scholes-like example from Section 27.2.5.2. For simplicity we consider a Markovian driver without drift, i.e.,

$$x(t_{i+1}) = x(t_i) + \sigma_i \Delta W(t_i).$$

together with functionals

$$S(t_i, \xi) = S(0) \exp\left(r t_i + \frac{1}{2}\bar{\sigma}_{\mathrm{BS}}(t_i)^2 t + \frac{\bar{\sigma}_{\mathrm{BS}}(t_i)}{\bar{\sigma}_i} \xi \right)$$

calibrating to European options with implied volatility $\bar{\sigma}_{\mathrm{BS}}(t_i)$.[5]

Then we have that the standard deviation of the increment $x(t_k) - x(t_i)$ is $\bar{\sigma}_{t_i, t_k} \sqrt{(t_k - t_i)}$, where $\frac{1}{t_k - t_i} \sum_{j=i}^{k-1} \sigma_j^2 \Delta t_j$. It follows that the implied volatility of an option with maturity t_k, given we are in state (t_i, ξ), is

$$\bar{\sigma}_{\mathrm{BS}}(t_k) \frac{\bar{\sigma}_{t_i, t_k}}{\bar{\sigma}_{t_k}}.$$

Thus a decay in the instantaneous volatility of the driver process will result in a forward volatility decaying with simulation time t_i (for fixed maturity t_k).

Example: Exponential Decaying Instantaneous Volatility

Consider the case of a time continuous Markovian driver $dx = \sigma(t)\, dW$ with decaying instantaneous volatility

$$\sigma(t) = \exp(-a\,t), \qquad a \neq 0.$$

Then

$$\bar{\sigma}_{t_1, t_2} := \sqrt{\frac{1}{t_2 - t_1} \int_{t_1}^{t_2} \sigma(t)\, dt} = \sqrt{\frac{-1}{2a(t_2 - t_1)}(\exp(-2a\,t_2) - \exp(-2a\,t_1))}.$$

Assuming functionals

$$S(t, \xi) = S(0) \exp\left(r \cdot t + \frac{1}{2}\bar{\sigma}_{\mathrm{BS}}(t)^2 t + \frac{\bar{\sigma}_{\mathrm{BS}}(t)}{\bar{\sigma}(0, t)} \xi \right)$$

the forward volatility for an option with maturity T, given we are in t, is

$$\bar{\sigma}_{\mathrm{BS}}(T) \frac{\bar{\sigma}(t, T)}{\bar{\sigma}(0, T)} = \bar{\sigma}_{\mathrm{BS}}(T) \sqrt{\frac{T}{T-t} \frac{\exp(-2at) - \exp(-2aT)}{1 - \exp(-2aT)}}.$$

---

[5] As before we use the notation $\bar{\sigma}_i^2 := \frac{1}{t_i} \sum_{j=0}^{i-1} \sigma_j^2 \Delta t_j$.

27.2.6 Implementation

The model may be implemented in the same way as is done for a one-dimensional Markov functional LIBOR model. The basic steps for a numerical implementation are (Figure 27.1)

- Choose a suitable discretization of $(t, x(t))$, i.e., set up a grid $(t_i, x_{i,j})$.
- For each $x^* = x_{i,j}$:
  - Calculate $V^{\text{model}}(T, x^*; 0)$.
  - Find K^* such that $V^{\text{market}}(T, K^*; 0) = V^{\text{model}}(T, x^*; 0)$.
  - Set $S(x_{i,j}) := K^*$.

See Section 27.4.

27.3 LIBOR Markov Functional Model

We postulate that the forward rate viewed on its reset date (fixing) may be given as a function of the realization of the underlying Markov process x:

$$
\boxed{
\begin{array}{c}
\text{The forward rate (LIBOR)} \\[2mm]
L(T_k) = L_k(T_k) = \dfrac{1 - P(T_{k+1}; T_k)}{P(T_{k+1}; T_k)(T_{k+1} - T_k)} \\[2mm]
\text{(seen on its reset date } T_k\text{) is a (deterministic) function of} \\
x(T_k), \text{ where } x \text{ is a Markov process of the form} \\[2mm]
\mathrm{d}x = \sigma(t)\,\mathrm{d}W \quad \text{under } \mathbb{Q}^N, \qquad x(0) = x_0.
\end{array}
}
\tag{27.12}
$$

At that point we leave the choice of the numéraires N and the corresponding martingale measure $\mathbb{Q}$ open. We will make this choice now, and depending on this we obtain a Markov functional model in *terminal measure* (Section 27.3.1) or in *spot measure* (Section 27.3.2).

27.3.1 LIBOR Markov Functional Model in Terminal Measure

We choose as numéraire the T_n-bond $N := P(T_n)$. The measure $\mathbb{Q}$ should denote the corresponding martingale measure (terminal measure). From assumption (27.12) we have:

Lemma 233 (Numéraire of the Markov Functional Model under Terminal Measure): The numéraire $N(T_i) = P(T_n; T_i)$ is a (deterministic) function of $x(T_i)$, i.e., $N(T_i) = N(T_i, x(T_i))$. For the functional $\xi \mapsto N(T_i, \xi)$ we have the recursion

$$N(T_n, \xi) = 1$$

$$N(T_i, \xi) = \left[E^Q\left(\frac{1}{N(T_{i+1}, x(T_{i+1}))} \mid \{x(T_i) = \xi\} \right) (1 + L_i(T_i, \xi)(T_{i+1} - T_i)) \right]^{-1}.$$

$$(27.13)$$

Remark 234 (Notation for the Functionals): Here and in the following we denote the functionals by the same symbol as the corresponding random variables they are representing. In the equation

$$N(T_i) = N(T_i, x(T_i))$$

the $N(T_i)$ on the left-hand side denotes the random variable; on the right-hand side $\xi \mapsto N(T_i, \xi)$ denotes a function. If we used different symbols it would reduce the readability of the following text. The difference between the two will be obvious from the additional argument ξ or $x(T_i)$ in its place.

Proof: Since Q is the corresponding martingale measure we can use the *pricing theorem*. Thus for the zero-coupon bond $P(T_k)$ (maturity in T_k)

$$\frac{P(T_k; T_i)}{N(T_i)} = E^{Q^N}\left(\frac{1}{N(T_k)} \mid \mathcal{F}_{T_i} \right), \text{ i.e.}$$

$$P(T_k; T_i) = N(T_i) E^{Q^N}\left(\frac{1}{N(T_k)} \mid \mathcal{F}_{T_i} \right).$$

Since the process x generates the filtration $\{\mathcal{F}_{T_i}\}$ and since x is Markovian, it is sufficient to know the $\mathcal{F}_{T_i}$-measurable random variable $P(T_k; T_i)$ on the set $\{x(T_i) = \xi\}$. Thus the bond $P(T_k)$ seen at time T_i may be given as a function of $x(T_i)$, namely as $P(T_k; T_i) = P(T_k; T_i, x(T_i))$ with

$$P(T_k; T_i, \xi) := N(T_i, \xi) E^Q\left(\frac{1}{N(T_k, x(T_k))} \mid \{x(T_i) = \xi\} \right).$$

$$(27.14)$$

Since (compare Definition 99)

$$1 + L_i(T_i)(T_{i+1} - T_i) := \frac{P(T_i; T_i)}{P(T_{i+1}; T_i)} = \frac{1}{P(T_{i+1}; T_i)},$$

we have

$$1 + L_i(T_i, \xi)(T_{i+1} - T_i) = \left[N(T_i, \xi) \, \mathrm{E}^Q \left(\frac{1}{N(T_{i+1}, x(T_{i+1}))} \, \Big| \, \{x(T_i) = \xi\} \right) \right]^{-1},$$

and thus (27.13). □|

With $N(T_n) = P(T_n; T_n) = 1$ we have from (27.13) a recursion to determine the functionals $N(T_i, \xi)$ from the functionals $L_i(T_i, \xi)$.

With the specification of the numéraire the model is fully described. The functionals $\xi \mapsto L_i(T_i, \xi)$ are the free quantities that may be used to calibrate the model.

27.3.1.1 Evaluation within the LIBOR Markov Functional Model

As preparation for the discussion of the calibration of the model we consider the evaluation of a caplet and a digital caplet using the LIBOR Markov functional model.

Valuation of Caplets within the LIBOR Markov Functional Model

Let T_i denote the fixing and T_{i+1} the payment date of caplet with notional 1 and strike K. Then for its value $V_{\text{caplet}}(T_0)$ in the LIBOR Markov functional model

$$
\begin{aligned}
V_{\text{caplet}}(T_0) &= N(T_0) \, \mathrm{E}^Q \left(\frac{\max(L(T_i, x(T_i)) - K, \, 0) \, (T_{i+1} - T_i)}{N(T_{i+1}, x(T_{i+1}))} \, \Big| \, \{x(T_0) = x_0\} \right) \\
&= N(T_0) \, \mathrm{E}^Q \big(\max(L(T_i, x(T_i)) - K, \, 0) \, (T_{i+1} - T_i) \\
&\qquad \cdot \bar{P}(T_{i+1}; T_i, x(T_i)) \, | \, \{x(T_0) = x_0\} \big),
\end{aligned}
$$

(27.15)

where

$$\bar{P}(T_{i+1}; T_i, \xi) = \mathrm{E}^Q \left(\frac{1}{N(T_{i+1}, x(T_{i+1}))} \, \Big| \, \{x(T_i) = \xi\} \right).$$

Valuation of Digital Caplets within the LIBOR Markov Functional Model

Let T_i denote the fixing and T_{i+1} the payment date of digital caplet with notional 1 and strike K. Then for its value $V_{\text{digital}}(T_0)$ in the LIBOR Markov functional model

$$
\begin{aligned}
V_{\text{digital}}(T_0) &= N(T_0) \, \mathrm{E}^Q \left(\frac{\mathbf{1}(L(T_i, x(T_i)) - K) \, (T_{i+1} - T_i)}{N(T_{i+1}, x(T_{i+1}))} \, \Big| \, \{x(T_0) = x_0\} \right) \\
&= N(T_0) \, \mathrm{E}^Q \big(\mathbf{1}(L(T_i, x(T_i)) - K) \, (T_{i+1} - T_i) \\
&\qquad \cdot \bar{P}(T_{i+1}; T_i, x(T_i)) \, | \, \{x(T_0) = x_0\} \big),
\end{aligned}
$$

(27.16)

where

$$\bar{P}(T_{i+1}; T_i, \xi) = \mathrm{E}^Q(\frac{1}{N(T_{i+1}, x(T_{i+1}))} \mid \{x(T_i) = \xi\}).$$

A detailed description of how the given expectations may be numerically calculated, i.e., how to implement the model, is given in Section 27.4.

27.3.1.2 Calibration of the LIBOR Functional

The LIBOR functionals may be derived from the market prices of caplets. One possibility is to give a parametrization of the functional and optimize the parameters by comparing the model prices to given market prices. Another possibility is to derive the functional pointwise from a continuum of given market prices. For this we have to know the market prices of the caplets with periods $[T_i, T_{i+1}]$ and strikes $K \in [0, \infty)$. If the prices are known only for a finite number of strikes K_j, then we require a corresponding interpolation of the caplet prices. This is nontrivial; see Chapter 6.

Calibration of Parametrized LIBOR Functionals

If we assume a parametrized functional form for $L(T_i)$, then we can calculate the model prices of caplet on $L(T_i)$ from a known functional for $N(T_{i+1})$. This allows us to optimize the parameters of a parametrized functional form to achieve the replication of given market prices. Thus we have the following backward induction:

Induction start:

- $N(T_n) = 1$.

Induction step ($T_{i+1} \to T_i$):

- Calculate $\bar{P}(T_{i+1}; T_i, \xi) = \mathrm{E}^Q\left(\frac{1}{N(T_{i+1}, x(T_{i+1}))} \mid \{x(T_i) = \xi\}\right)$.

- "Optimize" $L(T_i)$ by comparing (27.15) with given market prices.

- Calculate $N(T_i)$ from $L(T_i)$ and $\bar{P}(T_{i+1}; T_i)$ via (27.13).

If $L(T_i)$ should be almost lognormally distributed, then a good starting point for the parametrization is an exponential. In [26] the following parametrization is discussed:

$$L_i(\xi) = \exp(a + b\xi + c(\xi - d)^2).$$

Calibration of Discretized LIBOR Functionals (Pointwise)

The evaluation of a digital caplet allows us to calculate the value $L(T_i, x^*)$ of the functional for an arbitrary point x^*. We make the following additional assumption:

> The LIBOR functional $\xi \mapsto L(T_i, \xi)$ is
>
> strictly monotone increasing in ξ. (27.17)

With this assumption we can provide a simple algorithm to derive the functional $L(T_i)$ from a given curve of caplet market prices.[6] We assume that for any T_i and any strike K we have a market caplet price $V_{\text{caplet}}(T_i, K; T_0)$.[7] Then we can calculate all digital caplet prices $V_{\text{digital}}^{\text{market}}(T_i, K^*; T_0)$. We have

$$V_{\text{digital}}^{\text{market}}(T_i, K; T_0) = \frac{\partial}{\partial K} V_{\text{caplet}}^{\text{market}}(T_i, K; T_0). \qquad (27.18)$$

Let x^* denote a given state point. We wish to derive the functional $\xi \mapsto L(T_i, \xi)$, i.e., $K^* := L(T_i, x^*)$. The model price $V_{\text{digital}}^{\text{model}}(T_i, K^*; T_0)$ for a digital caplet with strike K^* is given by

$$
\begin{aligned}
V_{\text{digital}}(T_0) &= N(T_0) \, \mathrm{E}^{\mathbb{Q}}(\mathbf{1}(L(T_i, x(T_i)) - K) \, (T_{i+1} - T_i) \\
&\qquad \cdot \bar{P}(T_{i+1}; T_i, x(T_i)) \mid \{x(T_0) = 0\}), \\
&= N(T_0) \, \mathrm{E}^{\mathbb{Q}}(\mathbf{1}(x(T_i) - x^*) \, (T_{i+1} - T_i) \\
&\qquad \bar{P}(T_{i+1}; T_i, x(T_i)) \mid \{x(T_0) = 0\}),
\end{aligned}
\qquad (27.19)
$$

thus, from the monotony assumption $(L(x(T_i)) > K \Leftrightarrow x(T_i) > x^*)$ it follows that the product may be evaluated without the knowledge of the LIBOR functional $L(T_i)$. We thus write in brief $V_{\text{digital}}^{\text{model}}(T_i, x^*; T_0)$. By solving the equation

$$V_{\text{digital}}^{\text{market}}(T_i, K; T_0) = V_{\text{digital}}^{\text{model}}(T_i, x^*; T_0) \qquad (27.20)$$

to K we obtain K^*, i.e., $L(x^*)$, for the given x^*. Compare Figure 27.1.

The LIBOR functional obtained from this procedure replicates the given market price curve of the digital caplets and thus the market price curve of the caplets. Using backward induction we can calibrate the model to caplets of different maturities:

---

[6] See also the considerations in Chapter 5, where we showed how to derive the probability density of the underlying from market prices of European options.

[7] A complete price curve is usually not available. It has to be constructed by interpolating on given market prices. This interpolation procedure has to be understood as part of the model; see Chapter 6.

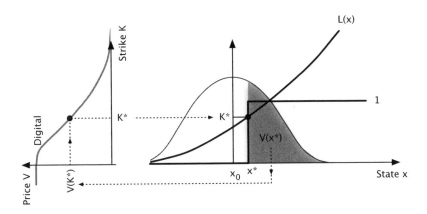

Figure 27.1. *Calibration of the LIBOR functional $L(x)$ within the Markov functional model: For a given x^* we calculate the model price of a digital caplet with strike $K^* = L(x^*)$ (payoff profile in black). This is possible without knowing the functional L. For the given model price $V(x^*)$ (gray surface) we find the corresponding strike K^* by looking it up on the (inverse) of the market price curve $K \mapsto V(K)$ (left graph). This determines the LIBOR functional $x \mapsto L(x)$ (right graph).*

Induction start:

- $N(T_n) = 1$.

Induction step ($T_{i+1} \rightarrow T_i$):

- Calculate $P(T_{i+1}; T_i, \xi) = E^Q \left(\frac{1}{N(T_{i+1}, x(T_{i+1}))} \,\middle|\, \{x(T_i) = \xi\} \right)$.
- For any given $\{x_k^*\}$:
  - Calculate the model price $V_{\text{digital}}^{\text{model}}(T_i, x^*; T_0)$ from (27.19).
  - Calculate $K^* = L(x^*)$ from (27.20) and (27.18).

- If required, calculate an interpolation from sample points $x^*, L(x^*)$ obtained in the previous step.

- Calculate $N(T_i)$ from $L(T_i)$ and $P(T_{i+1}; T_i)$ through (27.13).

27.3.2 LIBOR Markov Functional Model in Spot Measure

In this section we will discuss the Markov functional model under the spot measure, i.e., we choose the money market account as numéraire and present an efficient calibration method for this model. By money market account numéraire we mean (cf. [24, 81])

$$N(T_i) := \prod_{k=0}^{i-1} (1 + L(T_k))(T_{k+1} - T_k), \qquad (27.21)$$

which is the value of repeated reinvestments of the initial value $N(0) = 1$ in the shortest bond in our time discretization $\{T_0, \ldots, T_N\}$. As in Section 27.3 we make the assumption:

The forward rate (LIBOR)

$$L(T_k) = L_k(T_k) = \frac{1 - P(T_{k+1}; T_k)}{P(T_{k+1}; T_k)(T_{k+1} - T_k)}$$

(seen upon its maturity T_k) is a (deterministic) function of $x(T_k)$, where x is a Markovian process given by

$$dx = \sigma(t)\, dW \quad \text{under } \mathbb{Q}^N, \qquad x(0) = x_0.$$

(27.12)

Note that this does imply that the numéraire $N(T_k)$ given in (27.21) is *not* a function of $x(T_k)$ alone. Here the numéraire $N(T_i)$ is path-dependent, i.e., it is given as a function of $x(T_0), x(T_1), \ldots, x(T_{i-1})$:

$$N(T_i; x(T_0), x(T_1), \ldots, x(T_{i-1})) := \prod_{k=0}^{i-1} (1 + L(T_k; x(T_k))(T_{k+1} - T_k), \qquad (27.22)$$

and $\mathcal{F}_{T_{i-1}}$-measurable. In contrast to this, for the Markov functional model under terminal measure the numéraire $N(T_i)$ was a function of $x(T_i)$ alone (i.e., not path-dependent and $\mathcal{F}_{T_i}$-measurable, not $\mathcal{F}_{T_{i-1}}$-measurable).

27.3.2.1 Calibration of the Markov Functional Model under Spot Measure

The calibration procedure of the Markov Functional model under the terminal measure was presented in Section 27.3.1.2. It seems as if the feasibility of the calibration process is tied to the choice of the terminal measure as it induced a simple *backward induction* for the LIBOR functionals. The LIBOR functionals were calculated by the pricing of digital caplets which simply involved expectations of indicator functionals (i.e., half-integrals over given distributions).

We will show that the calibration procedure of the Markov functional model under the spot measure is given by a simple *forward induction* for the LIBOR functionals. They are calculated by the pricing of a portfolio of a caplet and digital caplets. This involves only a simple half integral over the given distribution and a known expectation step.

27.3.2.2 Forward Induction Step

We assume that the LIBORs $L(T_j)$ for $T_j < T_i$ and thus $N(T_i)$ have already been calculated and present the induction step $T_i \rightarrow T_{i+1}$.[8] Together with $N(0) := 1$ this gives the calibration procedure as a *forward-in-time* algorithm. Let $V_{T_i}(T_k)$ denote the time T_k value of a product with a time T_{i+1} value $V_{T_i}(T_{i+1}; L(T_i))$ depending on $L(T_i)$ only (e.g., the value of a caplet or digital caplet with fixing date T_i and payment date T_{i+1}). Then the value of this product is

$$V_{T_i}(0) = N(0) \, \mathrm{E}\left(\frac{V_{T_i}(T_{i+1}; L(T_i))}{(1 + L(T_i)(T_{i+1} - T_i))N(T_i)} \,\middle|\, \mathcal{F}_{T_0} \right).$$

On the right-hand side, we take the expectation of a function depending on $L(T_i)$ and $N(T_i)$. As the numéraire $N(T_i)$ is known from the previous induction step; the functional form of $L(T_i; x(T_i))$ is the only unknown in this equation and it may be used to calibrate the functional form $\xi \mapsto L(T_i; \xi)$ to given market prices.

---

[8] Note that $N(T_i)$ depends on $L(T_j)$ for $T_j < T_i$ only.

27.3.2.3 Dealing With the Path Dependency of the Numéraire

The path dependency of the numéraire (27.22) implies that (conditional) expectations have to be calculated time step by time step using

$$E\left(\frac{V_{T_i}(T_k)}{N(T_k)} \mid \mathcal{F}_{T_i}\right) = E\left(\frac{V_{T_i}(T_{k-1})}{N(T_{k-1})} \mid \mathcal{F}_{T_i}\right)$$

where

$$V_{T_i}(T_{k-1}) = \frac{E\left(V_{T_i}(T_k) \mid \mathcal{F}_{T_{k-1}}\right)}{1 + L(T_{k-1})(T_k - T_{k-1})}.$$

The need for the time step by time step calculation of conditional expectations (induced by the path dependency of the numéraire) seems to be a major computational bottleneck, when compared to the Markov functional model in terminal measure. However, we will discuss in Section 27.3.3 the fact that $\mathcal{F}_{T_0}$ conditioned expectations may be calculated fast using a single scalar product with precalculated projection vectors.

27.3.2.4 Efficient Calculation of the LIBOR Functional From Given Market Prices

The LIBOR functional are now derived from the model pricing formula of a portfolio of a caplet and digital caplets. Consider the following payout function:

$$V_{T_i,K}(T_{i+1}, L(T_i)) := \begin{cases} 1 + L(T_i)(T_{i+1} - T_i) & \text{if } L_i - K > 0 \\ 0 & \text{else} \end{cases} \quad \text{paid in } T_{i+1}.$$

$$(27.23)$$

This is a *digital caplet in arrears* or equivalently the portfolio of 1 strike K caplet and $K + \frac{1}{(T_{i+1}-T_i)}$ strike K digital caplets. Given market prices of caplets, we have market prices for the digital caplet in arrears for any strike K; see [71]. Its model price is given by

$$V_{T_i,K}^{\text{model}}(T_0) = \mathrm{E}\left(\frac{V_{T_i,K}(T_{i+1})}{N(T_{i+1})} \,\Big|\, \mathcal{F}_{T_0}\right)$$

$$= \mathrm{E}\left(\frac{V_{T_i,K}(T_{i+1})}{1 + L(T_i, x(T_i))(T_{i+1} - T_i)\, N(T_i)} \,\Big|\, \mathcal{F}_{T_0}\right)$$

$$= \mathrm{E}\left(\mathbf{1}(L(T_i, x(T_i)) - K)\, \underbrace{\frac{1}{N(T_i)}}_{\mathcal{F}_{T_{i-1}}\text{-measurable}} \,\Big|\, \mathcal{F}_{T_0}\right)$$

$$= \mathrm{E}\left(\mathrm{E}\left(\mathbf{1}(L(T_i, x(T_i)) - K)\,|\,\mathcal{F}_{T_{i-1}}\right)\frac{1}{N(T_i)} \,\Big|\, \mathcal{F}_{T_0}\right),$$

where $\mathbf{1}$ denotes the indicator function with $\mathbf{1}(R) = 0$ if $R \le 0$ and $\mathbf{1}(R) = 1$ if $R > 0$ and $\xi \mapsto L(T_i, \xi)$ denotes the functional form of the LIBOR, assumed to be increasing. If x^* is such that

$$L(T_i, x^*) = K \tag{27.24}$$

we have

$$V_{T_i,K}(T_{i-1}) = \mathrm{E}\left(\mathbf{1}(L(T_i, x(T_i)) - K)\,|\,(T_{i-1}, \xi)\right) = \mathrm{E}\left(\mathbf{1}(x(T_i) - x^*)\,|\,(T_{i-1}, \xi)\right)$$

$$= \int_{x^*}^{\infty} \phi(\eta - \xi; \sigma(T_{i-1}, T_i))\, d\eta.$$

This reduces the model price to an integral over the indicator function(al) and then taking the expectation $\mathrm{E}\left(\frac{V(T_{i-1})}{N(T_{i-1})}\,|\,\mathcal{F}_{T_0}\right)$. The latter is known from the previous calibration steps from T_{i-1} back to T_0. It is implemented efficiently as a scalar product with a precalculated projection vector.[9]

The calculation of the functional form $L(T_i; \xi)$ thus involves the calculation of model prices as outlined above for suitable discretization points x^* and calculating the corresponding strikes K by inverting the market price function. This determines $L(T_i, x^*)$ using (27.24).

The calibration step is as simple as it was under the terminal measure: Model prices of calibration products are evaluated by a half-integral together with a known expectation step and matched with the market price function. Here, the half-integral only represents a slightly different product.

Often a certain measure is chosen to simplify the pricing of a given product (e.g., the Black '76 caplet pricing formula (10.2) is best derived under the terminal measure associated with the caplet's payment date). Here this technique is reversed by considering a certain product with a simple (model) pricing formula under a given

---

[9] We will discuss this aspect of the implementation in the next section.

measure. The suitable product for the terminal measure is the digital caplet while the digital caplet in arrears seems the best choice for the spot measure.

27.3.3 Remark on Implementation

Given a certain functional $\xi \mapsto f(\xi)$ and a lattice time and state discretization

$$\{x_{T_j,k} \mid k = 1, \ldots, m_i\} \subset x(T_j, \Omega) = \mathbb{R}, \qquad 0 \le j \le n,$$

where $m_0 = 1$, $x_{T_0,1} = x_0$. The expectation of $f(x(T_{i+1}))$ conditional on state $x(T_i) = x_{T_i,k}$ is given by

$$\int_{-\infty}^{\infty} f(\xi)\, \phi(\xi - x_{T_i,k}; \bar{\sigma}^2)\, d\xi, \tag{27.25}$$

where $\phi(\cdot; \bar{\sigma})$ is the density of the normal distribution with variance $\bar{\sigma}^2 = \int_{T_i}^{T_{i+1}} \sigma^2(\tau)\, d\tau$. The approximation of this integral within the lattice is given by a numerical integration based on sampled values $f_k := f(x_{T_i,k})$. We represent this integration by

$$A_{T_i}^{T_{i+1}} \cdot (f_1, \ldots, f_{m_{i+1}})^{\top}, \tag{27.26}$$

where $A_{T_i}^{T_{i+1}}$ is a linear operator given by a $m_i \times m_{i+1}$ matrix. Defining

$$A_{T_0}^{T_{i+1}} := A_{T_0}^{T_i} \cdot A_{T_i}^{T_{i+1}}, \tag{27.27}$$

the large time expectation step

$$E(f(x(T_{i+1})) \mid \{x(T_0) = x_0\})$$

is represented numerically by $A_{T_0}^{T_{i+1}}$. The matrix multiplication with $A_{T_0}^{T_{i+1}}$ is fast as $A_{T_0}^{T_{i+1}}$ is a row vector.

27.3.3.1 Fast Calculation of Price Functionals

In the model calibration and the application of the model to derivative pricing expectations of numéraire relative prices have to be calculated. For a given time T_{i+1} functional V we have to calculate

$$I_{T_i}^{T_{i+1}}[V](x_{T_i,k}) := \int_{-\infty}^{\infty} \frac{V(\xi)}{N(T_{i+1}, \xi)}\, \phi(\xi - x_{T_i,k}; \bar{\sigma}^2)\, d\xi. \tag{27.28}$$

It is advantageous to view $\xi \mapsto \frac{1}{N(T_{i+1}, \xi)} \cdot \phi(\xi - x_{T_i,k}; \bar{\sigma}^2)$ as a convolution kernel and directly precalculate the numerical approximation of the (linear) operator $V \mapsto I[V]$.

Redefining the $A_{T_i}^{T_{i+1}}$ in this sense, we are able to numerically calculate large time-step expectations

$$\mathrm{E}\left(\frac{V(T_{i+1})}{N(T_{i+1})} \,\Big|\, \mathcal{F}_{T_0}\right)$$

even for the path-dependent numéraire (27.22) by a single scalar product of the projection vector $A_{T_0}^{T_{i+1}}$ with the sample vector $(V(x_{T_{i+1},1}), \dots, V(x_{T_{i+1},n_{i+1}}))$. The vectors $A_{T_0}^{T_{i+1}}$ may be precalculated iteratively in each forward induction step.

The elements of projection vector $A_{T_0}^{T_{i+1}}$ are Arrow-Debreu like prices.

27.3.3.2 Discussion on the Implementation of the Markov Functional Model under Terminal and Spot Measure

It appears that the precalculation of the large time expectation step is only necessary to cope with the path-dependent numéraire in the spot measure Markov functional model. However, in our experience the precalculation of projection vectors by means of the iteration (27.27) is advantageous even for the terminal measure variant as it will prevent numerically inconsistent ways of calculating the large time expectation. Numerical approximation errors will lead to significant differences between iterated expectation and single, large time-step expectations, thus violating the tower law[10]. By enforcing the calculation of large time-step expectations by iterated expectations the tower law will by definition be valid in the model implementation. It might seem as if the iteration (27.27) will then lead to a propagation of numerical errors. Indeed the terminal distributions are much less close to a normal distribution, but exact sampling of the terminal distribution is not crucial and the calibration quality of the discrete model will not suffer.

27.3.4 Change of Numéraire in a Markov Functional Model

Having presented Markov functional models under different measures, it is natural to ask how the functionals relate, i.e. under what conditions a functional calibrated in one measure may be reused in the other.

Let N, M be two numéraires. Then for any traded asset V:

$$\frac{V(T_i)}{N(T_i)} = \mathrm{E}^{\mathbb{Q}^N}\left(\frac{V(T_{i+1})}{N(T_{i+1})} \,\Big|\, \mathcal{F}_{T_i}\right)$$

$$\frac{V(T_i)}{M(T_i)} = \mathrm{E}^{\mathbb{Q}^M}\left(\frac{V(T_{i+1})}{M(T_{i+1})} \,\Big|\, \mathcal{F}_{T_i}\right) = \mathrm{E}^{\mathbb{Q}^M}\left(\frac{V(T_{i+1})}{N(T_{i+1})}\frac{N(T_{i+1})}{M(T_{i+1})} \,\Big|\, \mathcal{F}_{T_i}\right).$$

---

[10] The tower law is the equation of iterated expectation, i.e., $\mathrm{E}\big(\mathrm{E}(Z \mid \mathcal{F}_{T_j}) \mid \mathcal{F}_{T_i}\big) = \mathrm{E}(Z \mid \mathcal{F}_{T_i})$ for $T_i < T_j$.

Thus

$$\mathrm{E}^{\mathbb{Q}^M}\left(\frac{V(T_{i+1})}{N(T_{i+1})} \underbrace{\frac{N(T_{i+1})}{M(T_{i+1})}\frac{M(T_i)}{N(T_i)}}_{=:C(T_i,T_{i+1})}\Big|\mathcal{F}_{T_i}\right) = \mathrm{E}^{\mathbb{Q}^N}\left(\frac{V(T_{i+1})}{N(T_{i+1})}\Big|\mathcal{F}_{T_i}\right)$$

i.e.,

$$\mathrm{E}^{\mathbb{Q}^M}\left(\frac{V(T_{i+1})}{N(T_{i+1})}\, C(T_i, T_{i+1})\Big|\mathcal{F}_{T_i}\right) = \mathrm{E}^{\mathbb{Q}^N}\left(\frac{V(T_{i+1})}{N(T_{i+1})}\Big|\mathcal{F}_{T_i}\right). \qquad (27.29)$$

We want to see this in the light of a Markov functional model and thus require that all three quantities V, N, and M are functions of a (scalar) time-discrete Markovian stochastic process $x(t)$. For illustration purposes we additionally assume that the functional of V under $\mathbb{Q}^M$ is the same as V under $\mathbb{Q}^N$ and that

$$x(T_{i+1}) = x(T_i) + \sigma(T_i)\Delta W(T_i, T_{i+1}) \qquad\qquad \text{under } \mathbb{Q}^N,$$
$$x(T_{i+1}) = x(T_i) + \mu(T_i, x(T_i))\Delta T_i + \sigma(T_i)\Delta W(T_i, T_{i+1}) \qquad\qquad \text{under } \mathbb{Q}^M,$$

where $\Delta T_i := (T_{i+1} - T_i)$—it will become clear below that this assumption cannot hold in general. Under these assumptions we find from (27.29)[11]

$$\mathrm{E}^{\mathbb{Q}^N}\left(\frac{V(T_{i+1})}{N(T_{i+1})}\, C(T_i, T_{i+1})\,\Big|\, x(T_i) + \mu(x(T_i))\,\Delta T_i\right) = \mathrm{E}^{\mathbb{Q}^N}\left(\frac{V(T_{i+1})}{N(T_{i+1})}\,\Big|\, x(T_i)\right),$$

i.e.,

$$\mathrm{E}^{\mathbb{Q}^N}\left(\frac{V(T_{i+1})}{N(T_{i+1})}\cdot C(T_i, T_{i+1})\mid x(T_i)\right) = \mathrm{E}^{\mathbb{Q}^N}\left(\frac{V(T_{i+1})}{N(T_{i+1})}\,\Big|\, x(T_i) + \mu(x(T_i))\,\Delta T_i\right).$$
$$(27.30)$$

Equation (27.30) is valid for all traded assets V.[12] Choosing $V(T_2) \equiv 1$ (a bond), we see that Equation (27.30) determines $\mu(x(T_1))$ from the change of numéraire integration kernel C.[13] With μ fixed we see that (27.30) cannot hold for general functionals V. This is clear from Girsanov's theorem: Over a discrete time step a change of numéraire will introduce a change in the conditional probability density, which cannot be just a shift of the mean as in general $\int_t^{t+\Delta t}\mu(t)\,\mathrm{d}t$ is not $\mathcal{F}_t$-previsible. (Girsanov's theorem states that the conditional probability density changes by an infinitesimal shift of the mean (the drift adjustment) over an *infinitesimal* time step.

---

[11] We assume here that the two measures are identical on $\mathcal{F}_{T_i}$, i.e., $[T_i, T_{i+1}]$ is the first time interval where the change of numéraire applies. This is not a restriction, for example, the argument applies to the first time step $[T_0, T_1]$.

[12] Equation (27.30) is just a discrete version of Girsanov's theorem.

[13] Note that $C(T_i, T_{i+1})$ is $\mathcal{F}_{T_{i+1}}$-measurable but not $\mathcal{F}_{T_i}$-measurable.

Therefore, in a discrete time model it is usually not possible to perform a change of numéraire by means of an adapted change of the drift (if it is done, it is an approximation).

Thus we have to relax our assumptions to either

- the drift μ is path-dependent, i.e., we consider

$$dx = \mu(t,x)dt + \sigma(t)\,dW \qquad \text{resulting in}$$

$$x(T_i) = x(T_{i-1}) + \int_{T_{i-1}}^{T_i} \mu(t,x(t))\,dt + \sigma(T_{i-1})\,\Delta W(T_{i-1},T_i) \quad \text{under } \mathbb{Q}^M, \text{ or}$$

- the functional V is different under $\mathbb{Q}^N$ and $\mathbb{Q}^M$.

The first will work because it is simply the proposition of Girsanov's theorem.

As we do not want such a path-dependent drift in the driving process, we choose the second approach. Then we can fit *terminal* (!) distributions of V by means of a change of the functional form. Under the changed numéraire one has to recalculate the functional form V^M.

Note that a recalculation of the functional forms *is not* a change of numéraire in the strict sense. The functional forms may be used to match the relevant terminal, but not the transition, distributions under $\mathbb{Q}^N$ and $\mathbb{Q}^M$. As a result prices of some products, e.g., Bermudans, may differ. The two models are not "equivalent".[14]

This can already be seen for the Markov functional model under the terminal measure. Two such models with different time horizons $T_m < T_n$ are not equivalent over the common time interval $[0, T_m]$.[15]

27.4 Implementation: Lattice

Consider an implementation which relies on functionals discretized in the state space (in contrast to parametrized functionals). In addition to the time discretization $\{0 = T_0 < \cdots < T_n\}$ we consider a discretization of the state space

$$x_{i,j} \in x(T_i)(\Omega) = \mathbb{R} \quad j = 1,\ldots,m_i, \quad i = 0,\ldots,n. \tag{27.31}$$

---

[14] This problem also exists in Monte Carlo simulations. For example, the Euler discretization of the LIBOR market model's SDE $dL_i(t) = \mu_i(t)L_i(t)\,dt + \sigma_i(t)L_i(t)\,dW(t)$ exhibits different discretization errors for different measures; see also Section 13.1.2. For the Monte Carlo simulation this problem can be solved by arbitrage-free discretization techniques [73] or by reducing the size of the discrete time step Δt.

[15] It is a charming aspect of the spot measure Markov functional model that it does not exhibit this dependence on the time horizon (since there is no time horizon at all).

For any given time discretization point T_i and any given space discretization point $x^* = x_{i,j}$, we calculate the model price of the digital caplet $V_{\text{digital}}^{\text{model}}(T_i, x^*; T_0)$ and calculate from it the value of the LIBOR functional $L(T_i, x_{i,j})$ and from there the numéraire functional $N(T_i, x_{i,j})$. It remains to specify how we interpolate the functionals $\xi \mapsto L(T_i, \xi)$, $\xi \mapsto N(T_i, \xi)$ for $\xi \neq x_{i,j}$ and calculate the conditional expectations, i.e., how we perform the numerical integration

$$\mathrm{E}^Q\left(\frac{1}{N(T_{i+1}, x(T_{i+1}))} \mid \{x(T_i) = x_{i,j}\}\right) = \int_{-\infty}^{\infty} \frac{1}{N(T_{i+1}, \xi)} \phi(\xi - x_{i,j}, \sigma_i) \, \mathrm{d}\xi$$

where $\phi(\xi - \mu, \sigma_i) = \frac{1}{\sqrt{2\pi}\sigma_i} \exp(-\frac{(\xi-\mu)^2}{2\sigma_i^2})$ denotes the (transition) probability density of $x(T_{i+1}) - x(T_i)|_{x(T_i)=\mu}$ with $\sigma_i := (\int_{T_i}^{T_{i+1}} \sigma^2(t) \, \mathrm{d}t)^{1/2}$.

27.4.1 Convolution with the Normal Probability Density

A main part of the implementation of the model is the numerical integration, i.e., the convolution of a functional with the density of the normal distribution:

$$\mathrm{E}^Q\left(f(x(T_{i+1})) \mid \{x(T_i) = x_{i,j}\}\right) = \int_{-\infty}^{\infty} f(\xi) \, \phi(\xi - x_{i,j}, \sigma_i) \, \mathrm{d}\xi =: I \qquad (27.32)$$

with $\phi(\xi - \mu, \sigma_i) = \frac{1}{\sqrt{2\pi}\sigma_i} \exp(-\frac{(\xi-\mu)^2}{2\sigma_i^2})$. In the literature one can find many methods for numerical integration. Since f is given only at the state space sample points (lattice) $\{x_{i+1,k} | j = 1, \ldots, m_i\}$, but the density ϕ is given analytically, it is natural to make use of this.

27.4.1.1 Piecewise Constant Approximation

Let $x_{i+1,k+\frac{1}{2}} = \frac{x_{i+1,k+1}+x_{i+1,k}}{2}$ denote the center point of the interval $[x_{i+1,k}, x_{i+1,k+1}]$ and let $f_k := f(x_{i+1,k})$. Then an approximation of the integral I from (27.32) is given by

$$I \approx \sum_{k=1}^{k=n} f_k \left(\Phi_{x_{i,j}, \sigma_i}(x_{i+1,k+\frac{1}{2}}) - \Phi_{x_{i,j}, \sigma_i}(x_{i+1,k-\frac{1}{2}})\right),$$

where $\Phi_{\mu,\sigma}(x) = \int_{-\infty}^{x} \phi(\xi - \mu, \sigma) \, \mathrm{d}\xi$. For the cumulative normal distribution function $\Phi_{\mu,\sigma}$ there are very accurate approximations using rational functions.

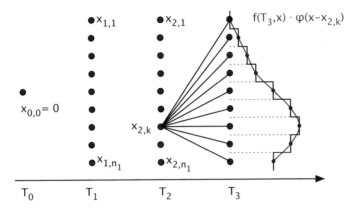

Figure 27.2. *Calculation of the conditional expectation (numerical integration) within a "lattice".*

27.4.1.2 Piecewise Polynomial Approximation

If we approximate the integrand f by a piecewise polynomial, then we have to calculate integrals of the form

$$\int_{h_1}^{h_2} \xi^m \, \phi(\xi - x_{i,j}, \sigma_i) \, d\xi.$$

These may be given recursively in terms of the cumulative distribution function Φ. We have

Lemma 235 (Convolution of a Monomial with the Normal Distribution Function[16]):

$$\int_{h_1}^{h_2} \xi^m \, \phi(\xi - \mu, \sigma) \, d\xi = \mu \int_{h_1}^{h_2} \xi^{m-1} \, \phi(\xi - \mu, \sigma) \, d\xi$$

$$+ (m - 1) \int_{h_1}^{h_2} \xi^{m-2} \, \phi(\xi - \mu, \sigma) \, d\xi$$

$$- \sigma^2 \left[\xi^{m-1} \, \phi(\xi - \mu, \sigma) \right]_{h_1}^{h_2}$$

---

[16] Compare [26].

405

Proof: For $m \in \mathbb{N}$

$$\frac{\partial}{\partial \xi} \xi^{m-1} \exp\left(-\frac{1}{2}\left(\frac{\xi - \mu}{\sigma}\right)^2\right)$$

$$= (m-1) \xi^{m-2} \exp\left(-\frac{1}{2}(\frac{\xi - \mu}{\sigma})^2\right) + \xi^{m-1} \exp\left(-\frac{1}{2}\left(\frac{\xi - \mu}{\sigma}\right)^2\right) \frac{-(\xi - \mu)}{\sigma^2}$$

$$= (m-1) \xi^{m-2} \exp\left(-\frac{1}{2}\left(\frac{\xi - \mu}{\sigma}\right)^2\right) - \frac{\xi^m}{\sigma^2} \exp\left(-\frac{1}{2}\left(\frac{\xi - \mu}{\sigma}\right)^2\right)$$

$$+ \frac{\mu}{\sigma^2} \xi^{m-1} \exp\left(-\frac{1}{2}\left(\frac{\xi - \mu}{\sigma}\right)^2\right),$$

i.e.

$$\xi^m \exp\left(-\frac{1}{2}\left(\frac{\xi - \mu}{\sigma}\right)^2\right)$$

$$= (m-1)\sigma^2 \xi^{m-2} \exp\left(-\frac{1}{2}\left(\frac{\xi - \mu}{\sigma}\right)^2\right) + \mu \xi^{m-1} \exp\left(-\frac{1}{2}\left(\frac{\xi - \mu}{\sigma}\right)^2\right)$$

$$- \frac{\partial}{\partial \xi} \sigma^2 \xi^{m-1} \exp\left(-\frac{1}{2}\left(\frac{\xi - \mu}{\sigma}\right)^2\right).$$

The claim follows after multiplication with $\frac{1}{\sqrt{2\pi}\sigma}$ and integration $\int_{h_1}^{h_2} d\xi$. $\quad\square|$

To calculate the interpolating polynomials many different methods may be used (e.g. *cubic spline interpolation*). The Neville algorithm gives a simple recursion by which we can calculate a piecewise polynomial interpolation function.

Lemma 236 (Neville Algorithm—Piecewise Polynomial Interpolation Function):
Let x_i denote given sample points and f_i the corresponding values. If $F_{i-k_1, i+k_2}$ is a polynomial of degree $k_1 + k_2$ interpolating the points (x_j, f_j), $j = i - k_1, \ldots i + k_2$, i.e.,

$$F_{i-k_1, i+k_2}(x_j) = f_j \quad \text{for } j = i - k_1, \ldots i + k_2,$$

then

$$F_{i-k_1, i+k_2+1}(\xi) = \frac{(\xi - x_{i-k_1})\, F_{i-k_1+1, i+k_2+1}(\xi) + (x_{i+k_2+1} - \xi)\, F_{i-k_1, i+k_2}}{x_{i+k_2+1} - x_{i-k_1}}$$

defines an interpolating polynomial of degree $k_1 + k_2 + 1$ for the points (x_j, f_j), $j = i - k_1, \ldots, i + k_2 + 1$. With the trivial interpolating polynomial $F_{i,i} \equiv f_i$ having degree 0, induction gives a construction of the desired interpolating polynomial.

Proof: A polynomial of degree $k_1 + k_2 + 1$ is uniquely determined by $k_1 + k_2 + 2$ points. The claim follows then by induction, evaluating the approximations polynomial

$F_{i-k_1,i+k_2+1}$ at the sample points x_{i+k_2+1}, x_{i-k_1} and the sample points which are common to $F_{i-k_1+1,i+k_2+1}$ and $F_{i-k_1,i+k_2}$. □|

The algorithm is called the *Neville algorithm*. The polynomial $F_{i-k_1,i+k_2}(x_j)$ will then be used as an interpolating polynomial on the interval x_i, x_{i+1}.[17] Using Lemma 235 the corresponding integral part is calculated.

The interpolation function is (for degree > 0) continuous, but is not differentiable at the interval bounds, as would be the case with a cubic spline interpolation.

27.4.2 State Space Discretization

The rule for choosing the state space discretization points $x_{i,j}$, $j = 1, \ldots, m_i$ (the lattice) for the realization of the Markovian driver $x(T_i)$ has a major impact on the accuracy of the model.

27.4.2.1 Equidistant Discretization

A simple rule gives an equidistant discretization of the interval $[x_{i,\min}, x_{i,\max}]$, where the interval has been chosen such that the neglected area $(-\infty, x_{i,\min}) \cup (x_{i,\max}, \infty)$ supports only a small probability measure $\mathbb{Q}(x(T_i) \notin [x_{i,\min}, x_{i,\max}]) = \epsilon$; see Figure 27.3. The bounds $x_{i,\min}$, $x_{i,\max}$ may be derived from the definition of the process x, e.g., in terms of the standard deviation of $x(T_i)$:

$$x_{i,\max} = -x_{i,\min} = k\,\bar{\sigma}_i, \qquad \bar{\sigma}_i := \left(\int_0^{T_i} \sigma^2(t)\,dt \right)^{1/2}.$$

The factor k has to be chosen sufficiently large (e.g. $k > 3$).

Instead of an equidistant discretization one may choose the interval $[x_{i,j}, x_{i,j+1}]$ such that all intervals support the same probability mass, i.e.,

$$\mathbb{Q}(x_{i,j} \leq x(T_i) \leq x_{i,j+1}) = \frac{1 - \epsilon}{m_i - 1}.$$

 Further Reading: The LIBOR and swap rate Markov functional model in terminal measure is discussed in [13, 26] as well as in the original article [79]. A LIBOR Markov functional model in spot measure is discussed in [71]. A hybrid cross currency Markov functional model is discussed in [63, 68, 71]. ◁|

---

[17] It is natural to choose the sample point symmetrically, i.e., use $k_2 = k_1 + 1$.

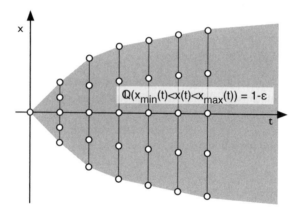

Figure 27.3. *State space discretization for the Markov functional model: A simple rule to choose the discretization points $x_{i,j}$.*

Part VI

Extended Models

CHAPTER 28

Credit Spreads

28.1 Introduction—Different Types of Spreads

First, we wish to clarify that there are essentially two different concepts of "spreads".

28.1.1 Spread on a Coupon

We have already encountered the term *spread* in the definition of some products. There, a spread was an additional fixed rate added to the floating rate to form the coupon, e.g., as in

$$C_i = L(T_i, T_{i+1}) + s_i,$$

where s_i is a constant—the spread; see Section 12.2. The spread is paid as part of the floating leg. It may be interpreted as a periodic fee. Usually the spread is chosen such that the deal is at par when issued, i.e., for (structured) swaps the spread is chosen such that the swap's value is 0, for (structured) bonds the spread is chosen such that the bond's value is 1, which would relate the spread to market parameters. Nevertheless, this type of a spread is a static feature of the product. It is determined when the product is issued and does not change over the lifetime of the product.

28.1.2 Credit Spread

The credit spread has a different origin than the spread on a coupon: It is a market parameter derived from market prices (like interest rates). Apart from the fact that it is derived (e.g., bootstrapped) from (specific) market products, it is product independent. To introduce the credit spread we will consider the defaultable zero-coupon bond in the following section.

411

However, in some situations it is possible to express a credit spread in terms of a spread on a coupon. Actually, for zero-coupon bonds, there is a transformation which is almost linear. Thus, the two notions of a spread may be related, which, unfortunately, may give rise to misunderstandings. It is important to understand that the original definition of a credit spread (as a market parameter) is different from a spread on a coupon (as a static product feature).

28.2 Defaultable Bonds

We started our consideration of interest rates with the definition of the zero-coupon bond. We viewed the zero-coupon bond as the fundamental (atomic) interest rate product and all other interest rate quantities (forward rate, short rate, swap rate, coupon bond, floater, etc.) could be expressed in terms of zero-coupon bonds. The continuum of zero-coupon bonds gives a complete representation of the interest rate curve; see Section 8.2.

The zero coupon bond $P(T; t)$ represents the value of a guaranteed payment of unit 1 in time T, seen in time t. However, in reality the guarantee of the payment of 1 may be limited. For example, if the issuer of the bond goes bankrupt, he does not repay the value or only a fraction of it. The financial term used for this case is that the issuer *defaults*. Let $P^d(T; t)$ denote the time t-value of a zero coupon bond that carries a default risk, paying 1 in T in case of nondefault. The default event may accure any time in (t, T).

Remark 237: Since each issuer (obligor) might carry his individual default risk, each issuer defines a unique continuum of defaultable bonds, e.g., $P^{d,US-GOV}$, $P^{d,IBM}$, etc. In the following, P^d denotes a defaultable bond of some fixed (yet unspecified) issuer. In reality all bonds are defaultable and the (nondefaultable) zero-coupon bond is an idealization.

A continuum of defaultable bonds $T \mapsto P^d(T; t)$ allows us to define all the associated rates, as we considered for the nondefaultable case in Section 8.2. This will give rise to the definition of the defaultable forward rate, the defaultable short rate, the defaultable instantaneous forward rate, etc.

Definition 238 (Defaultable Instantaneous Forward Rate): ⌐
For fixed t let $T \mapsto P^d(T; t)$ denote a family of defaultable bonds with maturities $T \geq t$. Assume that $T \mapsto P^d(T; t)$ is differentiable in T. We define

$$f^d(t, T) := -\frac{\partial}{\partial T} \log(P^d(T; t))$$

as the *defaultable instantaneous forward rate*. From the definition we have

$$P^{\mathrm{d}}(T;t) = \exp\left(-\int_t^T f^{\mathrm{d}}(t,\tau)\,\mathrm{d}\tau\right).$$

We are interested in the relation of the defaultable quantities to the nondefaultable quantities. We start by defining the default intensity. For an interpretation see below.

Definition 239 (Default Intensity, Credit Spread):
For fixed t let $T \mapsto P^{\mathrm{d}}(T;t)$ denote a family of defaultable bonds with maturities $T \geq t$. Assume that $T \mapsto P^{\mathrm{d}}(T;t)$ is differentiable in T. We define

$$\lambda(t,T) := f^{\mathrm{d}}(t,T) - f(t,T)$$

as the *default intensity* or *credit spread*. From the definition

$$P^{\mathrm{d}}(T;t) = P(T;t)\,\exp\left(-\int_t^T \lambda(t,\tau)\,\mathrm{d}\tau\right),$$

i.e.

$$\frac{P^{\mathrm{d}}(T;t)}{P(T;t)} = \exp\left(-\int_t^T \lambda(t,\tau)\,\mathrm{d}\tau\right).$$

Interpretation (Default Probability, Survival Probability, Intensity): Obviously the value of a bond which carries a default risk is less than or equal to the value of a (nondefaultable) zero-coupon bond, i.e., we have

$$P^{\mathrm{d}}(T;t) \leq P(T;t), \qquad \text{i.e.,} \qquad 0 \leq \frac{P^{\mathrm{d}}(T;t)}{P(T;t)} \leq 1.$$

The term $\frac{P^{\mathrm{d}}(T;t)}{P(T;t)}$ has an intuitive interpretation: It may be viewed as the probability of survival in (t,T). Consequently $\lambda(t,\cdot)$ is just the intensity of an exponential distribution. If $\exp\left(-\int_t^T \lambda(t,\tau)\,\mathrm{d}\tau\right)$ is the survival probability, then $1 - \exp\left(-\int_t^T \lambda(t,\tau)\,\mathrm{d}\tau\right)$ is the default probability, where—in this interpretation—default means that the bond pays 0. The parameter λ is called the *default intensity*. The default intensity λ has the (physical) unit $\frac{1}{\text{time}}$. It is the rate at which defaults occur. The inverse $\frac{1}{\lambda}$ has the (physical) unit time. It is the average survival time. Note that the expectation of a random variable having exponential distribution is $\frac{1}{\lambda}$.

413

There is one important aspect here which one may easily overlook. The survival probability

$$\frac{P^{\mathrm{d}}(T;t)}{P(T;t)} = \exp\left(-\int_t^T \lambda(t,\tau)\,\mathrm{d}\tau\right)$$

is not the *real* survival probability. It is the market implied probability from the price of a defaultable bond. It is the probability we should use as a pricing measure if we hedge default risk by trading in defaultable bonds. We will comment again on this point later. ◁|

28.3 Integrating Deterministic Credit Spread into a Pricing Model

A model which models one or more additional credit spread curves may become quite complex. Note that so far it is possible that $t \mapsto \lambda(\cdot, T)$ is a stochastic process. Because the credibility of an issuer varies uncertainly, it is natural to assume stochastic intensities.

However, if we make the simplifying assumption that the intensity is a deterministic function, then it is straightforward to endow all models discussed so far with the ability to handle defaultable quantities.

Assume a general (nondefaultable) pricing model with numéraire N and equivalent martingale measure $\mathbb{Q} = \mathbb{Q}^N$, modeled over the probability space $(\Omega, \mathcal{F})$. Then the value of a nondefaultable zero coupon bond is given by

$$P(T;t) = N(t)\,\mathrm{E}^{\mathbb{Q}}\left(\frac{1}{N(T)}\,\Big|\,\mathcal{F}_t\right),$$

and the time t-value of a payoff $V(T)$ paid in T is

$$P(T;t) = N(t)\,\mathrm{E}^{\mathbb{Q}}\left(\frac{V(T)}{N(T)}\,\Big|\,\mathcal{F}_t\right).$$

If $\mathbb{Q}$ is the pricing measure of a pure nondefaultable model (i.e., default events are not considered under $\mathbb{Q}$, i.e., falsely have probability zero), then we may simply correct for it by multiplying by the survival probability. The value of a defaultable bond is given by

$$P^{\mathrm{d}}(T;t) = N(t)\,\mathrm{E}^{\mathbb{Q}}\left(\frac{1}{N(T)}\exp\left(-\int_t^T \lambda(t,\tau)\,\mathrm{d}\tau\right)\Big|\,\mathcal{F}_t\right),$$

(check!) and the time t-value of a defaultable payoff $V^d(T)$ paid in T thus is

$$V^d(t) = N(t) \, \mathrm{E}^{\mathbb{Q}} \left(\frac{V^d(T)}{N(T)} \, \exp\left(- \int_t^T \lambda(t, \tau) \, \mathrm{d}\tau \right) \Big| \mathcal{F}_t \right).$$

Essentially we split the probability space into the nondefaultable part (pricing measure $\mathbb{Q}$) and the default part (survival probability $\exp\left(- \int_t^T \lambda(t, \tau) \, \mathrm{d}\tau \right)$). In a more rigorous treatment we would have introduced the product measure.

28.3.1 Deterministic Credit Spread

If the credit spread is nonstochastic, to be precise, an Itô process with zero volatility, then this implies that λ does not depend on t.

Lemma 240 (Nonstochastic Credit Spread): If

$$\mathrm{d}\left(\frac{P^d(T; t)}{P(T; t)} \right) = \mathrm{d}\left(\exp\left(- \int_t^T \lambda(t, \tau) \, \mathrm{d}\tau \right) \right) = \frac{\partial}{\partial t} \left(\exp\left(- \int_t^T \lambda(t, \tau) \, \mathrm{d}\tau \right) \right) \mathrm{d}t, \quad (28.1)$$

then λ does not depend on t, i.e.,

$$\lambda(t, T) = \lambda(T).$$

Proof: The proof starts with a useful little trick: By our assumption the stochastic process $t \mapsto \exp\left(- \int_t^T \lambda(t, \tau) \, \mathrm{d}\tau \right)$ has zero volatility. Thus, the theorem of Girsanov implies that a corresponding change of measure does not change the process—there is no change of drift since there is no diffusion. So considering (28.1) under the T-terminal measure $\mathbb{Q}^{P(T;t)}$ we immediately obtain

$$\frac{\partial}{\partial t} \left(\exp\left(- \int_t^T \lambda(t, \tau) \, \mathrm{d}\tau \right) \right) = 0.$$

From that, with

$$\frac{\partial}{\partial t} \left(\exp\left(- \int_t^T \lambda(t, \tau) \, \mathrm{d}\tau \right) \right) = \exp\left(- \int_t^T \lambda(t, \tau) \, \mathrm{d}\tau \right) \left(\lambda(t, t) - \int_t^T \frac{\partial}{\partial t} \lambda(t, \tau) \, \mathrm{d}\tau \right)$$

we find

$$\lambda(t, t) - \int_t^T \frac{\partial}{\partial t} \lambda(t, \tau) \, \mathrm{d}\tau = 0,$$

i.e.,

$$\lambda(t, t) = \int_t^T \frac{\partial}{\partial t} \lambda(t, \tau) \, \mathrm{d}\tau \quad \forall \, T.$$

The right-hand side does not depend on T. Consequently the left-hand side does not depend on T. Differentiating with respect to T, we find the integrand on the right-hand side to be zero:

$$\frac{\partial}{\partial t}\lambda(t,T) = 0 \quad \forall T.$$

In other words, λ does not depend on t.   □|

 Interpretation (The Default Event—Recovery Rates): There is a somewhat surprising aspect of our consideration of the default risk so far: We did not specify the nature of the credit event itself. Usually, when a (risky) bond defaults, then it pays some fraction of the original notional called the recovery rate. So far we did not consider if and how much recovery a defaultable assets pays. We did not even consider the nature of the credit event itself. Is it a total default at a single point in time or a gradual process?

If we only want to evaluate payments made by counterparties with credit risk, i.e., where the corresponding zero-coupon bond has a value different from the nonrisky zero-coupon bond, then these aspects are irrelevant. The credit event itself (e.g., the recovery rate) has to be modeled only if a financial product relies on the nature of the credit event (e.g., a guaranteed compensation of the loss (as for a *Credit Default Obligation (CDO)*)).

In this sense, we once again stress that the intensity λ is not the intensity of the true default probability (whatever the true default is). It is the equivalent intensity under the assumption of defaults (modeled as poisson events) with zero recovery, implied (!) by market prices. Like the equivalent martingale measure, it is a calculational tool rather than a real quantity.   ◁|

28.3.2 Implementation

Implementing deterministic credit spread into a given pricing model turns out to be a minor modification of the discounting. Given two curves of zero-coupon bond prices:

$$T \mapsto P(T;0), \qquad\qquad \text{nondefaultable zero-coupon bonds,}$$

$$T \mapsto P^{\text{d}}(T;0), \qquad\qquad \text{defaultable zero-coupon bonds,}$$

calculate the credit spread curve

$$s(T) := -\frac{1}{T}\log\left(\frac{P^{\text{d}}(T;0)}{P(T;0)}\right).$$

From this, calculate the *forward credit spread* as

$$s(T_1,T_2) := \frac{1}{T_2 - T_1}(s(T_2)\,T_2 - s(T_1)\,T_1)$$

for $T_1 < T_2$. A defaultable cash flow to be received in T_2, evaluated as seen in $T_1 < T_2$ is then adjusted by the factor

$$\exp\left(-s(T_1, T_2)\,(T_2 - T_1)\right).$$

In other words, while the time t relative value of nondefaultable cash flow X_i to be made in T_i is given by

$$E^{Q^N}\left(\frac{X_i}{N(T_i)} \mid \mathcal{F}_t\right),$$

the time t relative value of a defaultable cash flow X_i to be made in T_i is

$$E^{Q^N}\left(\frac{X_i}{N(T_i)}\,\exp\left(-s(t, T_i)\,(T_i - t)\right) \mid \mathcal{F}_t\right).$$

This can be implemented by replacing $N(T_i)$ by $N(T_i)\,\exp\left(+s(T_i)\,T_i\right)$, i.e.,

$$N(T_i) \quad \text{is replaced by} \quad N(T_i)\,\frac{P^d(T_i; 0)}{P(T_i; 0)}. \tag{28.2}$$

in the application of the universal pricing theorem (i.e., discounting) for all defaultable cash flows.

 Interpretation (Seeing (28.2) **as an Exchange Rate or as an Interest Rate):** The modification in (28.2) is similar to an exchange rate. For example, if one needs to price foreign cash flows using a model with a domestic numéraire, then one does so by exchanging the foreign cash flow into domestic currency using the exchange rate at payment date T_i, then discounting in the domestic model and converting back to foreign currency with the inverse of the exchange rate.

The same is happening here. At payment date a defaultable cash flow is being interpreted as nondefaultable. The corresponding conversion factor is 1, since at payment date the amount of the cash flow is indeed the same. Then the cash flow is discounted in the nondefaultable model and converted back to a defaultable quantity. The exchange rate is $\frac{P^d(T_i; t)}{P(T_i; t)}$, the value of a defaultable time T_i cash flow in terms of a nondefaultable time T_i cash flow.

This "exchange rate" depends on the maturity of the cash flow. So actually it is more similar to an interest rate and the interpretation of $\lambda(t)$ as an (additional) interest rate is even more striking: Assume that we are under the risk-neutral measure, i.e. the numéraire is $N(t) := \exp(\int_0^t r(\tau)d\tau)$. Then (28.2) will replace $N(t)$ by $\exp(\int_0^t r(\tau) + \lambda\tau d\tau)$. ◁

28.4 Receiver's and Payer's Credit Spreads

All future cash flows of a (nonforward starting) bond have the same direction: The holder of the bonds receives coupons until maturity and notional upon maturity. If the bond is forward starting, then the holder pays notional to the issuer upon start, in which case there is also a future cash flow from the holder to the issuer. For a swap the direction of the cash flow depends on the fixing and may vary for each period. If the cash flow is positive, the holder receives the cash flow from the issuer. If the cash flow is negative the holder pays its absolute value to the issuer.

Obviously, receiver and payer may have different credit spreads, i.e., different default risks. Let λ^r denote the receiver's credit spread and λ^p the payer's credit spread. Let X denote an $\mathcal{F}_T$-measurable random variable modeling a defaultable cash flow in T. If $\mathbb{Q}^N$ denotes the nondefaultable pricing measure corresponding to the nondefaultable numéraire N, then the time t-value of the defaultable time T-cash flow X is

$$N(t)\, \mathrm{E}^{\mathbb{Q}^N}\left(\frac{\max(X,0)}{N(T)}\, e^{-\int_t^T \lambda^r(\tau)\, d\tau} + \frac{\min(X,0)}{N(T)}\, e^{-\int_t^T \lambda^p(\tau)\, d\tau} \,\middle|\, \mathcal{F}_t \right).$$

For a financial product consisting of multiple (potential) cash flow we have to calculate the net exposure consisting of the present cash flow and the value of the future product. Let X_i denote an $\mathcal{F}_{T_i}$-measurable random variable modeling a defaultable cash flow in T_i. Then $V_{i+1}^d(T_i)$ denotes the time T_i value of the defaultable cash flow X_j for $j \geq i+1$. Then we have the backward induction

$$
\begin{aligned}
\frac{V_i^d(T_{i-1})}{N(T_{i-1})} = \mathrm{E}^{\mathbb{Q}^N}\Bigg(& \frac{\max(X_i + V_{i+1}^d(T_i), 0)}{N(T_i)}\, \exp\left(-\int_{T_{i-1}}^{T_i} \lambda^r(\tau)\, d\tau\right) \\
& + \frac{\min(X_i + V_{i+1}^d(T_i), 0)}{N(T_i)}\, \exp\left(-\int_{T_{i-1}}^{T_i} \lambda^p(\tau)\, d\tau\right) \,\bigg|\, \mathcal{F}_{T_{i-1}} \Bigg).
\end{aligned}
\tag{28.3}
$$

 Interpretation: For products with multiple cash flow (i.e., periods) with possibly different signs, the pricing is thus given as a backward algorithm like that for the pricing of Bermudan options. Using two different spread curves on a swap immediately results in a Bermudan pricing algorithm. ◁

Remark 241 (Netting): There are usually netting agreements between counterparties such that upon default the outstanding debt is calculated across a portfolio and only the net debt is under default. When this is done, the pricing Equation (28.3) needs to be considered on a portfolio level, which requires that all products are evaluated using the same model and within the same simulation.

28.4.1 Example: Defaultable Forward Starting Coupon Bond

Recall that the value of a forward starting coupon bond with an initial notional payment 1 in T_1 is

$$V_{\text{fwdCpnBnd}}(t) = \sum_{i=1}^{n-1} C_i \, (T_{i+1} - T_i) \, P(T_{i+1}; t) + P(T_n; t) - P(T_1; t),$$

which is identical to a swap

$$= \sum_{i=1}^{n-1} (C_i - L(T_i, T_{i+1}; t)) \, (T_{i+1} - T_i) \, P(T_{i+1}; t);$$

see Section 9.2.1.1.

If we consider defaults, then the time T_1 value of the cash flows received after T_1 is

$$V_{\text{CpnBnd}}^{\text{d}}(T_1) = \sum_{i=1}^{n-1} C_i \, (T_{i+1} - T_i) \, P(T_{i+1}; t) \, \exp\left(-\int_{T_1}^{T_{i+1}} \lambda^{\text{p}}(\tau) \, d\tau\right)$$

$$+ P(T_n; t) \, \exp\left(-\int_{T_1}^{T_N} \lambda^{\text{p}}(\tau) \, d\tau\right).$$

Together with the initial cash flow at time T_1, the time t-value of the defaultable forward starting coupon bond is

$$V_{\text{fwdCpnBnd}}^{\text{d}}(t) = (V_{\text{CpnBnd}}^{\text{d}}(T_1)-1)P(T_1; t) \begin{cases} \exp\left(-\int_t^{T_1} \lambda^{\text{p}}(\tau) \, d\tau\right) & \text{if } V_{\text{CpnBnd}}^{\text{d}}(T_1) > 1, \\ \exp\left(-\int_t^{T_1} \lambda^{\text{r}}(\tau) \, d\tau\right) & \text{else.} \end{cases}$$

For $V_{\text{fwdCpnBnd}}^{\text{d}}(T_1)$ we have

$$V_{\text{fwdCpnBnd}}^{\text{d}}(T_1) = \sum_{i=1}^{n-1} C_i \, (T_{i+1} - T_i) \, P^{\text{d,p}}(T_{i+1}; T_1) + P^{\text{d,p}}(T_n; t) - 1,$$

which is identical to

$$= \sum_{i=1}^{n-1} (C_i - L^{\text{d,p}}(T_i, T_{i+1}; T_1)) \, (T_{i+1} - T_i) \, P^{\text{d,p}}(T_{i+1}; T_1), \quad (28.4)$$

where

$$P^{\mathrm{d,p}}(T_i; T_1) = P(T_i; T_1) \exp(-\int_{T_1}^{T_i} \lambda^{\mathrm{p}}(\tau)\,\mathrm{d}\tau) \quad \text{and}$$

$$L^{\mathrm{d,p}}(T_i, T_{i+1}; T_1)\,(T_{i+1} - T_i) = \frac{P^{\mathrm{d,p}}(T_i; T_1)}{P^{\mathrm{d,p}}(T_{i+1}; T_1)} - 1.$$

Note that (28.4) is a swap where both counterparties share the *same* default probabilities.

28.4.2 Example: Option on a Defaultable Coupon Bond

Consider an option on the defaultable coupon bond with exercise date T_1. Since the option is exercised if and only if $V^{\mathrm{d}}_{\mathrm{CpnBnd}}(T_1) - 1 > 0$ its value is

$$V^{\mathrm{d}}_{\mathrm{CpnBndOpt}}(t) = N(t)\,\mathrm{E}^{\mathbb{Q}^N}\left(\frac{\max(V^{\mathrm{d}}_{\mathrm{CpnBnd}}(T_1) - 1, 0)}{N(T_1)} \exp\left(-\int_{t}^{T_1} \lambda^{\mathrm{p}}(\tau)\,\mathrm{d}\tau\right)\right).$$

Consequently, the value of an option on a defaultable coupon bond corresponds to $\exp\left(-\int_{t}^{T_1} \lambda^{\mathrm{p}}(\tau)\,\mathrm{d}\tau\right)$ times the value of a (nondefaultable) option on the defaultable swap. Note that due to the optionality λ^{r} does not enter the valuation.

It appear as if this allows to derive an adjusted Black formula using only implied volatilities of the (nondefaultable) swap rates. However, the implied volatilities refer to swaptions paying a constant coupon $C_i = C$ in each period. If these coupons are weighted by the survival probability, then, effectively, we have an option on a weighted sum of different swap rates.[1] Thus, the pricing of an option on a defaultable swap (or bond) requires additional information on the correlation of the swap rates.

 Further Reading: The setup of two spread curves is identical to considering a market where the interest rate for borrowing is different from the interest rate for lending. For a thorough treatment of the underlying theory see [12]. ◁|

---

[1] See Exercise 11.

CHAPTER 29

Hybrid Models

In this chapter we introduce several kinds of hybrid models. A hybrid model is a model that models multiple (different) assets in a single unified model. In general one combines several well-known models into a single unified model. Since different models usually come under different pricing measures, the essence of a hybrid model is the question "How do these models look under a common pricing measure?". So apart from the prerequisite that the models be compatible, the construction of a hybrid model is just a change of measure.

In this sense, we have already encountered the basic technique required for a hybrid model in the discussion of the LIBOR market model: There we wrote down multiple individual Black models (Chapter 10) and asked ourself how they look under a common pricing measure. The result was the LIBOR market model.

29.1 Cross-Currency LIBOR Market Model

For two currencies, "domestic" and "foreign", we model the interest rate curves, each with a *LIBOR market model*, as was discussed for an interest rate curve in a single currency in Section 19.

In addition we model the foreign exchange rate $FX(t)$. In Chapter 11 we presented the pricing of a *quanto caplet* by modeling the *FX forward* as a lognormal process. Here the spot exchange rate $FX(t)$ is modeled directly, also as a lognormal process. Let $FX(t)$ denote the amount (in domestic currency) that has to be paid by a *domestic* investor at time t for one unit of foreign currency (for). Thus, $FX(t)$ has the (physical) unit $[FX(t)] = \frac{1 \text{ dom}}{1 \text{ for}}$.

We assume that for the chosen numéraire N that there exists a corresponding equivalent martingale measure $\mathbb{Q}^N$. By the change of measure theorem (Girsanov theorem, 59) the modeled quantities are again lognormal processes under $\mathbb{Q}^N$, i.e.,

$$dL_i(t) = L_i(t)\,\mu_i(t)\,dt \;+\; L_i(t)\sigma_i(t)\,dW_i^{\mathbb{Q}^N}(t), \qquad (0 \le i \le n-1)$$

$$dFX(t) = FX(t)\,\mu^{FX}(t)\,dt \;+\; FX(t)\sigma^{FX}(t)\,dW_{FX}^{\mathbb{Q}^N}(t)$$

$$d\tilde{L}_i(t) = \tilde{L}_i(t)\,\tilde{\mu}_i(t)\,dt \;+\; \tilde{L}_i(t)\tilde{\sigma}_i(t)\,d\tilde{W}_i^{\mathbb{Q}^N}(t) \qquad (0 \le i \le n-1),$$

with initial conditions

$$L_i(0) = L_{i,0}, \qquad FX(0) = FX_0, \qquad \tilde{L}_i(0) = \tilde{L}_{i,0}.$$

As before, this is the starting point for

- Determination of the drift terms μ_i, μ^{FX}, and $\tilde{\mu}_i$ for a chosen numéraire $N(t)$ using the $\mathbb{Q}^N$-martingale property of N-relative prices.

- Determination of the initial conditions $L_{i,0}$, FX_0, $\tilde{L}_{i,0}$ using the bond and foreign bond prices observed at time $t = 0$.

- Determination/choice of the volatility and correlation to reproduce given option prices.

29.1.1 Derivation of the Drift Term under Spot Measure

As numéraire we chose the *rolled over one period bond* we had already made use of in Section 19.1.2:

$$N(t) := P(T_{m(t)+1}; t) \prod_{j=0}^{m(t)} (1 + L_j(T_j)\,\delta_j),$$

where $m(t) := \max\{i \,:\, T_i \le t\}$, $\delta_j := T_{j+1} - T_j$. We will now derive the corresponding processes (i.e., the drifts) under the corresponding equivalent martingale measure $\mathbb{Q}^N$, the *spot measure*.

29.1.1.1 Dynamic of the *Domestic* LIBOR under Spot Measure

For the drift μ_i we have, exactly as in Section 19.1.2

$$\mu_i(t) = \sum_{l=m(t)+1}^{i} \frac{\delta_l L_l(t)}{1 + \delta_l L_l(t)}\,\sigma_i(t)\sigma_l(t)\rho_{i,l}(t).$$

This is the already known LIBOR market model in *domestic* currency.

29.1.1.2 Dynamic of the *Foreign* LIBOR under Spot Measure

We derive the drift $\tilde{\mu}_i(t)$ of the foreign LIBOR by considering financial product from the foreign market. The foreign bond $\tilde{P}(T_{i+1})$, converted to domestic currency, i.e. $\tilde{P}(T_{i+1})\,FX$ is a traded asset for the domestic investor. Thus, $\frac{\tilde{P}(T_{i+1};t)\,FX(t)}{N(t)}$ is a $\mathbb{Q}^N$-martingale:[1]

$$\operatorname*{Drift}_{\mathbb{Q}^N}\left(\frac{\tilde{P}(T_{i+1};t)\,FX(t)}{N(t)}\right) = 0. \tag{29.1}$$

Likewise

$$\tilde{L}_i\,\tilde{P}(T_{i+1})\,FX = \frac{\tilde{P}(T_i) - \tilde{P}(T_{i+1})}{T_{i+1} - T_i}\,FX \tag{29.2}$$

is a traded asset for the domestic investor, because it is a portfolio of (foreign) bonds converted to domestic currency. Thus,

$$\operatorname*{Drift}_{\mathbb{Q}^N}\left(\tilde{L}_i\,\frac{\tilde{P}(T_{i+1};t)\,FX(t)}{N(t)}\right) = 0.$$

From the product rule we find

$$
\begin{aligned}
& d\left(\tilde{L}_i(t)\,\frac{\tilde{P}(T_{i+1};t)\,FX(t)}{N(t)}\right) \\
&= \frac{\tilde{P}(T_{i+1};t)\,FX(t)}{N(t)}\,d\tilde{L}_i + \tilde{L}_i\,d\left(\frac{\tilde{P}(T_{i+1};t)\,FX(t)}{N(t)}\right) \\
&\quad + d\tilde{L}_i\,d\left(\frac{\tilde{P}(T_{i+1};t)\,FX(t)}{N(t)}\right)
\end{aligned}
\tag{29.3}
$$

from which we derive $\tilde{\mu}_i$ after calculation of $d\frac{\tilde{P}(T_{i+1};t)\,FX(t)}{N(t)}$ by comparing drift terms. We have

$$d\left(\frac{\tilde{P}(T_{i+1};t)\,FX(t)}{N(t)}\right) = d\left(\frac{\tilde{P}(T_{m(t)+1};t)}{P(T_{m(t)+1};t)}\,\frac{\prod_{j=m(t)+1}^{i}\left(1 + \tilde{L}_j(t)\delta_j\right)^{-1}}{\prod_{j=0}^{m(t)}\left(1 + L_j(T_j)\delta_j\right)}\,FX(t)\right).$$

Remark 242 (Interpolation of Bond Prices): At this point we encounter an interesting difference to the single currency LIBOR market model (see Section 19): While for the case of the domestic currency the corresponding term $\frac{P(T_{m(t)+1};t)}{P(T_{m(t)+1};t)}$ vanishes, it is necessary to provide an interpretation of $\frac{\tilde{P}(T_{m(t)+1};t)}{P(T_{m(t)+1};t)}$. See also Section 19.2.3.

---

[1] In the view of the domestic investor the foreign bond is not a traded asset. Only after conversion to the domestic currency by the applicable conversion rate $FX(t)$ does the product become tradable for the domestic investor. This is also apparent from the fact that relative prices are dimensionless: While $\frac{\tilde{P}(T_{i+1};t)\,FX(t)}{N(t)}$ is dimensionless, we find that $\frac{\tilde{P}(T_{i+1};t)}{N(t)}$ has the (physical) unit $\frac{1\text{ for}}{1\text{ dom}}$.

We have not yet defined the value of the short period bonds $P(T_{i+1};t)$ and $\tilde{P}(T_{i+1};t)$ for $t \neq T_j$, $j = 0, 1, 2, \ldots$. We do this now and define for $T_j < t \leq T_{j+1}$:

$$P(T_{m(t)+1};t) := \left(1 + L_{m(t)}(T_{m(t)}) \, (T_{m(t)+1} - t)\right)^{-1},$$

$$\tilde{P}(T_{m(t)+1};t) := \left(1 + \tilde{L}_{m(t)}(T_{m(t)}) \, (T_{m(t)+1} - t)\right)^{-1}.$$

This concludes the definition of the numéraire.

From this we find

$$d\frac{\tilde{P}(T_{m(t)+1};t)}{P(T_{m(t)+1};t)} = \frac{\partial}{\partial t}\left(\frac{1 + L_{m(t)}(T_{m(t)}) \, (T_{m(t)+1} - t)}{1 + \tilde{L}_{m(t)}(T_{m(t)}) \, (T_{m(t)+1} - t)}\right) dt,$$

i.e., the term $\frac{\tilde{P}(T_{i+1};t)}{P(T_{i+1};t)}$ has no diffusion part dW for $T_j < t \leq T_{j+1}$. This is sufficient for the following derivation: The specific form of the drift does not need to be known.

Indeed, it would be sufficient to require that $P(T_{i+1};t)$ and $\tilde{P}(T_{i+1};t)$ have zero volatility, i.e., no diffusion part, in the short period $t \in (T_i, T_{i+1}]$. No specific interpolation is required. The zero volatility assumption for $P(T_{i+1};t)$ and $\tilde{P}(T_{i+1};t)$ on $t \in (T_i, T_{i+1}]$ closes the definition of the LIBOR market model for all $t \in \{T_0, T_1, \ldots\}$.[2]

Continuing with the derivation of the drift, we have

$$d\left(\frac{\tilde{P}(T_{i+1};t)\,FX(t)}{N(t)}\right)$$

$$= d\left(\frac{\tilde{P}(T_{i+1};t)}{P(T_{i+1};t)} \frac{\prod_{j=}^{i}(1 + \tilde{L}_j(t)\delta_j)^{-1}}{\prod_{j=0}^{-1}(1 + L_j(T_j)\delta_j)} FX(t)\right)$$

$$= (\ldots)\,dt - \frac{\tilde{P}(T_{i+1};t)\,FX(t)}{N(t)} \sum_{j=m(t)+1}^{i} \frac{\delta_j\tilde{L}_j}{1 + \delta_j\tilde{L}_j}\tilde{\sigma}_j\,d\tilde{W}_j^{Q^N}(t)$$

$$+ \frac{\tilde{P}(T_{i+1};t)\,FX(t)}{N(t)} \sigma^{FX}(t)\,dW_{FX}^{Q^N}(t).$$

If we plug this into (29.3) and compare drift terms (the coefficients of dt) we get, together with (29.1) and (29.2),

$$0 = \frac{\tilde{P}(T_{i+1};t)\,FX(t)}{N(t)} \tilde{L}_i(t)\,\tilde{\mu}_i(t) + 0$$

$$+ \frac{\tilde{P}(T_{i+1};t)\,FX(t)}{N(t)} \tilde{L}_i(t)\left(-\sum_{j=m(t)+1}^{i} \frac{\delta_j\tilde{L}_j}{1 + \delta_j\tilde{L}_j}\tilde{\sigma}_i\tilde{\sigma}_j\,d\tilde{W}_i^{Q^N}(t)\,d\tilde{W}_j^{Q^N}(t)\right.$$

$$\left. + \tilde{\sigma}_i\sigma^{FX}(t)\,d\tilde{W}_j^{Q^N}(t)\,dW_{FX}^{Q^N}(t)\right).$$

---

[2] See [24].

Denoting the interest rate correlation within the foreign currency by $\tilde{\rho}_{i,j}$, i.e.,

$$d\tilde{W}_i^{Q^N}(t)\,d\tilde{W}_j^{Q^N}(t) \;=\; \tilde{\rho}_{i,j}(t)\,dt$$

and denoting the correlation of the foreign currency and the foreign exchange rate by $\rho_{\tilde{i},FX}$, i.e.,

$$d\tilde{W}_i^{Q^N}(t)\,dW_{FX}^{Q^N}(t) \;=\; \rho_{\tilde{i},FX}(t)dt,$$

gives

$$\tilde{\mu}_i(t) \;=\; \sum_{j=m(t)+1}^{i} \frac{\delta_j\tilde{L}_j(t)}{(1+\delta_j\tilde{L}_j)(t)}\tilde{\sigma}_i(t)\tilde{\sigma}_j(t)\tilde{\rho}_{i,j}(t) \;-\; \tilde{\sigma}_i(t)\sigma^{FX}(t)\rho_{\tilde{i},FX}.$$

29.1.1.3 Dynamic of the FX Rate under Spot Measure

For a given $t \in [T_{m(t)}, T_{m(t)+1})$ consider the foreign bond with maturity $T_{m(t)+1}$—the next possible maturity in our tenor discretization. Converting it to domestic currency its N-relative value $\frac{FX(t)\,\tilde{P}(T_{m(t)+1};t)}{N(t)}$ is a martingale. Furthermore we have

$$d\left(FX(t)\,\frac{\tilde{P}(T_{m(t)+1};t)}{N(t)}\right)$$

$$\stackrel{\text{Def. } N}{=} \prod_{j=0}^{m(t)}(1+L_j(T_j)\delta_j)\,d\left(FX(t)\,\frac{\tilde{P}(T_{m(t)+1};t)}{P(T_{m(t)+1};t)}\right)$$

$$=\; dFX(t)\,\frac{\tilde{P}(T_{m(t)+1};t)}{N(t)}$$

$$+\; FX(t)\prod_{j=0}^{m(t)}(1+L_j(T_j)\delta_j)\,\frac{\partial}{\partial t}\left(\frac{\tilde{P}(T_{m(t)+1};t)}{P(T_{m(t)+1};t)}\right)\,dt$$

$$\stackrel{\text{Def. } N}{=} dFX(t)\,\frac{\tilde{P}(T_{m(t)+1};t)}{N(t)} + FX(t)\,\frac{\tilde{P}(T_{m(t)+1};t)}{N(t)}\,\frac{\frac{\partial}{\partial t}\left(\frac{\tilde{P}(T_{m(t)+1};t)}{P(T_{m(t)+1};t)}\right)}{\frac{\tilde{P}(T_{m(t)+1};t)}{P(T_{m(t)+1};t)}}\,dt$$

and thus

$$\mu^{FX}(t) \;=\; -\frac{\frac{\partial}{\partial t}\left(\frac{\tilde{P}(T_{m(t)+1};t)}{P(T_{m(t)+1};t)}\right)}{\frac{\tilde{P}(T_{m(t)+1};t)}{P(T_{m(t)+1};t)}} \;=\; -\frac{\partial}{\partial t}\log\left(\frac{\tilde{P}(T_{m(t)+1};t)}{P(T_{m(t)+1};t)}\right)$$

$$=\; \frac{\partial}{\partial t}\log\left(\frac{P(T_{m(t)+1};t)}{\tilde{P}(T_{m(t)+1};t)}\right).$$

The drift $\mu^{FX}(t)$ thus depends on the chosen interpolation of the bond prices; see Remark 242. However, for its use in an implementation via a time discretization scheme it is not necessary to calculate the corresponding derivative after t, since only the integrated drift enters the time discretization scheme. For the integral $\int_{T_i}^{T_{i+1}} \mu^{FX}(t) \, dt$ we have

$$
\int_{T_i}^{T_{i+1}} \mu^{FX}(t) \, dt = \log\left(\frac{P(T_{i+1}; T_{i+1})}{\tilde{P}(T_{i+1}; T_{i+1})}\right) - \log\left(\frac{P(T_{i+1}; T_i)}{\tilde{P}(T_{i+1}; T_i)}\right) = -\log\left(\frac{P(T_{i+1}; T_i)}{\tilde{P}(T_{i+1}; T_i)}\right)
$$

$$
= \log\left(\frac{1 + \tilde{L}(T_i, T_{i+1}; T_i)}{1 + L(T_i, T_{i+1}; T_i)}\right),
$$

i.e., it is independent of the interpolation of the bond prices. See also Section 29.3.3.

29.1.2 Implementation

We will discuss the implementation together with the *equity hybrid LIBOR market model* in Section 29.3.3.

29.2 Equity Hybrid LIBOR Market Model

In Chapter 4 we introduced the Black-Scholes model for a (single) stock. The stock S was modeled as a lognormal process:

$$
dS(t) = \mu^{S,\mathbb{P}}(t)S(t) \, dt + \sigma^S(t)S(t) \, dW^{S,\mathbb{P}}(t) \qquad \text{under the real measure } \mathbb{P}.
$$

We had assumed interest rates to be nonstochastic and constant and the chosen numéraire was then $B(t) := \exp(r\,t)$. Under the corresponding martingale measure $\mathbb{Q}^B$ we could then derive an analytic formula for the price of a European stock option.

The pricing of products exhibiting optionalities related to both stock and interest rates requires joint stochastic modeling of the stock and the interest rates. If we chose as the interest rate model the LIBOR market model and as the stock process model the Black-Scholes model, then the construction of the joint model is simply the derivation of the drifts under a common measure.

29.2.1 Derivation of the Drift Term under Spot-Measure

As before, we choose as numéraire the *rolled over one period bond*:

$$
N(t) := P(T_{m(t)+1}; t) \prod_{j=0}^{m(t)} (1 + L_j(T_j)\, \delta_j),
$$

where $m(t) := \max\{i : T_i \le t\}$ and $\delta_j := T_{j+1} - T_j$.

29.2.1.1 Dynamic of the Stock Process under Spot Measure

The stock S is a traded asset, thus $\frac{S}{N}$ is a martingale under the equivalent martingale measure $\mathbb{Q}^N$. From the product rule we have

$$d\left(\frac{S}{N}\right) = dS \, \frac{1}{N} + S \, d\left(\frac{1}{N}\right) + dS \, d\left(\frac{1}{N}\right).$$

For $T_i < t < T_{i+1}$ let the bond price $P(T_{i+1}; t)$ defined (interpolated) as in Remark 242, i.e.,

$$P(T_{i+1}; t) := (1 + L_i(T_i) \, (T_{i+1} - t))^{-1} \quad \text{for } T_i < t < T_{i+1}.$$

In this case

$$
\begin{aligned}
d\left(\frac{1}{N}\right) &= \left(\frac{1}{P(T_{m(t)+1}; t)} \cdot \prod_{j=0}^{m(t)} (1 + L_j(T_j)\delta_j)\right) \\
&= d\left(\frac{1}{P(T_{m(t)+1}; t)} \prod_{j=0}^{m(t)} (1 + L_j(T_j)\delta_j)\right) \\
&= d\left((1 + L_{m(t)}(T_{m(t)}) \, (T_{m(t)+1} - t)) \cdot \prod_{j=0}^{m(t)} (1 + L_j(T_j)\delta_j)\right) \\
&= -L_{m(t)}(T_{m(t)}) \, dt \prod_{j=0}^{m(t)} (1 + L_j(T_j)\delta_j) \\
&= -\frac{L_{m(t)}(T_{m(t)})}{1 + L_{m(t)}(T_{m(t)}) \cdot (T_{m(t)+1} - t)} \, dt \, \frac{1}{N} \\
&= \frac{\partial}{\partial t} \log\left(1 + L_{m(t)}(T_{m(t)}) \cdot (T_{m(t)+1} - t)\right) \, dt \, \frac{1}{N}.
\end{aligned}
$$

Remark 243 (On the Numéraire Process $\frac{1}{N}$): Under the assumption that $N(t)$ does not have a diffusion (i.e., dW) term, which is the case for the above definition of the short period bond, then $d\left(\frac{1}{N}\right) = \left(\frac{\partial}{\partial t}\frac{1}{N}\right) dt$. Nothing else has been calculated above. See also Remark 242.

Assuming that N does not have a diffusion (i.e., dW) term, $dS \, d\left(\frac{1}{N}\right) = 0$ and from $\text{Drift}_{\mathbb{Q}^N}\left(\frac{S}{N}\right) = 0$ we find

$$0 = \mu^S \frac{S}{N} + \left[\frac{\partial}{\partial t} \log\left(1 + L_{m(t)}(T_{m(t)}) \, (T_{m(t)+1} - t)\right)\right] \frac{S}{N} + 0,$$

thus

$$\mu^S(t) = -\frac{\partial}{\partial t} \log\left(1 + L_{m(t)}(T_{m(t)}) (T_{m(t)+1} - t)\right)$$

$$= \frac{L_{m(t)}(T_{m(t)})}{1 + L_{m(t)}(T_{m(t)}) (T_{m(t)+1} - t)}. \qquad (29.4)$$

 Interpretation (Comparison to the Dynamic of the Black-Scholes Model): Using the numéraire $B(t) := \exp(r \cdot t)$ used in the Black-Scholes model (see Chapter 4) we had derived the drift of the stock process under the martingale measure $\mathbb{Q}^B$ as $\mu^{S,\mathbb{Q}^B} = r$. Equation (29.4) is simply a discrete (and stochastic) version of this drift: For the average drift over one period $[T_i, T_{i+1}]$

$$\frac{1}{T_{i+1} - T_i} \int_{T_i}^{T_{i+1}} \mu^{S,\mathbb{Q}^N}(t)\, dt = \log(1) - \log(1 + L_i(T_i)) (T_{i+1} - T_i))$$

$$= -\frac{\log(1 + L_i(T_i)) (T_{i+1} - T_i))}{T_{i+1} - T_i}$$

and for infinitesimal period lengths $T_{i+1} \to T_i$ we find—see Definition 103—that

$$\frac{-\log(1 + L_i(T_i)) (T_{i+1} - T_i))}{T_{i+1} - T_i} \xrightarrow[T_{i+1} \to T_i]{} r(T_i) = \mu^{S,\mathbb{Q}^B},$$

i.e.,

$$\mu^{S,\mathbb{Q}^N}(t) \xrightarrow[T_{i+1} \to T_i]{} \mu^{S,\mathbb{Q}^B}.$$

◁|

29.2.2 Implementation

We discuss the implementation together with the previously discussed *cross-currency LIBOR market model* in Section 29.3.3.

29.3 Equity Hybrid Cross-Currency LIBOR Market Model

The models given in Sections 29.1 and 29.2 may be combined. This is now trivial since the numéraire and thus the martingale measure are the same in both models. Thus we have a unified model for interest rates, foreign exchange rates, foreign interest rates, and equity. We will now add the model of a foreign stock. Let $\tilde{S}$ denote another stock process, modeling a stock from the foreign market, i.e., the process $\tilde{S}$ has the dimension (currency unit) 1for.

29.3.1 Dynamic of the Foreign Stock under Spot Measure

We assume that the foreign stock $\tilde{S}$ follows a lognormal process

$$d\tilde{S}(t) = \mu^{\tilde{S},\mathbb{P}}(t)\tilde{S}(t)\,dt + \sigma^{\tilde{S}}(t)\tilde{S}(t)\,dW^{\tilde{S},\mathbb{P}}(t) \qquad \text{under the real measure } \mathbb{P}.$$

As in Sections 29.1 and 29.2, we chose as numéraire $N(t) := P(T_{m(t)+1};t)\prod_{j=0}^{m(t)}(1 + L_j(T_j)\cdot\delta_j)$. As for the cross-currency LIBOR market model, the foreign stock $\tilde{S}$ has to be converted to domestic currency to be a traded asset for the domestic investor. That is, $FX \cdot \tilde{S}$ is a traded asset and the N-relative price $\frac{FX\cdot\tilde{S}}{N}$ a $\mathbb{Q}^N$-martingale. From the product rule

$$d\left(\frac{FX\cdot\tilde{S}}{N}\right) = d(FX\,\tilde{S})\frac{1}{N} + FX\,\tilde{S}\cdot d\left(\frac{1}{N}\right) + d(FX\,\tilde{S})\,d\left(\frac{1}{N}\right)$$

and

$$d(FX\cdot\tilde{S}) = dFX\cdot\tilde{S} + FX\,d\tilde{S} + dFX\,d\tilde{S}.$$

With the definition of the numéraire from Remarks 242 and 243

$$d\left(\frac{1}{N}\right) = \left(\frac{\partial}{\partial t}\frac{1}{N}\right)dt = \left(\frac{1}{N}\cdot\frac{\partial}{\partial t}\log\left(\frac{1}{N}\right)\right)dt$$

and thus

$$\mathop{\mathrm{Drift}}_{\mathbb{Q}^N}\left(\frac{FX\cdot\tilde{S}}{N}\right)$$
$$= \frac{FX\,\tilde{S}}{N}\left(\mu^{FX} + \mu^{\tilde{S}} + \sigma^{FX}(t)\,dW_{FX}^{\mathbb{Q}^N}(t)\,\sigma^{\tilde{S}}(t)\,dW_{\tilde{S}}^{\mathbb{Q}^N}(t) + \frac{\partial}{\partial t}\log\left(\frac{1}{N}\right)\right).$$

If we denote the instantaneous correlation of the stock process $\tilde{S}$ and the foreign exchange rate process FX by $\rho_{FX,\tilde{S}}$, i.e., we have $dW_{FX}^{\mathbb{Q}^N}(t)\,dW_{\tilde{S}}^{\mathbb{Q}^N}(t) = \rho_{FX,\tilde{S}}\,dt$, then we get with $\mathrm{Drift}_{\mathbb{Q}^N}\left(\frac{FX\cdot\tilde{S}}{N}\right) = 0$:

$$\mu^{\tilde{S}} = -\mu^{FX} - \frac{\partial}{\partial t}\log\left(\frac{1}{N}\right) - \sigma^{FX}(t)\sigma^{\tilde{S}}(t)\rho_{FX,\tilde{S}}(t).$$

With

$$\mu^{FX}(t) = \frac{\partial}{\partial t}\log\left(\frac{P(T_{m(t)+1};t)}{\tilde{P}(T_{m(t)+1};t)}\right)$$

and

$$\frac{\partial}{\partial t}\log\left(\frac{1}{N}\right) = \frac{\partial}{\partial t}\log\left(\frac{1}{P(T_{m(t)+1};t)}\right)$$

we get

$$\mu^{\tilde{S}} = \frac{\partial}{\partial t}\log(\tilde{P}(T_{m(t)+1};t)) - \sigma^{FX}(t)\sigma^{\tilde{S}}(t)\rho_{FX,\tilde{S}}(t)$$

$$= -\frac{\partial}{\partial t}\log\left(1 + \tilde{L}_{m(t)}(T_{m(t)})\,(T_{m(t)+1} - t)\right) - \sigma^{FX}(t)\sigma^{\tilde{S}}(t)\rho_{FX,\tilde{S}}(t).$$

29.3.2 Summary

Under the numéraire

$$N(t) := P(T_{m(t)+1};t)\prod_{j=0}^{m(t)}(1 + L_j(T_j)\cdot\delta_j),$$

where $m(t) := \max\{i : T_i \le t\}$, and the assumption that $N(t)$ does not have a diffusion part, as would be the case for

$$\left.\begin{array}{l} P(T_{i+1};t) := (1 + L_i(T_i)\,(T_{i+1} - t))^{-1} \\ \tilde{P}(T_{i+1};t) := (1 + \tilde{L}_i(T_i)\,(T_{i+1} - t))^{-1} \end{array}\right\} \quad \text{for } T_i < t < T_{i+1}, \tag{29.5}$$

the dynamic of the *equity hybrid cross-currency LIBOR market model* under the corresponding martingale measure $\mathbb{Q}^N$ (*spot measure*) is given by

$$\begin{aligned} dL_i(t) &= L_i(t)\cdot\mu_i(t)\,dt + L_i(t)\sigma_i(t)\,dW_i^{\mathbb{Q}^N}(t), & \text{for } i = 0,\ldots,n-1,\\ dFX(t) &= FX(t)\cdot\mu^{FX}(t)\,dt + FX(t)\sigma^{FX}(t)\,dW_{FX}^{\mathbb{Q}^N}(t),\\ d\tilde{L}_i(t) &= \tilde{L}_i(t)\cdot\tilde{\mu}_i(t)\,dt + \tilde{L}_i(t)\tilde{\sigma}_i(t)\,d\tilde{W}_i^{\mathbb{Q}^N}(t), & \text{for } i = 0,\ldots,n-1,\\ dS(t) &= S(t)\mu^S(t)\,dt + S(t)\sigma^S(t)\,dW_S^{\mathbb{Q}^N}(t),\\ d\tilde{S}(t) &= \tilde{S}(t)\mu^{\tilde{S}}(t)\,dt + \tilde{S}(t)\sigma^{\tilde{S}}(t)\,dW_{\tilde{S}}^{\mathbb{Q}^N}(t), \end{aligned}$$

where

$$\mu_i = \sum_{l=m(t)+1}^{i} \frac{\delta_l L_l}{1 + \delta_l L_l} \sigma_i \sigma_l \rho_{i,l},$$

$$\tilde{\mu}_i(t) = \sum_{j=m(t)+1}^{i} \frac{\delta_j \tilde{L}_j}{1 + \delta_j \tilde{L}_j} \tilde{\sigma}_i \tilde{\sigma}_j \tilde{\rho}_{i,j} - \tilde{\sigma}_i \sigma^{FX}(t) \rho_{\tilde{i},FX},$$

$$\mu^{FX}(t) = \frac{\partial}{\partial t} \log\left(\frac{P(T_{m(t)+1}; t)}{\tilde{P}(T_{m(t)+1}; t)} \right),$$

$$\mu^S(t) = \frac{\partial}{\partial t} \log(P(T_{m(t)+1}; t)),$$

$$\mu^{\tilde{S}}(t) = \frac{\partial}{\partial t} \log(\tilde{P}(T_{m(t)+1}; t)) - \sigma^{FX}(t) \sigma^{\tilde{S}}(t) \rho_{FX,\tilde{S}}(t).$$

If we choose for $P(T_{m(t)+1}; t)$ and $\tilde{P}(T_{m(t)+1}; t)$ the interpolation given in (29.5), then it follows that

$$\mu^{FX}(t) = \frac{\partial}{\partial t} \log\left(\frac{1 + \tilde{L}_{m(t)}(T_{m(t)}) (T_{m(t)+1} - t)}{1 + L_{m(t)}(T_{m(t)}) (T_{m(t)+1} - t)} \right),$$

$$\mu^S(t) = \frac{\partial}{\partial t} \log\left(\frac{1}{1 + L_{m(t)}(T_{m(t)}) (T_{m(t)+1} - t)} \right),$$

$$\mu^{\tilde{S}}(t) = \frac{\partial}{\partial t} \log\left(\frac{1}{1 + \tilde{L}_{m(t)}(T_{m(t)}) (T_{m(t)+1} - t)} \right) - \sigma^{FX}(t) \sigma^{\tilde{S}}(t) \rho_{FX,\tilde{S}}(t).$$

29.3.3 Implementation

Due to the many state variables, i.e., the high Markov dimension, it is natural to consider an implementation to be path simulation (Monte Carlo simulation). The first step toward an implementation is the discretization of the simulation time t by suitable discretization scheme. Since the processes are lognormal processes, we use the Euler scheme for the log process; see Sections 13.1 and 13.1.2.3.

The discretization of the interest rate processes L_i has been presented in Section 19.3, the interest rate processes of the foreign currency rates $\tilde{L}_i$ are discretized likewise. For FX we find

$$FX(t + \Delta t)$$

$$= FX(t) \exp\left(\int_t^{t+\Delta t} \mu^{FX}(\tau) - \frac{1}{2} \sigma^{FX}(\tau)^2 \, d\tau + \int_t^{t+\Delta t} \sigma^{FX}(\tau) \, dW^{FX}(\tau) \right)$$

$$= FX(t) \exp\left((\bar{\mu}^{FX}(t) - \frac{1}{2} \bar{\sigma}^{FX}(t)^2) \Delta t + \bar{\sigma}^{FX}(t) \Delta W^{FX}(t) \right),$$

with $\Delta W^{FX}(t) := W^{FX}(t + \Delta t) - W^{FX}(t)$ and

$$\bar{\mu}^{FX}(t) := \frac{1}{\Delta t} \int_t^{t+\Delta t} \mu^{FX}(\tau)\, d\tau$$

$$\bar{\sigma}^{FX}(t) := \sqrt{\frac{1}{\Delta t} \int_t^{t+\Delta t} \sigma^{FX}(\tau)^2\, d\tau}.$$

If we especially choose the time discretization of the Monte Carlo simulation to match the *tenor* structure $T_0, T_1, T_2, \ldots$, then we have

$$FX(T_{i+1}) = FX(T_i)\, \exp\left(\bar{\mu}^{FX}(T_i)\, \Delta T_i - \frac{1}{2}\bar{\sigma}^{FX}(T_i)^2\, \Delta T_i + \bar{\sigma}^{FX}(T_i)\, \Delta W^{FX}(T_i)\right),$$

with $\Delta T_i := T_{i+1} - T_i$ and $\Delta W^{FX}(T_i) := W^{FX}(T_{i+1}) - W^{FX}(T_i)$ and

$$
\begin{aligned}
\bar{\mu}^{FX}(T_i) &:= \frac{1}{\Delta T_i} \int_{T_i}^{T_{i+1}} \mu^{FX}(\tau)\, d\tau = \frac{1}{\Delta T_i} \log\left(\frac{1 + \tilde{L}_i(T_i) \cdot (T_{i+1} - t)}{1 + L_i(T_i) \cdot (T_{i+1} - t)}\right)\Bigg|_{t=T_i}^{T_{i+1}} \\
&= \frac{1}{\Delta T_i} \log\left(\frac{1 + L_i(T_i) \cdot (T_{i+1} - T_i)}{1 + \tilde{L}_i(T_i) \cdot (T_{i+1} - T_i)}\right), \\
\bar{\sigma}^{FX}(T_i) &:= \sqrt{\frac{1}{\Delta T_i} \int_{T_i}^{T_{i+1}} \sigma^{FX}(\tau)^2\, d\tau}.
\end{aligned}
$$

This Euler discretization is exact. The discretization of the processes S and $\tilde{S}$ follows likewise. With "discrete drift term"

$$\bar{\mu}^S(T_i) := \frac{1}{\Delta T_i} \log(1 + L_i(T_i)\, (T_{i+1} - T_i)),$$

$$\bar{\mu}^{\tilde{S}}(T_i) = \frac{1}{\Delta T_i} \log(1 + \tilde{L}_i(T_i)\, (T_{i+1} - T_i)) - \int_{T_i}^{T_{i+1}} \sigma^{FX}(t)\sigma^{\tilde{S}}(t)\rho_{FX,\tilde{S}}(t)\, dt.$$

Part VII

Implementation

CHAPTER 30

Object-Oriented Implementation in Java™

> From early on, we wanted a product that would seem so natural and so inevitable and so simple, you almost wouldn't think of it as having been designed.
>
> *Jonathan Ive*
> *iPod™ Design Team, Apple Computer [37]*

30.1 Elements of Object-Oriented Programming: Class and Objects

First we define the two concepts *class* and *object*.

Definition 244 (Class):
A class consists of

- A description of a *data structure*.

- A description of a set of functions, the *methods* that act on the data structures and other data (given as arguments). The description of the methods consists of

  - A description of the calling convention of the methods, the *interface*; see Section 30.1.3 and Definition 247.

  - A description of how the method (function) actually acts on the data, the *implementation*.

Definition 245 (Object): ⌐

We say that X *is an object of the class* K, if

- X provides memory to store data according to the structure (layout) described by K and

- the methods described in K may be applied to the data in X.

In this case, K is also called the *type* of X. ⌐

Interpretation: A class is the blueprint of an object, while an object is a real *instance* of the class.

Considering how classes and objects are realized in a computer, it becomes apparent that an object X of class K merely stores the data according to the storage layout in K, while the algorithms (code) that operate on X are given by K. The class is a description of the storage layout and the functionality, while the object represents the corresponding data record. The definition of a class exists only once, while the object of a class (data records) may exist multiple times. *Class and object distinguish between logic (code) and data.* Obviously, the logic, i.e., the class, has to know the layout (structure) of the data.

To illustrate the relation of classes and objects some authers use an analogy like, e.g., *human* is a class while the *specific individual "Christian Fries"* is an object of the class *human*. Such analogies do not hold very far. For example, it does not become apparent that an object is just a data storage, disfunctional without the class and that the code, i.e.; the algorithm that acts on the data exists only once, namely inside the class. However, each individual has its own experiences (data) and patterns of (re-)action (code processing experiences).

That the code is stored only inside the class also becomes apparent in the memory requirements: If we add a data field to a class and create 100 objects (instances) of this class, then, of course, this consumes 100 times the memory of the new data field. If a new method is added to a class, then its code is stored only once and the memory requirement is totally independent of the number of objects created.

As well as the data described in the class, an object carries another data item, namely its type. This *type* specifies the class of the objects. Thus there is a link back from the object to the class and thus to the methods that may be invoked on the objects' data. ◁|

30.1.1 Example: Class of a Binomial Distributed Random Variable

Let B denote a binomial distributed random variable defined over a probability space $(\Omega, \mathcal{F}, P)$ with $\Omega = \{\omega_1, \omega_2\}$. Probability space and random variable may be characterized by three values $b_1, b_2, p \in \mathbb{R}$:

$$B(\omega_1) = b_1, \qquad P(\omega_1) = p,$$
$$B(\omega_2) = b_2, \qquad P(\omega_2) = 1 - p. \tag{30.1}$$

Equation (30.1) describes the class of "binomial distributed random variables" while the random variable C with

$$C(\omega_1) = 1, \qquad P(\omega_1) = 0.5,$$
$$C(\omega_2) = -1, \qquad P(\omega_2) = 0.5, \tag{30.2}$$

is a (specific) object, i.e., an instance of the class " binomial distributed random variable". Of course, operators that may be applied to this random variable have to be defined only on the class level. For example, the calculation of the mean is defined by

$$E^{\mathbb{P}}(B) \;=\; \int_{\Omega} B \, dP \;=\; p \, B(\omega_1) + (1 - p) \, B(\omega_2) \;=\; p \, b_1 + (1 - p) \, b_2. \tag{30.3}$$

That $E(C) = 0.5 \times 1.0 + 0.5 \times (-1.0) = 0.0$ follows from (30.3).

In Java™ a corresponding class could look as follows: The class definition starts (after a comment) with the description of the data layout, here `value1`, `value2` and `pobabilityOfState1` for b_1, b_2, and p, respectively.

Listing 30.1. `BinomialDistributedRandomVariable`: *A class for binomial distributed random variables*

```
package com.christianfries.finmath.tutorial.example1;

/**
 * @author Christian Fries
 * @version 1.0
 */
class BinomialDistributedRandomVariable {

    private double value1;
    private double value2;
    private double probabilityOfState1;
```

followed by the constructor

```
    /**
     * This class implements a binominal distributed random variable
     * @param value1 The value in state 1.
     * @param value2 The value in state 2.
     * @param probabilityOfState1 The probability of state 1.
     */
    public BinomialDistributedRandomVariable(double value1, double value2,
        double probabilityOfState1) {
        this.value1 = value1;
        this.value2 = value2;
        this.probabilityOfState1 = probabilityOfState1;
    }
```

and the description of the method `getExpectation`.

```
/**
 * @return The expectation of the random variable
 */
public double getExpectation() {
    double expectation = probabilityOfState1 * value1
        + (1-probabilityOfState1) * value2;
    return expectation;
}
}
```

30.1.2 Constructor

The constructor of a class is a (special) method that is called upon the instantiation (construction) of an object (there may be many different constructors and then it is possible to choose which constructor is called). With a constructor it is possible to do additional initializations beyond the allocation of the memory—in our case the initializations are setting the value of b_1, b_2, and p.

The code of the above constructor of the class `BinomialDistributedRandomVariable` may be confusing: The arguments of the constructor have the same names as the data fields of the class. This is allowed and is often used for data initialization in constructors, but it is dangerously confusing in other methods with longer code. In this case the corresponding name always denotes the argument of the constructor (or method). To access the data field with the corresponding name the prefix `this.` has to be added. So the constructor above sets the data fields of the object to the values given by the arguments.

30.1.3 Methods: Getter, Setter, and Static Methods

30.1.3.1 Calling Convention, Signatures

The calling convention of a method is the name of the methods together with the list of its argument types, i.e., the calling convention defines which name and argument type have to be used to call a method. The list of argument types is called the *signature* of a method. Two methods of the same name but with different signatures are seen as different methods. Providing another method with the same name but a different signature is called *overloading* the method.[1]

---

[1] Within a class there cannot be two methods with the same name and the same signature. The return value may not be changed by overloading.

30.1.3.2 Getter, Setter

If the data fields of an object are made accessible, then they may be accessed through objectName.dataFieldName, i.e., they may be read or modified. The access to the data fields of an object may be allowed or denied; see also *data hiding* in Section 30.2.1.

After an object has been constructed data fields may still be changed by means of methods. We may set the data or get the data. Methods that do this are called *setter* or *getter*. It is a convention that all getter methods start with the prefix get and all setter methods start with the prefix set, both followed by a name of the entity they modify, starting with a capital letter.

We add a setter for the value of b_1 to the class definition:

```
/**                                            45
 * @param value1 The value of state 1.         46
 */                                            47
public void setValue1(double value1) {         48
  this.value1 = value1;                        49
}                                              50
```

The method only changes the state of object and thus does not return a value. This is indicated by the keyword void.

30.1.3.3 Static Methods

Methods that do not require knowledge of the data fields of an object, i.e., that do not read or modify data from an object, are called *static methods*. Put differently, the method does not need an object; it is sufficient to have the class definition.

Definition 246 (Static Method):
A method of class K which keeps objects of the class K invariant and is independent of its data is called *static*. A static method is also called a *class method*.

In Java™ method is declared static by the keyword static.

To apply the (nonstatic) method to data we (have to) create objects. A corresponding code, demonstrating how to work with object of the toy class above by doing some tests is given in the main method[2].

```
public static void main( String args[] ) {              26
  /*                                                    27
   * Test of class BinominalDistributedRandomVariable   28
   */                                                   29
                                                        30
  System.out.println("Creating random variable.");      31
```

---

[2] The main method may be called from outside without requiring a corresponding object. It is *static*. Thus (especially since it does not require the existence of an object) it may function as a possible entry point to a program.

```
BinomialDistributedRandomVariable randomVariable =       32
new BinomialDistributedRandomVariable(1.0,-1.0,0.5);   // Create object   33
                                                         34
double expectation = randomVariable.getExpectation();    35
System.out.println("Expectation is: " + expectation);    36
                                                         37
System.out.println("Changing value of state 1 to 0.4."); 38
randomVariable.setValue1(0.4);                           39
                                                         40
expectation = randomVariable.getExpectation();           41
System.out.println("Expectation is: " + expectation);    42
}                                                        43
```

In line 33 we create a new object of the type
BinomialDistributedRandomVariable by using the keyword new (reserv-
ing the memory corresponding to the data layout) followed by the specification of the
constructor to use (note that the constructor is essentially a method having the same
name as the class) (right side of =). The result is stored in an object reference of type
BinomialDistributedRandomVariable (left side of =).

 Further Reading: In [11, 36]: primitive types, object references,
static methods (the keyword static), return values (the keyword void),
the main method, comments, and the JavaDoc standard. ◁|

30.2 Principles of Object Oriented Programming: Data Hiding, Abstraction, Inheritance, and Polymorphism

30.2.1 Encapsulation and Interfaces

To access the data of an object there are two possible ways: One is to provide two
(or more) methods that allow us to read and modify the data, i.e. getters and setters
are implemented. In our example class BinomialDistributedRandomVariable
we provide an example of this for the data field probabilityOfState1:

Listing 30.2. *Getter and setter*

```
/**                                                      52
 * @return Returns the probability of state 1.            53
 */                                                      54
public double getProbabilityOfState1() {                 55
  return probabilityOfState1;                            56
}                                                        57
                                                         58
/**                                                      59
 * @param probabilityOfState1 The probability of state 1. 60
 */                                                      61
```

440

```
public void setProbabilityOfState1(double probabilityOfState1) {    62
  this.probabilityOfState1 = probabilityOfState1;                    63
}                                                                    64
```

The use of these methods could look as follows:

```
BinomialDistributedRandomVariable randomVariable =
  new BinomialDistributedRandomVariable(1.0,-1.0,0.5);

// Get probability of state 1
double p = randomVariable.getProbabilityOfState1();
System.out.println( "Current value of p is: " + p);

// Change probability of state 1 to 0.3
randomVariable.setProbabilityOfState1(0.3);
```

Another possiblily is to use the direct access to the corresponding data field:

```
BinomialDistributedRandomVariable randomVariable =
  new BinomialDistributedRandomVariable(1.0,-1.0,0.5);

// Get probability of state 1
double p = randomVariable.probabilityOfState1;
System.out.println( "Current value of p is: " + p);

// Change probability of state 1 to 0.3
randomVariable.probabilityOfState1 = 0.3;
```

The last variant works without special methods.[3] It is the direct access to the internal data of the object.

This kind of access to the data structure appears to be more convenient for both the developer who does not need to implement special getter and setter methods as well as for the user of the class. Direct access to the internal data structure of a class has to be allowed explicitly. To allow direct access to a data field the keyword `public` has to be used:

```
class BinomialDistributedRandomVariable {

  public double value1;
  public double value2;
  public double probabilityOfState1;
```

30.2.1.1 Encapsulation

Hiding the internal data structure and implementation and thus denying direct access to the data structure is called *encapsulation*. The fundamental advantage of encapsulation is that the data structure and the way the methods process that data may be changed. Users of the class, having access to methods on the objects only, may be left untouched by such changes. From "outside" the class behaves as before.

The advantage of encapsulation may be illustrated with the very simple example of a binomial distributed random variable. We give two examples.

---

[3] As long as access to the data is allowed, in Java™ this is done by adding the keyword `public` before the data.

Example of Encapsulation: Offering Alternative Methods: Like the getter and setter for the probability $P(\{\omega_1\})$ of the state ω_1 we offer a getter and setter for the probability $P(\{\omega_2\})$ of the state ω_2:

```
/**                                                              66
 * @return Returns the probability of state 2.                   67
 */                                                              68
public double getProbabilityOfState2() {                         69
  return 1.0 - probabilityOfState1;                              70
}                                                                71
                                                                 72
/**                                                              73
 * @param probabilityOfState2 The probability of state 2.        74
 */                                                              75
public void setProbabilityOfState2(double probabilityOfState2) { 76
  this.probabilityOfState1 = 1.0 - probabilityOfState2;          77
}                                                                78
```

Previously, the state ω_1 was distinguished by the methods available (only its probability p could be read) and the probability of the state ω_2 was a derived quantity $P(\{\omega_2\}) = 1 - p$. Now both states are equally represented. How the properties of a binomial distributed random variable are represented internally, i.e., how the data is stored, cannot be inferred from outside. It is possible to change the data layout and keep the specification (behavior) of the methods unchanged by adapting their implementation.

Example of Encapsulation: Performance Improvement by Adding a Cache to the Internal Data Modell: The data model described in Listing 30.1 consists of $B(\omega_1)$ (value1), $B(\omega_2)$ (value2) and $P(\{\omega_1\})$ (probabilityOfState1). As a consequence we have to calculate the expectation as

$$\mathrm{E}(B) := p\,(B(\omega_1) + (1 - p)\,B(\omega_2)) = p\,(B(\omega_1) - B(\omega_2)) + B(\omega_2).$$

This is done by the method `getExpectation()`. If this method is called very often, we may improve performance by calculating the result once and storing it in a cache. We add a data field mean as *cache*

```
class BinomialDistributedRandomVariable {

  private double value1;
  private double value2;
  private double probabilityOfState1;

  private double mean;              // Cache for the mean
```

which is updated to the mean by the method `updateMean`

```
private void updateMean() {
  mean = probabilityOfState1 * value1 + (1-probabilityOfState1) * value2;
}
```

In addition we add a call to `updateMean()` at the end of the constructor as well as at the end of any setter modifying the state of the object, i.e., modifying the data. This ensures that the field `mean` contains the valid mean. The method `getExpectation` does not do any calculations but merely returns the value from the cache `mean`:

```
/**
 * @return The expectation of the random variable
 */
public double getExpectation() {
  return mean;
}
```

A multiply call to `getExpectation` does not result in a multiple calculation of the (same) mean.

Obviously, the user must not gain access to the field `mean` of a corresponding mean. This would be fatal. The lines

```
BinomialDistributedRandomVariable randomVariable =
    new BinomialDistributedRandomVariable(1.0,-1.0,0.5);   // Create object

randomVariable.mean = 0.2;
```

would put the object `randomVariable` into an inconsistent state. The user must neither assume the existence of a cache nor manipulate it. Thus both `mean` and `updateMean` are declared private. All other data fields also have to be declared *private*. If they are changed, then `mean` has to be recalculated. This is ensured by adding a call to `updateMean` to any setter. A direct manipulation of the data fields would disable this.

30.2.1.2 Interfaces

The advantage of encapsulation is that the internal data layout may be changed if required. By adapting the implementation of the methods which is also hidden it is ensured that the methods offer the same functionality as before. For the user of (the objects of) a class it is only relevant to know the calling convention of the methods, the interface.

Definition 247 (Interface):
The description of the calling convention of methods is called the *interface*. Similar to Definition 244 an interface consists of

- A description of the calling convention of a set of functions, the *methods*.

Definition 248 (Encapsulation):
If a class offers its functionality only through an *interface*, then we call the class *encapsulated*. This is called *encapsulation*.

An example of an interface for an discrete real-valued random variable, i.e., a real-valued random variable defined over a space $\Omega = \{\omega_1, \ldots, \omega_n\}$, is given by:

Listing 30.3. `DiscreteRandomVariableInterface`: *Interface description of a discrete random variable*

```
/*
 * Created on 04.12.2004
 *
 * (c) Copyright Christian P. Fries, Germany.
 * Contact: email@christian-fries.de.
 */
package com.christianfries.finmath.tutorial.randomVariables;

/**
 * This is the interface for a discrete real valued random variable.
 *
 * @author Christian Fries
 */
public interface DiscreteRandomVariableInterface {
  int getNumberOfStates();
  double getProbabilityOfState(int stateIndex);
  double getValueOfState(int stateIndex);

  double getExpectation();
  double getVariance();
}
```

 Further Reading: In [11, 36]: The keywords `public`, `private`, and `protected` for data fields and interfaces.   ◁|

30.2.2 Abstraction and Inheritance

Interface and class are two extremes and something in between may be considered, namely classes in which some methods have an implementation, while others are given only through their calling convention (i.e., as an interface). Methods for which the implementation is not yet specified are called *abstract methods*.

An example may be given by considering the interface `DiscreteRandomVariableInterface` above: The implementation of expectation and variance may be added without the knowledge of the internal data layout of the class. It is possible to add a partial implementation.[4]

Listing 30.4. `DiscreteRandomVariable`: *Abstract base class for a discrete random variable*

```
/*
 * Created on 05.12.2004
 *
 * (c) Copyright Christian P. Fries, Germany.
 * Contact: email@christian-fries.de.
 */
```

---

[4] An abstract class does not need to have any data layout.

```
package com.christianfries.finmath.tutorial.randomVariables;

/**
 * @author Christian Fries
 */
public abstract class DiscreteRandomVariable
    implements DiscreteRandomVariableInterface
{
    public double getExpectation() {
        double expectation = 0.0;
        for(int stateIndex=0; stateIndex<this.getNumberOfStates(); stateIndex++) {
            expectation += this.getValueOfState(stateIndex)
                * this.getProbabilityOfState(stateIndex);
        }

        return expectation;
    }

    public double getVariance() {
        // Calculate second moment
        double secondMoment = 0.0;
        for(int stateIndex=0; stateIndex<this.getNumberOfStates(); stateIndex++) {
            double value = this.getValueOfState(stateIndex);
            secondMoment += value * value * this.getProbabilityOfState(stateIndex);
        }

        // Calculate expectation
        double expectation = this.getExpectation();

        // Return variance
        return secondMoment - expectation*expectation;
    }
}
```

To define a class which provides an implementation to the interface DiscreteRandomVariableInterface is is only necessary to *extend* the class DiscreteRandomVariable with the implementations of the remaining abstract methods. This is possible in an elegant way by specifying that the new class should *inherit* the already defined properties from DiscreteRandomVariable. By doing so we may define a (new) class for the binomial distributed random variable:

Listing 30.5. *BinomialDistributedRandomVariable:*   *derived from DiscreteRandomVariable*

```
/*
 * Created on 05.12.2004
 *
 * (c) Copyright Christian P. Fries, Germany.
 * Contact: email@christian-fries.de.
 */
package com.christianfries.finmath.tutorial.randomVariables;

/**
 * @author Christian Fries
 * @version 1.0
 */
class BinomialDistributedRandomVariable extends DiscreteRandomVariable {
```

```java
double value1;
double value2;
double probabilityOfState1;

class ChangeOfMeasureException extends Exception {
  String cause;

  /**
   * @param cause The reason for this exception.
   */
  public ChangeOfMeasureException(String cause) {
    super(cause);
    this.cause = cause;
  }
}

/**
 * This class implements a binominal distributed random variable
 *
 * @param value1 The value in state 1.
 * @param value2 The value in state 2.
 * @param probabilityOfState1 The probability of state 1.
 */
public BinomialDistributedRandomVariable(
  double value1, double value2, double probabilityOfState1) {

  this.value1 = value1;
  this.value2 = value2;
  this.probabilityOfState1 = probabilityOfState1;
}

public int getNumberOfStates() {
  return 2;
}

public double getProbabilityOfState(int stateIndex) {
  if (stateIndex == 0)  return probabilityOfState1;
  else           return 1.0 - probabilityOfState1;
}

public double getValueOfState(int stateIndex) {
  if (stateIndex == 0)  return value1;
  else           return value2;
}

public void changeMeasureToMatchGivenExpecation(double expectation)
  throws ArithmeticException, ChangeOfMeasureException {

  // Check if we have anything to do
  if (expectation == this.getExpectation()) return; // Noting to do here

  // Check if change of measure is possible (random variable is stochastic)
  if (value1 == value2)
    throw new ArithmeticException("Random variable is not stochastic.");

  // Calculate candidate for new measure
  double quasiProbability = (expectation - value2) / (value1 - value2);

  if (quasiProbability < 0 || quasiProbability > 1)
```

446

```
    throw new ChangeOfMeasureException("Given␣expectation␣out␣of␣range.");

    // Check if change of measure is possible
    probabilityOfState1 = quasiProbability;
  }
}
```

Inheritance is not limited to the implementation of abstract methods, i.e., inheriting from abstract classes. It is also possible to inherit from a class (not necessarily abstract) and to extend this class by a new data layout, new methods, or new implementation of existing methods.

Definition 249 (Inherited Class):   ⌐
Let A and B denote classes. B is called inherited from A, if B implements (at least) the interface of A. B is also called *derived class*. A is called *base class*, also *superclass*.

If class B is inherited from A, then all objects of type B are simultaneously objects of the type A; they are polymorph; see Definition 251.   ⌐

A convenient element of inheritance is the possibility of using the implementation of the base class by default. If the derived class does not provide an implementation for a base class method, then the implementation is inherited from the base class. To be precise, a call to a method on an object of the derived class is routed automatically to the base class object if the derived class does not provide an implementation. The method then works on the data fields of the base class object.

Definition 250 (Overwriting (a Method)):   ⌐
Supplying a new implementation to a method of a base class in a derived class is called *overwriting the method*.   ⌐

30.2.3 Polymorphism

The property that objects of a derived class are of multiple types is very important. Since the derived class implements the interface defined by the superclass, objects of the derived class's type may be used equally well in all applications of base class objects. This is possible and meaningful because these objects may simultaneously be seen as objects of type A (type of the super class) and as objects of type B (type of the derived class). We say that these objects are *polymorph*.[5] If a base class is itself a derived class, then objects of the derived class have all types of all base classes.

Definition 251 (Polymorph):   ⌐
An object is called *polymorph* if it is of multiple type and behaves according to its derived type, even if it is used in a context (originally) expecting a base class.   ⌐

---

[5] Objects are polymorph, i.e., of multiple types, not classes.

The importance of polymorphism becomes apparent in the method call of polymorphic objects. Method calls on polymorphic objects use what is called *late binding*. There a method call on a polymorphic object is routed to the implementation of the derived class even if the call is invoked in a context originally expecting a base class.

Remark 252 (Interface, The Message Paradigm): The concept of an interface is a central concept of object-oriented programming. Inheritance is to some extent only a short way of saying that a class offers a superset of the interface of another class, where the shortening is that for methods that do not have an implementation in the derived class the implementation in the base class is used as a proxy.

For inheritance (in Java™) there is also the concept of the type of an object (and thus the concept of polymorphism). In Java™ it is possible that two objects of two different classes providing methods with identical calling conventions, i.e., providing the same interfaces, are not interchangeable in their use since they are of different types. If this additional restriction (*type safety*) is left out, then the only characteristic of a class is the interface provided. The calling convention of a method is often interpreted as a "message which may be received by an object". Some programming languages do not have the concept of type safety and distinguish objects only by the messages they may receive. Nice examples are Smalltalk and Objective-C.

 Further Reading: The Java™ keywords `private`, `protected`, `public`, `void`, `static`, `final`, `implements`, `extends`, `package`, and `import` in [11, 36]. ◁|

30.3 Example: A Class Structure for One-Dimensional Root Finders

We consider the problem of finding a root[6] of a function $f : \mathbb{R} \to \mathbb{R}$. The algorithm for seeking the root is realized in a class that does not know the special shape of the function. Instead we realize a question-answer pattern: In each iteration the class proposes a value x (through a getter) for which it awaits the function value to be set, i.e., the class questions the function value $f(x)$ for a (chosen) x and develops a strategy for approaching the solution from the answers.

30.3.1 Root Finder for General Functions

30.3.1.1 Interface

Such a class has to provide a method that returns the suggested point x (double getNextPoint()) and a method that receives the corresponding value $f(x)$ (void setValue(double)). Together with some methods for controling the iteration (counting, accuracy achieved) we have to provide the following interface:

Listing 30.6. *RootFinder: Interface for a one-dimensional root finder*

```
/*
 * Created on 30.05.2004
 *
 * (c) Copyright Christian P. Fries, Germany.
 * Contact: email@christian-fries.de.
 */
package net.finmath.rootFinder;

/**
 * This is the interface for a one dimensional root finder
 * implemented as an question-and-answer algorithm.
 *
 * @author Christian Fries
 */
public interface RootFinder {

    /**
     * @return Next point for which a value should be set
     * using <code>setValue</code>.
     */
    public double getNextPoint();

    /**
     * @param value Value corresponding to point returned
     * by previous <code>getNextPoint</code> call.
     */
    public void setValue(double value);

    /**
```

---

[6] x is a root of f if $f(x) = 0$.

```
    * @return Returns the numberOfIterations.
    */
   public int getNumberOfIterations();

   /**
    * @return Best point optained so far
    */
   public double getBestPoint();

   /**
    * @return Returns the accuracy.
    */
   public double getAccuracy();

   /**
    * @return Returns the isDone.
    */
   public boolean isDone();

}
```

We still have no data layout and no specific implementation. RootFinder only describes the *interface*. Obviously, a class implementing a root finder according to this interface has to have some storage on the current state of the search (say a data field for the current x) to derive a strategy for seeking the root. Which strategy is used (the implementation) and which information is needed for the strategy (the data layout) is not required in order to use it. It is sufficient to know the interface. Thus we may write a method that tests a given RootFinder against some test function f without actually having a specific class implementing the RootFinder:

Listing 30.7. *Test for RootFinder classes*

```
public static void testRootFinder(RootFinder rootFinder) {
  System.out.println("Testing " + rootFinder.getClass().getName() + ":");

  // Find a solution to x^3 + x^2 + x + 1 = 0
  while(rootFinder.getAccuracy() > 1E-11 && !rootFinder.isDone()) {
    double x = rootFinder.getNextPoint();

    // Our test function. Analytic solution is -1.0.
    double y = x*x*x + x*x + x + 1;

    rootFinder.setValue(y);
  }

  // Print result:
  DecimalFormat formatter = new DecimalFormat("0.00E00");
  System.out.print("Root......: "+formatter.format(rootFinder.getBestPoint())
     +"\t");
  System.out.print("Accuracy..: "+formatter.format(rootFinder.getAccuracy() )
     +"\t");
  System.out.print("Iterations: "+rootFinder.getNumberOfIterations() +"\n");
}
```

30.3.1.2 Bisection Search

A simple root finding algorithm is the *bisection search*.

Definition 253 (Bisection Search): ⌐
Given a continuous function $f : \mathbb{R} \mapsto \mathbb{R}$ and x_1, x_2 with $f(x_1)\, f(x_2) < 0$. The sequence

$$x_{i+1} = \begin{cases} \dfrac{x_i + x_{i-1}}{2}, & \text{for } f(x_{i-1}) \cdot f(x_i) < 0 \\[2mm] \dfrac{x_i + x_{i-2}}{2}, & \text{else.} \end{cases}$$

is called *bisection search*. ⌐

The class `BisectionSearch` realizes this algorithm by implementing the interface `RootFinder`. A corresponding code is given in Appendix D.1.

30.3.2 Root Finder for Functions with Analytic Derivative: Newton's Method

Some root finding methods, like the Newton method, require knowledge of the derivative $f' = \frac{\partial f}{\partial x}$. The "search strategy" of the Newton method is

$$x_{i+1} = x_i - \frac{f(x_i)}{f'(x_i)}.$$

30.3.2.1 Interface

Obviously a corresponding class has to implement a slightly modified interface. Instead of a method `setValue(double value)` the interface `RootFinderWithDerivative` provides a method `setValueAndDerivative(double value, double derivative)`.

30.3.2.2 Newton Method

The class `NewtonsMethod` implements the interface `RootFinderWithDerivative` using a Newton method. For the Newton method the corresponding implementation looks as follows:

Listing 30.8. *NewtonsMethod: Implementing the RootFinderWithDerivative interface*

```
/**
 * @param value
 *      The value corresponding to the point returned by previous
 *      <code>getNextPoint</code> call.
 */
```

```
 *  @param derivative
 *      The derivative corresponding to the point returned by previous
 *      <code>getNextPoint</code> call.
 */
public void setValueAndDerivative(double value, double derivative) {

  if(Math.abs(value) < accuracy)
  {
    accuracy  = Math.abs(value);
    bestPoint = nextPoint;
  }

  // Calculate next point
  nextPoint = nextPoint - value/derivative;

  numberOfIterations++;
  return;
}
```

.

30.3.3 Root Finder for Functions with Derivative Estimation: Secant Method

30.3.3.1 Inheritance

The power of inheritance and interfaces becomes apparent in the following realization of the *secant method*. The search strategy of the secant method is

$$x_{i+1} = x_i - \frac{f(x_i)}{\frac{f(x_i)-f(x_{i-1})}{x_i-x_{i-1}}}.$$

From that, two aspects become apparent:

- The secant method *is* a Newton method with an estimate for the derivative:
  $f'(x_i) \approx \frac{f(x_i)-f(x_{i-1})}{x_i-x_{i-1}}$.

- In each iteration the secant method only required knowledge of the function value $f(x_i)$ for the proposed point x_i.

For the class, these properties translate to:

- The secant method *extends* the class NewtonsMethod by an estimator for the derivative.

- The secant method *implements* the RootFinder interface.

Thus, a corresponding class would look as follows:

Listing 30.9. *SecantMethod: Implementing the RootFinder interface*

```java
/*
 * Created on 19.02.2004
 *
 * (c) Copyright Christian P. Fries, Germany.
 * Contact: email@christian-fries.de.
 */
package net.finmath.rootFinder;

/**
 * This class implements a root finder as auestion-and-answer algorithm using
 * the secant method.
 *
 * @author Christian Fries
 * @version 1.1
 */
public class SecantMethod extends NewtonsMethod implements RootFinder {

    // We need a second guess for the inital secant
    double secondGuess;

    // State of the solver
    double currentPoint;   // Actually the same as NewtonsMethod.nextPoint
    double lastPoint;
    double lastValue;

    /**
     * @param firstGuess
     *         The first guess for the solver to use.
     * @param secondGuess
     *         A second guess for the solver to use (different from first guess).
     */
    public SecantMethod(double firstGuess, double secondGuess) {
        super(firstGuess);
        this.secondGuess = secondGuess;
    }

    public double getNextPoint() {
        // Ask NewtonsMethods for next point and rember it as current point
        currentPoint = super.getNextPoint();
        return currentPoint;
    }

    /**
     * @param value
     *         The value corresponding to the point returned
     *         by previous <code>getNextPoint</code> call.
     */
    public void setValue(double value) {
        // Calculate approximation for derivative
        double derivative;
        if (getNumberOfIterations() == 0) {
            /* Trick: This derivative will let Newton's method
                 * propose the second guess as next point
                 */
            derivative = value / (secondGuess - currentPoint);
        } else {
            derivative = (value - lastValue) / (currentPoint - lastPoint);
        }
```

```java
    // Remember last point
    lastPoint = currentPoint;
    lastValue = value;

    super.setValueAndDerivative(value, derivative);

    return;
}

/**
 * @param value
 *      The value corresponding to the point returned by previous
 *      <code>getNextPoint</code> call.
 * @param derivative
 *      The derivative corresponding to the point returned by previous
 *      <code>getNextPoint</code> call.
 */
public void setValueAndDerivative(double value, double derivative) {
    // Remember last point
    lastPoint = nextPoint;
    lastValue = value;

    super.setValueAndDerivative(value, derivative);

    return;
}
}
```

Remark 254 (SecantMethod): Note that in our implementation of the secant method we stored the current point x of each iteration in a field currentPoint. This is not necessary, as we could have used the field nextPoint from the base class NewtonsMethod. However, then we have to make the field visible to the derived class.[7] Using the additional field currentPoint makes the derived class independent of the data model of the base class (but also a bit less efficient since the point is stored twice).

30.3.3.2 Polymorphism

The class SecantMethod shows how polymorphism works. Objects of the class SecantMethod are simultaneously objects of the class NewtonMethod, since SecantMethod inherits from NewtonMethod and thus offers the corresponding interface. Thus the class SecantMethod not only implements the interface RootFinder but also implements the interface RootFinderWithDerivative as a Newton method. It is truly polymorphic. With respect to the interface RootFinderWithDerivative behaves like a NewtonMethod (by routing calls to the base class); with respect to the interface RootFinder it implements the secant

---

[7] It would be sufficient to declare the field protected, a weaker form of public making it visible only to derived classes.

method. That the class `SecantMethod` may act like a `NewtonMethod` is not surprising: We did not change any method of the interface of the base class (no method has been overwritten). This is also apparent in out test program Listing 30.10, testing all the root finders; see Listing 30.11.

Remark 255 (Inheritance: Specialization and Extension): The construct "*B inherits from A*" is often interpreted as "*B is an A*". For example, "*a discrete random variable is a random variable*" or "*a binomially distributed random variable is a discrete random variable*". The motivation for this "mnemonic trick" is the conception that the derived class *B* is a specialization of the base class *A*. This interpretation may help us to design a class hierarchy, but it is not universal. For example "*the secant method is a Newton method*" appears to be wrong. The use of "*... extends ...*" in place of "*... is a(n) ...*" is much more universal. For example: "*The secant method extends the Newton method by an approximation for the derivative*" makes sense. And in Java™ the corresponding keyword is `extends`.

We test the implementation of our root finders with the class `TestRootFinders`. Since there are only two different interfaces we have to write only two different test routines:

Listing 30.10. *Test for* `RootFinder` *and* `RootFinderWithDerivative` *classes*

```
/*
 * Created on 02.12.2004
 *
 * (c) Copyright Christian P. Fries, Germany.
 * Contact: email@christian-fries.de.
 */
package net.finmath.rootFinder;

import java.text.DecimalFormat;

/**
 * @author Christian Fries
 */
public class TestRootFinders {
  public static void main(String[] args) {
    System.out.println("Applying root finders to x^3 + 2*y^2 + x + 1 = 0\n");

    System.out.println("Root finders without derivative:");
    System.out.println("----------------------------");

    RootFinder rootFinder;

    rootFinder = new BisectionSearch(-10.0,10.0);
    testRootFinder(rootFinder);

    rootFinder = new RiddersMethod(-10.0,10.0);
    testRootFinder(rootFinder);

    rootFinder = new SecantMethod(2.0,10.0);
    testRootFinder(rootFinder);
```

```java
        System.out.println("");

        System.out.println("Root finders with    derivative:");
        System.out.println("-------------------------------");

        RootFinderWithDerivative rootFinderWithDerivative;

        rootFinderWithDerivative = new NewtonsMethod(2.0);
        testRootFinderWithDerivative(rootFinderWithDerivative);

        rootFinderWithDerivative = new SecantMethod(2.0,10.0);
        testRootFinderWithDerivative(rootFinderWithDerivative);
    }
    public static void testRootFinder(RootFinder rootFinder) {
        System.out.println("Testing " + rootFinder.getClass().getName() + ":");

        // Find a solution to x^3 + x^2 + x + 1 = 0
        while(rootFinder.getAccuracy() > 1E-11 && !rootFinder.isDone()) {
            double x = rootFinder.getNextPoint();

            // Our test function. Analytic solution is -1.0.
            double y = x*x*x + x*x + x + 1;

            rootFinder.setValue(y);
        }

        // Print result:
        DecimalFormat formatter = new DecimalFormat("0.00E00");
        System.out.print("Root......: "+formatter.format(rootFinder.getBestPoint())
            +"\t");
        System.out.print("Accuracy..: "+formatter.format(rootFinder.getAccuracy() )
            +"\t");
        System.out.print("Iterations: "+rootFinder.getNumberOfIterations() +"\n");
    }

    public static void testRootFinderWithDerivative(
                RootFinderWithDerivative rootFinder) {
        System.out.println("Testing " + rootFinder.getClass().getName() + ":");

        // Find a solution to x^3 + x^2 + x + 1 = 0
        while(rootFinder.getAccuracy() > 1E-11 && !rootFinder.isDone()) {
            double x = rootFinder.getNextPoint();

            double y = x*x*x + x*x + x + 1;
            double p = 3*x*x + 2*x + 1;

            rootFinder.setValueAndDerivative(y,p);
        }

        // Print result:
        DecimalFormat formatter = new DecimalFormat("0.00E00");
        System.out.print("Root......: "+formatter.format(rootFinder.getBestPoint())
            +"\t");
        System.out.print("Accuracy..: "+formatter.format(rootFinder.getAccuracy() )
            +"\t");
        System.out.print("Iterations: "+rootFinder.getNumberOfIterations() +"\n");
    }
}
```

Listing 30.11. *Output of the test 30.10*

```
Applying root finders to x^3 + 2*y^2 + x + 1 = 0

Root finders without derivative:
-------------------------------
Testing net.finmath.rootFinder.BisectionSearch:
Root......: -1,00E00  Accuracy..: 9,09E-12  Iterations: 43
Testing net.finmath.rootFinder.RiddersMethod:
Root......: -1,00E00  Accuracy..: 3,33E-16  Iterations: 22
Testing net.finmath.rootFinder.SecantMethod:
Root......: -1,00E00  Accuracy..: 0,00E00 Iterations: 15

Root finders with    derivative:
-------------------------------
Testing net.finmath.rootFinder.NewtonsMethod:
Root......: -1,00E00  Accuracy..: 1,33E-15  Iterations: 10
Testing net.finmath.rootFinder.SecantMethod:
Root......: -1,00E00  Accuracy..: 1,33E-15  Iterations: 10
```

30.4 Anatomy of a Java™ Class

In Figure 30.1 we show (part of) a Java™ class with the most important elements.

Before the declaration of the class the name of the packet to which the class belongs is specified. The full class name is the concatenation of the packet name and the class name and it should be unique. To achieve this, the packet name is often derived from the Internet address of its creator.

This is followed by the specification of other classes used in the declaration of this class by means of the keyword `import` followed by the full class name.

The declaration of the class starts with the keyword `class` followed by the class name and introduced by the keywords `extends` and `implements` the optional specification of a base class and the implemented interfaces. If the specification of a base class is missing, then `java.lang.Object` is used as a base class. Thus all objects inherit directly or indirectly from `java.lang.Object`.

Following is the declaration of the data layout by a list of data fields, also called attributes. An *attribute* is defined by the specification of its type (primitive types, like `double`, `int`, etc., or a class) and its name. To determine its visibility (encapsulation) it may be preceded by the keywords `private`, `protected`, or `public`. Without such a keyword the visibility `private` is assumed.

The remainder of the class declaration consists of the declaration and implementation of the constructors and methods. The method name is preceded by the type of the return value (or `void` for a method without return value). This may be preceded by further keywords (visibility: `public`, `private`; declaration as class method: `static`; prevention of overwriting: `final`). A constructor is a method for which the name corresponds to the class name. It is `public` and has no return value (the keyword `void` is missing, however).[8]

---

[8] Actually, the object created should be viewed as the return value of the constructor.

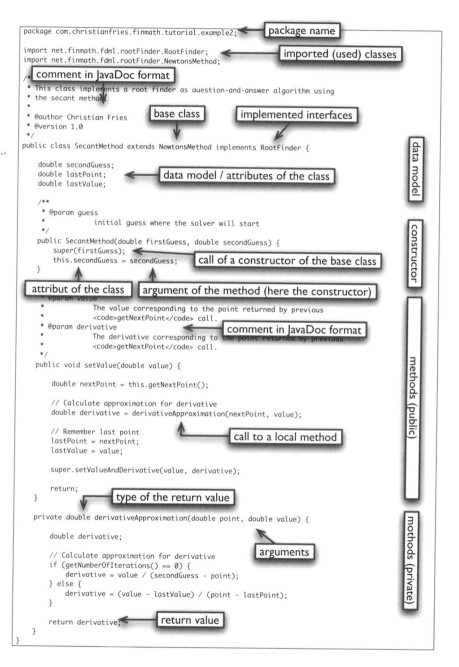

Figure 30.1. *Anatomy of Java™ class.*

30.5 Libraries

A major advantage of Java™ is its rich set of class libraries, which may easily be incorporated due to their unique package name and clear interfaces and often coming with a JavaDoc documentation. Not only are basic data management classes like *collections* available but also numerical libraries with algorithms from linear algebra, statistic and stochastic.

30.5.1 Java™ 2 Platform, Standard Edition (j2se)

The packets java.* of the *Java™ 2 Platform, Standard Edition*, offer basic functionalities, especially for the management of lists, strings, and files.

Packet(s) ... : java.*
Manufacturer : Sun Microsystems
Source ...... : http://java.sun.com/j2se/
Licence ..... : free, see http://java.sun.com/j2se/

30.5.2 Java™ 2 Platform, Enterprise Edition (j2ee)

The packets javax.* of the *Java™ 2 Platform, Enterprise Edition*, provide functionalities for the graphical user interface Swing and Internet communication.

Packet(s) ... : javax.*
Manufacturer : Sun Microsystems
Source dotfill: http://java.sun.com/j2ee/
Licence ..... : free, see http://java.sun.com/j2ee/

30.5.3 Colt

The Colt library offers in the packets cern.colt.* functionalities from linear algebra (matrix multiplication, matrix inversion, eigenvector decomposition) and in the packets cern.jet.* functionalities from stochastics (random number generators, distribution functions).

Packet(s) ... : cern.colt.*, cern.jet.*, cern.clhep
Manufacturer : Wolfgang Hoschek.
              Copyright (c) 1999 CERN - European Organization for Nuclear Research.
Source ...... : http://dsd.lbl.gov/~hoschek/colt/
Licence ..... : The packets cern.colt*, cern.jet*, cern.clhep are free for commercial use, see http://dsd.lbl.gov/~hoschek/colt/license.html.

30.5.4 Commons-Math: The Jakarta Mathematics Library

Manufacturer :   Various (Open Source), The Apache Software Foundation.
Source ......:   `http://jakarta.apache.org/commons/math/`
Licence .....:   Apache License, `http://jakarta.apache.org/commons/`
                 `license.html`.

30.6 Some Final Remarks

30.6.1 Object-Oriented Design (OOD)/Unified Modeling Language (UML)

Two key advantages of object-oriented programming are the modularization of the solution of a problem by encapsulation and abstraction and the reuse and extensibility of the solution of a problem by inheritance and polymorphism. Clean interfaces allow the independent development, refinement, and optimization of parts, independent both in time and personal.

Working out an object-oriented solution starts with the definition of the interfaces. These should provide an efficient communication of the objects. The design of the interfaces (and from that the classes) is called object-oriented design (*OOD*) [14]. For the design of more complex solutions a graphical language may be used (a convention of symbols): the *unified modeling language (UML)* [28].

 Further Reading: On the object-oriented design: Design patterns in [14], and UML in [28]. ◁|

461

Part VIII

Appendices

APPENDIX A

A Small Collection of Common Misconceptions

In a one-factor model a flat interest rate curve stays flat (a steep curve stays steep, an inverse curve stays inverse)

This assumption is wrong with respect to multiple aspects. If the diffusion part ($\sigma \, dW$) only allows a parallel movement of the interest rate curve, then the shape of the interest rate curve at a future time is given by the initial interest rate curve, the parallel movements of the interest rate curve, *and* the drift. The drift will change the shape of the interest rate curve. For example, a flat interest curve becomes steep under a one-factor LIBOR market model. In addition, a time structure of volatility allows parts of the movements of the interest rate curve to be independent. See also Chapter 25.

Specifying an interest rate model as short-rate model imposes a restriction. A short-rate model is incomplete since it models the short rate *only*

This is wrong. Under the martingale measure $\mathbb{Q}^B$ with numéraire $B(t) = \exp\left(\int_0^t r(\tau) \, d\tau\right)$ all bonds are given by $P(T;t) = \mathrm{E}^{\mathbb{Q}^B}(B(T)^{-1} \mid \mathcal{F}_t)$. Thus, in theory, the bond price curve $T \mapsto P(T;t)$, i.e., the interest rate curve, may be derived from the short-rate dynamics under the measure $\mathbb{Q}^B$. The short-rate dynamic gives a complete description of the interest rate curve dynamic. Conversely, any HJM model may be written as a short-rate model (this holds also for the LIBOR Market Model). However, the drift may then be path-dependent. The possible shapes of the interest rate curve are restricted, imposing special requirements on the model (e.g.,

465

the assumption of a Markov property of the short-rate process). See also Section 22.1.

An n factor model may be implemented in a lattice with n space dimensions

This is not necessarily the case. The amount of state space dimension necessary is the Markov dimension of the model, i.e., the number of state variables that are required to give the model as a Markov process. The Markov dimension may be significantly higher than the number of driving Brownian motions. Examples are given by the LIBOR market model and the Cheyette model.

In an n-factor (Monte Carlo) model the option value at time $t > 0$ can be described by an n-dimensional state vector (e.g., when pricing a Bermudan option by regression methods)

This is not necessarily the case. The reasoning corresponds to the previous considerations regarding the meaning of the number of factors. Also consider the counterexample from Figure 21.2, that a one-factor model may generate at maturity forward rates that are completely independent.

The LIBOR market modell exhibits no mean reversion

It is not reasonable to expect a mean-reversion term in the process for the forward rates, since the drift of the forward rates is given by the no-arbitrage requirement (martingale property). In this context, the property of being *mean reverting* makes sense only for the short rate. In a LIBOR market model the short rate may indeed exhibit a mean reversion. This is determined by the specific volatility structure of the forward rates. See also Section 25.3.

APPENDIX B

Tools (Selection)

B.1 Generation of Random Numbers

This section will consider the generation of (pseudo-)random numbers and shows how to construct a Monte Carlo simulation from these. There are numerous methods to generate random numbers and Monte Carlo simulations and a discussion of the various aspects of the quality of random numbers will not be discussed. We will give only an example based on the Mersenne twister. However, the methods presented are sufficient for most applications.

B.1.1 Uniform Distributed Random Variables

B.1.1.1 Mersenne Twister

A very popular (and also very good) random number generator for $[0, 1]$-equidistributed random numbers is the Mersenne twister (MT 19937). The random number generator has a period length of $2^{19937} - 1$, i.e., the random numbers generated repeat for the first time after 2^{19937} samples. The random numbers are also equidistributed in high dimensions (up to 623); see [87]. Based on the MT 19937 we may thus generate an n-dimensional stochastic process by drawing n sequential random numbers in each time step to calculate the increments of the stochastic process.[1]

Many libraries contain an implementation of the MT 19937. For the most popular languages it is available as source code.

---

[1] See also the remark at the end of Section B.1.5.

B.1.2 Transformation of the Random Number Distribution via the Inverse Distribution Function

If Z is an $[0,1]$-equidistributed random variable and Φ a cumulative distribution function, then $X := \Phi^{-1}(Z)$ is a random variable with a distribution given by Φ. If a random number generator for equidistributed random numbers is given (e.g., the Mersenne twister), then we draw realizations $Z(\omega_i)$, $i = 1, 2, \ldots$ and obtain from $\Phi^{-1}(Z(\omega_i))$ realizations of X. Thus, in addition to an $[0,1]$-equidistributed random number generator, we only require an inverse distribution function.

B.1.3 Normal Distributed Random Variables

B.1.3.1 Inverse Distribution Function

The density of the standard normal distribution is $\phi(x) := \frac{1}{\sqrt{2\pi}} \exp\left(-\frac{x^2}{2}\right)$, the (cumulative) distribution function is $\Phi(x) := \int_{-\infty}^{x} \phi(\xi)\, d\xi$. The algorithm described in [97] gives an approximation $\tilde{\Phi}^{-1}$ of Φ^{-1} with a relative error of

$$\frac{|\tilde{\Phi}^{-1} - \Phi^{-1}|}{1 + |\Phi^{-1}|} < 10^{-15}.$$

B.1.3.2 Box-Muller Transformation

The Box-Muller transformation transforms two independent $[0,1]$-equidistributed random numbers into two independent normally distributed random numbers.

Lemma 256 (Box-Muller Transformation): If z_1 and z_2 are two independent $[0,1]$-equidistributed random variables, then

$$x_1 := \rho \cos(\theta), \quad x_2 := \rho \sin(\theta)$$

with $\rho := \sqrt{-2\log(z_1)}$ and $\theta := 2\pi z_2$ two independent normally distributed random variables with mean 0 and standard deviation 1.

B.1.4 Poisson Distributed Random Variables

B.1.4.1 Inverse Distribution Function

The cumulative distribution function of the Poisson distribution is $\Phi(\tau) := 1 - \exp\left(-\int_0^\tau \lambda(t)\, dt\right)$, where λ denotes the *intensity*. If λ is constant then

$$\Phi^{-1}(z) = -\frac{\log(1-z)}{\lambda}.$$

If q is the probability that an event will occur in the interval $[T_1, T_2]$ and if λ is constant, then

$$\lambda = -\frac{\log(1-q)}{T_2 - T_1}.$$

B.1.5 Generation of Paths of an n-Dimensional Brownian Motion

Let $T_0 < T_1 < \cdots < T_m$ denote a given time discretization. We wish to generate the realization of an N-dimensional Brownian motion $W := (W_1, \ldots, W_n)$ on sample paths $\omega_1, \ldots \omega_k$. For a single path we have to draw $n \cdot m$ random numbers. To generate the $n \cdot m$-tuples we use the Mersenne twister and apply a transformation.

Let $\{z_i\}_{i=1,2,\ldots}$ denote the sequence of $[0, 1]$-equidistributed random numbers drawn from the Mersenne twister. Then $\{\Phi^{-1}(z_i)\}_{i=1,2,\ldots}$ is a sequence of standard normally distributed random variables. If `normalDistribution.nextDouble()` is a method returning a new element of the sequence $\{\Phi^{-1}(z_i)\}_{i=1,2,\ldots}$ upon each call, then a time-discrete Brownian motion is generated by the code in Listing B.1.

 Tip (Generation of Paths of Time-Discrete Stochastic Processes): The Mersenne twister is equidistributed in 623 dimensions. How well this property is preserved in higher dimensions is not clear. For an n-dimensional Brownian motion with independent increments, we have first the requirement that the increments are derived through a transformation of independent equidistributed random variables. For this reason it is advisable to generate the random numbers of the n increments of the n-dimensional processes in a sequence (i.e., as an n-tuple). Furthermore we have the requirement of the temporal independence of the m increments of the stochastic process, here the increments $\Delta W(T_i) := W(T_{i+1}) - W(T_i)$ of the Brownian motion. Thus, a path ω corresponds to the realization of an $n \cdot m$-dimensional random variable, here

$$(\Delta W_j(T_i))_{j=1,\ldots,n, i=1,\ldots,m}.$$

In other words, if we have to generate paths of an n-dimensional process with m time steps, then we draw an $n \cdot m$-tuple for every paths. This requires the random number generator to create the desired distribution in $n \cdot m$ dimensions. Thus, the high dimension of the Mersenne twister is of interest to our application. With the Mersenne twister we may, e.g., generate a 7-dimensional process with 89 time steps $(7 \times 89 = 623)$.[2]

Thus, the order of the loops in Listing B.1 has been chosen deliberately. ◁|

---

[2] Not all 623 dimensions have to be used, although in this example this would be the case.

Listing B.1. *Generation of an n-dimensional Brownian motion*

```
/**
 * This class represents a multidimensional brownian motion W = (W(1),...,W(n))
 * where W(i),W(j) are uncorrelated for i not equal j.
 * Here the dimension n is called factors since this brownian motion is used to
 * generate multi-dimensional multi-factor Ito processes and there one might
 * use a different number of factors to generate Ito processes of different
 * dimension.
 *
 * @author Christian Fries
 * @version 1.1
 */
public class BrownianMotion {
  double[][][]   brownianIncrement;

  double[]   timeDiscretization;
  int        numberOfFactors;
  int        numberOfPaths;
  int        seed;
```

⋮

```
  /**
   * Lazy initizialization of browniaIncrement.
   * Synchronized to ensure thread safty of lazy init.
   */
  private synchronized void doGenerateBrownianMotion() {
    if(brownianIncrement != null) return; // Nothing to do

    // Create random number sequence generator
    AbstractDistribution normalDistribution
      = new Normal(0,1, new MersenneTwister64(seed));

    // Allocate memory
    brownianIncrement = new double[timeDiscretization.length-1][numberOfFactors
      ][numberOfPaths];

    // Precalculate square roots of deltaT
    double[] sqrtOfTimeStep = new double[timeDiscretization.length-1];
    for(int timeIndex=0; timeIndex<timeDiscretization.length-1; timeIndex++) {
      sqrtOfTimeStep[timeIndex] = Math.sqrt(timeDiscretization[timeIndex+1]-
          timeDiscretization[timeIndex]);
    }

    // Set increments
    for(int path=0; path<numberOfPaths; path++) {
      for(int timeIndex=0; timeIndex<timeDiscretization.length-1; timeIndex++)
        {
        double sqrtDeltaT = sqrtOfTimeStep[timeIndex];
        // Generate uncorrelated Brownian increments
        for(int factor=0; factor<numberOfFactors; factor++) {
          brownianIncrement[timeIndex][factor][path]
              = normalDistribution.nextDouble() * sqrtDeltaT;
        }
      }
    }
  }
}
```

Further Reading: For the generation of random numbers, especially in the context of Monte Carlo simulations and derivative pricing, see [18]. Background information on the Mersenne twister and references to its source code and libraries may be found in the Wikipedia article "Mersenne twister", http://en.wikipedia.org/wiki/Mersenne_twister. ◁|

B.2 Factor Decomposition—Generation of Correlated Brownian Motion

Lemma 257 (Factor Decomposition):   Let $R = (\rho_{i,j})_{i,j=1...n}$ denote a given correlation matrix. Thus R is symmetric and positive semidefinite. This implies that R has real eigenvalues $\lambda_1 \geq \cdots \geq \lambda_n \geq 0$ and that a corresponding orthonormal basis of eigenvectors $v_1, \ldots, v_n$ of R exists, i.e.,

$$\exists V: \quad V^\top R V = D := \begin{pmatrix} \lambda_1 & & 0 \\ 0 & \ddots & 0 \\ 0 & & \lambda_n \end{pmatrix}, \quad \text{where } V = (v_1, \ldots, v_n),$$

and $R = V D V^\top$ as well as $V^\top V = I$.

Let $U_1, \ldots, U_n$ denote independent Brownian motions, $U := (U_1, \ldots, U_n)^\top$. Then W with

$$dW := (dW_1, \ldots, dW_n)^\top := V \sqrt{D}\, dU$$

is an n-dimensional Brownian motion with

$$< dW_i, dW_j > = \rho_{i,j}\, dt.$$

With $F := (v_1 \sqrt{\lambda_1}, \ldots, v_n \sqrt{\lambda_n})$ we thus have $dW = F\, dU$.

Proof: Obviously, a correlation matrix is symmetric and thus its eigenvalues are all real. If R is the correlation matrix of the random variable vector $X = (X_1, \ldots, X_n)^\top$ with $\mathrm{Var}(X_i) = 1$ and $\mathrm{E}(X_i) = 0$, then $R = \mathrm{E}(X \cdot X^\top)$. If v_i denotes the eigenvector corresponding to the eigenvalue λ_i, then

$$\mathrm{E}(\|X^\top v_i\|_{l_2}^2) \;=\; \mathrm{E}(v_i^\top X \cdot X^\top v_i) \;=\; v_i^\top R v_i = \lambda_i \|v_i\|_{l_2}^2$$

and thus

$$\lambda_i = \frac{\mathrm{E}(\|X^\top v_i\|_{l_2}^2)}{\|v_i\|_{l_2}^2} \geq 0.$$

Thus, R is positive semidefinite and $dW := V \sqrt{D}\, dU$ is well defined.   □|

B.3 Factor Reduction

Using the construction of correlated Brownian motion discussed in Section B.2, we may reduce the number of relevant factors (i.e., the number of nonzero eigenvalues), while keeping the correlation structure close to the original correlation structure. Let R, V, D be as in Section B.2 and $m < n$. Using

$$(f_1, \ldots, f_n) = F = V \sqrt{D}, \qquad f_i = (f_{j,i})_{j=1}^{n}$$

define

$$F^r = (f_i^r)_{i=1,\ldots,m} = (f_{j,i}^r)_{\substack{j=1,\ldots,n \\ i=1,\ldots,m}}, \qquad f_{j,i}^r := \frac{f_{j,i}}{(\sum_k f_{j,k}^2)^{1/2}},$$

i.e. the $n \times m$ matrix F^r is calculate from the $n \times m$ matrix $(v_1 \sqrt{\lambda_1}, \ldots, v_m \sqrt{\lambda_m})$ by re-normalizing the n rows.

Let $U_1, \ldots, U_m$ denote independent Brownian motions, $U := (U_1, \ldots, U_m)^\top$. Then W defined by

$$dW := (dW_1, \ldots, dW_n)^\top := F^r \, dU$$

is an n-dimensional m-factorial Brownian motion.

The factor reduction corresponds to a *pricipal component analysis* followed by a renormalization of the components.

Remark 258 (Factor Reduction):   The magnitude of the absolute value of the eigenvalue of λ_i represents the importance of the corresponding factor f_i. It may be used to decide upon the number of factors to use. A simple example is given by the limit case of perfect correlation $\rho_{i,j} = 1$. The corresponding correlation matrix has one eigenvalue n corresponding to the eigenvector $(1, \ldots, 1)$ and an $n - 1$-fold eigenvalue 0 corresponding to the orthogonal space. This implies that the dynamic of the n-dimensional Brownian motion may be explained by a one-dimensional Brownian motion (one factor).

In Figure B.1 we depict a reduction to the first three factors for the case of a high correlation $\rho_{i,j} = \exp(-0.005 * |i - j|)$. However, if many factors with relatively high weight (eigenvalues) are neglected, then the factor reduction has a significant impact on the correlation structure (see Figure B.3) as well as on the shape of the remaining factors (see Figure B.2).

 Experiment: The impact of a factor reduction on the correlation matrix may be studied for different correlation structures at
`http://www.christian-fries.de/finmath/applets/`
`FactorReduction.html.`   ◁|

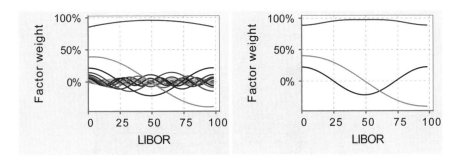

Figure B.1. *Factor reduction in the case of high correlation: The factors f_i (eigenvectors) of the correlation matrix $\rho_{i,j} = \exp(-0.005 * |i - j|)$ (left) and a reduction to the three factors having the largest eigenvalues (right).*

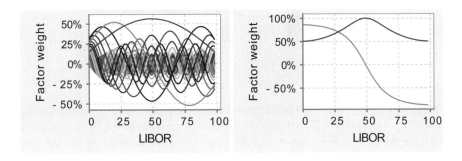

Figure B.2. *Factor reduction in the case of low correlation: The factors f_i (eigenvalues) of the correlation matrix $\rho_{i,j} = \exp(-0.1 * |i - j|)$ (left) and a reduction to the two factors having the largest eigenvalues (right).*

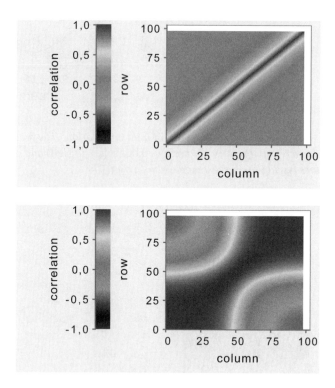

Figure B.3. *Factor reduction in the case of low correlation: The original correlation matrix $\rho_{i,j} = \exp(-0.1 * |T_i - T_j|)$ (top) and the correlation matrix corresponding to the reduction to two factors (bottom). This case corresponds to the factor reduction in Figure B.2.*

B.4 Optimization (One-Dimensional): Golden Section Search

Given a function $f : [a, b] \to \mathbb{R}$. Furthermore let $\lambda \in (0, 1)$ and $m_0 = \lambda a + (1 - \lambda)b$ such that $f(m_0) < \min\{f(a), f(b)\}$. Then the sequence $\{m_i\}_{i=0}^{\infty}$ defined by the following algorithm converges to a local minimum of f (and thus to a global minimum on $[a, b]$, if f is strictly convex on $[a, b]$):

Iteration start:

$$a_0 := a, \qquad m_0 := \lambda a + (1 - \lambda)b, \qquad b_0 := b.$$

Iteration step:

- If $b_i - m_i > m_i - a_i$, then set $z := \lambda m_i + (1 - \lambda)b_i$, and

$$\text{if } f(m_i) < f(z): \left\{ \begin{array}{ll} a_{i+1} & := a_i \\ m_{i+1} & := m_i \\ b_{i+1} & := z \end{array} \right\} \quad \text{else:} \left\{ \begin{array}{ll} a_{i+1} & := m_i \\ m_{i+1} & := z \\ b_{i+1} & := b_i \end{array} \right\}.$$

- If $b_i - m_i \le m_i - a_i$, then set $z := \lambda a_i + (1 - \lambda)m_i$, and

$$\text{if } f(z) < f(m_i): \left\{ \begin{array}{ll} a_{i+1} & := a_i \\ m_{i+1} & := z \\ b_{i+1} & := m_i \end{array} \right\} \quad \text{else:} \left\{ \begin{array}{ll} a_{i+1} & := z \\ m_{i+1} & := m_i \\ b_{i+1} & := b_i \end{array} \right\}.$$

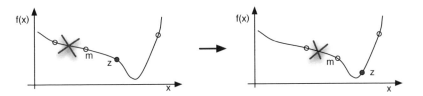

Figure B.4. *Golden section search.*

The algorithm places a point (z) into the larger of the two intervals $[a, m]$, $[m, b]$ and from the resulting three intervals it rejects the one that is adjacent to the larger value of $f(m)$, $f(z)$; see Figure B.4.

For the division ratio λ the value

$$\lambda = \frac{3 - \sqrt{5}}{2}$$

is optimal in the following sense: In the worst case, in which the algorithm rejects the smaller interval and retains the larger interval at every iteration step, then the value $\lambda = \frac{3-\sqrt{5}}{2}$ will result in the fastest convergence rate. Since this ratio is the *golden section*, the algorithm is called the *golden section search*.

B.5 Linear Regression

Lemma 259 (Linear Regression): Let $\Omega^* = \{\omega_1, \ldots, \omega_n\}$ be a given sample space, $V : \Omega^* \to \mathbb{R}$ and $Y := (Y_1, \ldots, Y_p) : \Omega^* \to \mathbb{R}^p$ given random variables. Furthermore let

$$f(y_1, \ldots, y_p, \alpha_1, \ldots, \alpha_p) := \sum \alpha_i y_i.$$

Then for any α^* with $X^T X \alpha^* = X^T v$

$$\|V - f(Y, \alpha^*)\|_{L_2(\Omega^*)} = \min_\alpha \|V - f(Y, \alpha)\|_{L_2(\Omega^*)},$$

where

$$X := \begin{pmatrix} Y_1(\omega_1) & \ldots & Y_p(\omega_1) \\ \vdots & & \vdots \\ Y_1(\omega_n) & \ldots & Y_p(\omega_n) \end{pmatrix}, \quad v := \begin{pmatrix} V(\omega_1) \\ \vdots \\ V(\omega_n) \end{pmatrix}.$$

If $(X^T X)^{-1}$ then $\alpha^* := (X^T X)^{-1} X^T v$. The $Y_1, \ldots, Y_p$ are called *basis functions* or *explanatory variables*.

Proof: We have to solve the minimization problem

$$g(\alpha) := \|V - f(Y, \alpha)\|_{L_2(\Omega^*)}^2 = (v - X \cdot \alpha)^T \cdot (v - X \cdot \alpha) \to \min.$$

The quadratic function on the right-hand side attains its minimum where the partial derivatives with respect to α_i are zero. We have

$$\frac{\partial g}{\partial \alpha} = 2X^T \cdot (v - X \cdot \alpha) = 2(X^T v - X^T X \cdot \alpha)$$

and thus

$$\frac{\partial g}{\partial \alpha} = 0 \iff X^T v = X^T X \cdot \alpha.$$

$\square$|

Further Reading: An extensive discussion of regression methods is given in [9]. ◁|

B.6 Convolution with Normal Density

Lemma 260 (Integration of $\exp(a \cdot X)$, X **Normally Distributed):** It is

$$\int_{h_1}^{h_2} \exp(a\,x)\,\phi(x; y, \sigma)\,dx$$

$$= \left[\Phi\left(\frac{h_2 - (y + a\sigma^2)}{\sigma} \right) - \Phi\left(\frac{h_1 - (y + a\sigma^2)}{\sigma} \right) \right] \exp\left(ay + \frac{a^2\sigma^2}{2} \right),$$

where $\phi(\xi; \mu, \sigma) = \frac{1}{\sqrt{2\pi}\sigma} \exp\left(-\frac{(\xi-\mu)^2}{2\sigma^2}\right)$ denotes the density and $\Phi(x) :=$ $\int_{-\infty}^{x} \phi(\xi; 0, 1)\,d\xi$ denotes the distribution function of the normal distribution. In particular:

$$\int_{-\infty}^{\infty} \exp(a\,x)\,\phi(x; y, \sigma)\,dx = \exp\left(ay + \frac{a^2\sigma^2}{2} \right).$$

Proof: It is

$$\int_{h_1}^{h_2} \exp(a\,x)\,\phi(x; y, \sigma)\,dx = \frac{1}{\sqrt{2\pi}\sigma} \int_{h_1}^{h_2} \exp(a\,x)\,\exp\left(-\frac{1}{2}(\frac{x-y}{\sigma})^2\right) dx$$

$$= \frac{1}{\sqrt{2\pi}\sigma} \int_{h_1}^{h_2} \exp\left(a\,x - \frac{1}{2}\left(\frac{x-y}{\sigma}\right)^2 \right) dx.$$

Since

$$\exp\left(a\,x - \frac{1}{2}\left(\frac{x-y}{\sigma}\right)^2 \right)$$

$$= \exp\left(a\,x - \frac{1}{2\sigma^2}\left(x^2 - 2x\,y + y^2\right) \right)$$

$$= \exp\left(-\frac{1}{2\sigma^2}\left(x^2 - 2x\,y + y^2 - 2a\sigma^2\,x\right) \right)$$

$$= \exp\left(-\frac{1}{2\sigma^2}\left(x^2 - 2x(y + a\sigma^2) + y^2\right) \right)$$

$$= \exp\left(-\frac{1}{2\sigma^2}\left(x^2 - 2x(y + a\sigma^2) + (y + a\sigma^2)^2 - 2a\sigma^2 y - a^2\sigma^4\right) \right)$$

$$= \exp\left(-\frac{1}{2\sigma^2}\left(x^2 - 2x(y + a\sigma^2) + (y + a\sigma^2)^2\right) \right) \exp\left(ay + \frac{a^2\sigma^2}{2} \right)$$

$$= \exp\left(-\frac{(x - (y + a\sigma^2))^2}{2\sigma^2} \right) \exp\left(ay + \frac{a^2\sigma^2}{2} \right),$$

it follows that

$$\int_{h_1}^{h_2} \exp(a\,x)\,\phi(x;y,\sigma)\,\mathrm{d}x$$

$$= \frac{1}{\sqrt{2\pi}\sigma} \int_{h_1}^{h_2} \exp\left(-\frac{(x-(y+a\sigma^2))^2}{2\sigma^2}\right)\,\mathrm{d}x \;\exp\left(ay+\frac{a^2\sigma^2}{2}\right)$$

$$= \int_{h_1}^{h_2} \phi(x;y+a\sigma^2,\sigma)\,\mathrm{d}x \;\exp\left(ay+\frac{a^2\sigma^2}{2}\right).$$

□

APPENDIX C

Exercises

In this appendix we give a small selection of exercises. The points are a rough indication of the complexity of the solution.

Exercise 1 (Probability Space, Random Variable [15 points]):   Let $(\Omega, \mathcal{F}, P)$ denote a probability space and $X : \Omega \to \mathbb{R}$ a random variable.

1. Give an example of a (false) modeling Ω, $\mathcal{F}$, X of some random experiment, such that X is not $(\mathcal{F}, \mathcal{B}(\mathbb{R}))$-measurable (give the definition of the mathematical objects and their interpretation).

2. Now let X be $(\mathcal{F}, \mathcal{B}(\mathbb{R}))$-measurable.

    a) Show that
    $$\{X^{-1}(A) \,|\, A \in \mathcal{B}(\mathbb{R})\}$$
    is a σ-algebra and a subset of $\mathcal{F}$ (sub-σ-algebra of $\mathcal{F}$). Give a possible an interpretation of the object?

    b) Show that
    $$P_X(A) := P(X^{-1}(A)) \qquad \forall A \in \mathcal{B}(\mathbb{R}))$$
    defines a probability measure (the image measure).

Exercise 2 (Conditional Expectation [20 points])[1]: Let X denote an $\mathcal{F}$-measurable (numerical) random variable and $\mathcal{G}$ a σ-algebra with $\mathcal{G} \subset \mathcal{F}$ (i.e., $\mathcal{G}$ is a sub-σ-algebra of $\mathcal{F}$). Prove the following properties of the conditional expectation:

1. If X is a $\mathcal{G}$-measurable random variable, then $E(X|\mathcal{G}) = X$ (P-almost surely).

2. If $X \geq 0$, then $E(X|\mathcal{G}) \geq 0$.

---

[1] See [41].

3. (*Tower Law*) If $\mathcal{H}$ is an σ-algebra with $\mathcal{H} \subset \mathcal{G}$, then $E(E(X|\mathcal{G})|\mathcal{H}) = E(X|\mathcal{H})$.

4. (*Taking out what is known*) If Z is a bounded $\mathcal{G}$-measurable random variable, then

$$E(ZX|\mathcal{G}) = ZE(X|\mathcal{G}). \tag{C.1}$$

Exercise 3 (Distribution Function [10 points])[2]: Let $X : \Omega \to \mathbb{R}$ denote a random variable and F the distribution function of X, i.e., $F(x) := P_X((-\infty, x))$. Show that:

1. F is left continuous, i.e., $F(x) = \lim_{h \nearrow 0} F(x + h)$.

2. If $g : \mathbb{R} \to \mathbb{R}$ is measurable with $E(|g(X)|) < \infty$, then

$$E(g(X)) = \int_{-\infty}^{\infty} g(x) \, dF(x),$$

where the integral is interpreted as the Lebesgue-Stieltjes integral (see [27]).

Exercise 4 (Brownian Scaling [10 points])[3]: Let W denote a Brownian motion and $c > 0$. Show that with

$$\tilde{W}(t) := \frac{1}{c} W(c^2 t)$$

$\tilde{W}$ is also a Brownian motion. Are the two processes W and $\tilde{W}$ equal in any sense (cf. Definition 20)?

Exercise 5 (Quadratic Variation [10 points])[4]: Let X denote a continuous stochastic process[5]. For $p > 0$ let the p-th variation process be defined as

$$< X, X >^{(p)} (t, \omega) := \lim_{\Delta t \to 0} \sum_{t_k \leq t} |X(t_{k+1}, \omega) - X(t_{k+1}, \omega)|^p,$$

where $\{t_i\}_{i=1}^{\infty}$ is a strictly monotone sequence with $t_0 = 0$, $\lim_{k \to \infty} t_k = \infty$, $\Delta t := \sup_k |t_{k+1} - t_k|$. The process $< X, X >^{(1)}$ is called the *total variation* of X and the process $< X, X >:=< X, X >^{(2)}$ is called the *quadratic variation* of X.

Let W denote a (one-dimensional) Brownian motion.

---

[2] See [27].
[3] See [27].
[4] See [27].
[5] A continuous stochastic processes is a stochastic process for which each path is a continuous function in time.

1. Show that
$$< W, W > (t, \omega) = t \quad P\text{-almost surely.}$$

2. Show that
$$< W, W >^1 (t, \omega) = \infty \quad P\text{-almost surely.}$$

Note that these properties hold pathwise for almost all paths and not only just in an averaged sense.

Exercise 6 (Itô Integral [10 points])[6]: Show by direct use of the definition of the Itô integrals that

1. $\int_0^T t \, dW(t) = T \, W(T) - \int_0^T W(t) \, dt,$

2. $\int_0^T W(t)^2 \, dW(t) = \frac{1}{3} W(T)^3 - \int_0^T W(t) \, dW(t).$

Exercise 7 (Stratonovich Integral [10 points])[7]: Let $T^{(n)} := \{t_0, \cdots, t_n\}$ with $0 = t_0 < t_1 < \cdots < t_n = T$ denote a *decomposition* of the interval $[0, T]$ and $\Delta T^{(n)} := \sup_i |t_i - t_{i-1}|$ its *fineness*. Furthermore let $f : [0, T] \times \Omega \to \mathbb{R}$ be of the *class of integrands of the Itô integral (on $[0, T]$)* and $t \mapsto f(t, \omega)$ continuous for (almost) all ω. Then

$$\int_0^T f(t, \omega) \, dW(t, \omega) = \lim_{\Delta T^{(n)} \to 0} \sum_{j=1}^n f(t_{j-1}, \omega)(W(t_j, \omega) - W(t_{j-1}, \omega))$$

is in $L_2(P)$ (proof?). The *Stratonovich integral* is defined correspondingly as

$$\int_0^T f(t, \omega) \circ dW(t, \omega) = \lim_{\Delta T^{(n)} \to 0} \sum_{j=1}^n f(t_{j-\frac{1}{2}}, \omega)(W(t_j, \omega) - W(t_{j-1}, \omega))$$

with $t_{j-\frac{1}{2}} := \frac{t_{i-1} + t_i}{2}$. Calculate

1. $\int_0^T W(t, \omega) \circ dW(t, \omega)$ and

2. $\int_0^T W(t, \omega) \circ dW(t, \omega) - \int_0^T W(t, \omega) \, dW(t, \omega).$

---

[6] See [27].
[7] See [27].

Solution: It is

$$\int_0^T W(t,\omega) \circ dW(t,\omega) = \lim_{\Delta T^{(n)} \to 0} \sum_{j=1}^n W(t_{j-\frac{1}{2}},\omega)(W(t_j,\omega) - W(t_{j-1},\omega))$$

and

$$
\begin{aligned}
&W(t_{j-\frac{1}{2}},\omega)(W(t_j,\omega) - W(t_{j-1},\omega)) \\
=\ & W(t_{j-\frac{1}{2}},\omega)(W(t_j,\omega) - W(t_{j-\frac{1}{2}},\omega)) + W(t_{j-\frac{1}{2}},\omega)(W(t_{j-\frac{1}{2}},\omega) - W(t_{j-1},\omega)) \\
=\ & \frac{1}{2}(W(t_j,\omega) + W(t_{j-\frac{1}{2}},\omega))(W(t_j,\omega) - W(t_{j-\frac{1}{2}},\omega)) \\
& -\frac{1}{2}(W(t_j,\omega) - W(t_{j-\frac{1}{2}},\omega))(W(t_j,\omega) - W(t_{j-\frac{1}{2}},\omega)) \\
& +\frac{1}{2}(W(t_{j-\frac{1}{2}},\omega) - W(t_{j-1},\omega))(W(t_{j-\frac{1}{2}},\omega) - W(t_{j-1},\omega)) \\
& +\frac{1}{2}(W(t_{j-\frac{1}{2}},\omega) + W(t_{j-1},\omega))(W(t_{j-\frac{1}{2}},\omega) - W(t_{j-1},\omega)) \\
=\ & \frac{1}{2}(W(t_j,\omega)^2 - \cancel{W(t_{j-\frac{1}{2}},\omega)^2}) + \frac{1}{2}(\cancel{W(t_{j-\frac{1}{2}},\omega)^2} - W(t_{j-1},\omega)^2) \\
& -\frac{1}{2}(W(t_j,\omega) - W(t_{j-\frac{1}{2}},\omega))(W(t_j,\omega) - W(t_{j-\frac{1}{2}},\omega)) \\
& +\frac{1}{2}(W(t_{j-\frac{1}{2}},\omega) - W(t_{j-1},\omega))(W(t_{j-\frac{1}{2}},\omega) - W(t_{j-1},\omega)) \\
=\ & \frac{1}{2}(W(t_j,\omega)^2 - W(t_{j-1},\omega)^2) \\
& -\frac{1}{2}(W(t_j,\omega) - W(t_{j-\frac{1}{2}},\omega))^2 + \frac{1}{2}(W(t_{j-\frac{1}{2}},\omega) - W(t_{j-1},\omega))^2.
\end{aligned}
$$

Thus, from

$$\lim_{\Delta T^{(n)} \to 0} \sum_{j=1}^n (W(t_j,\omega) - W(t_{j-\frac{1}{2}},\omega))^2 = \lim_{\Delta T^{(n)} \to 0} \sum_{j=1}^n (W(t_{j-\frac{1}{2}},\omega) - W(t_{j-1},\omega))^2$$

we have

$$\int_0^T W(t,\omega) \circ dW(t,\omega) = \frac{1}{2} W(t_n,\omega)^2.$$

Exercise 8 (Itô Product Rule, Itô Quotient Rule [15 points]): Use the Itô formula and prove

482

1. The product rule: Let X and Y denote Itô processes. Then

$$d(X\,Y) \;=\; Y\,dX + X\,dY + dX\,dY$$

2. The quotient rule: Let X and Y denote Itô processes, $Y > c$ for some $c \in (0, \infty)$. Then

$$d\left(\frac{X}{Y}\right) \;=\; \frac{X}{Y}\left(\frac{dX}{X} - \frac{dY}{Y} - \left(\frac{dX}{X}\right)\left(\frac{dY}{Y}\right) + \left(\frac{dY}{Y}\right)^2\right)$$

3. The drift adjustment of a lognormal process: Let $S(t) > 0$ denote an Itô process of the form

$$dS(t) \;=\; \mu(t)S(t)\,dt + \sigma(t)S(t)\,dW(t),$$

$$dY(t) \;=\; (\mu(t) - \frac{1}{2}\sigma^2(t))\,dt + \sigma(t)\,dW(t).$$

Exercise 9 (Martingale: Itô Formula [15 points])[8]:   Use the Itô formula and show that the following processes are $\mathcal{F}_t$-martingales:

1. $X(t) = \exp(\frac{1}{2}t)\cos(W(t))$

2. $X(t) = \exp(\frac{1}{2}t)\sin(W(t))$

3. $X(t) = (t + W(t))\exp(-\frac{1}{2}t - W(t))$,

where $W(t)$ denotes a one-dimensional Brownian motion.

Exercise 10 (Black-Scholes Partial Differential Equation in the Coordinates t, $N(t)$, $S(t)$ [20 points]):   Show that the function

$$V(t, n, s)) \;=\; s\Phi(d_+) - n\frac{K}{N(T)}\Phi(d_-)$$

with

$$d_\pm \;=\; \frac{1}{\bar\sigma\sqrt{(T-t)}}\left[\log\left(\frac{s}{n}\frac{N(T)}{K}\right) \pm \frac{\bar\sigma^2(T-t)}{2}\right].$$

and $\Phi'(x) = \phi(x) = C\exp(-x^2/2)$ solves the partial differential equation

$$\frac{\partial V(t)}{\partial t}\,dt + \frac{1}{2}\frac{\partial^2 V(t)}{\partial s\,\partial s}\,\sigma^2 \;=\; 0$$

---

[8] See [27].

with the *final time condition*

$$V(T, N(T), s) = \max(s - K, 0).$$

Exercise 11 (Black 76 Formula for Swaption [30 points]): Let $T_1 < \cdots < T_n$ denote given times. Let V_{swap} denote s swap as in Definition 117 with constant swap rates $S_i = K$, $i = 1, \ldots, n - 1$ ($T_1, \ldots, T_{n-1}$ are fixing dates and $T_2, \ldots, T_n$ are payment dates). Let S denote the corresponding *par swap rate* as in Definition 122 (cf. Remark 121). Derive the formula for the value V_{swaption} of an (*European*) option on V_{swap} (with exercise date T_1), assuming the S has lognormal dynamics

$$dS(t) = \mu(t)S(t)\, dt + \sigma(t)S(t)\, dW(t) \quad \text{under } \mathbb{P}.$$

Hint: First rewrite the value of the swap V_{swap} as a function of S and K by transforming cash flow: $L(T_i, T_{i+1}) - K = (L(T_i, T_{i+1}) - S) + (S - K)$; note the definition of S. Then consider the value of the swap at exercise date of the option, i.e. $V_{\text{swap}}(T_1)$ and try to choose a suitable numéraire (compare with the evaluation of a caplet). Say why the numéraire chosen is a traded product.

Solution (sketched): From Definition 117 a swap pays

$$(L(T_i, T_{i+1}; T_i) - S_i)\ (T_{i+1} - T_i) \quad \text{in } T_{i+1}. \tag{C.2}$$

Let $t \leq T_1$. The value of the payment (C.2) in t is

$$(L(T_i, T_{i+1}; T_i) - S_i)\ (T_{i+1} - T_i)\ P(T_{i+1}; t).$$

Thus, the value of the swap in t is given by

$$V_{\text{swap}}(t) = \sum_{i=1}^{n-1} (L(T_i, T_{i+1}; T_i) - S_i)\ (T_{i+1} - T_i)\ P(T_{i+1}; t).$$

By Definition 122 the *par swap rate* is given by

$$S_{\text{par}}(t) := \frac{P(T_1; t) - P(T_n; t)}{\sum_{i=1}^{n-1}(T_{i+1} - T_i)\ P(T_{i+1}; t)},$$

so that

$$\sum_{i=1}^{n-1} \left(L(T_i, T_{i+1}; T_i) - S_{\text{par}}(t) \right)\ (T_{i+1} - T_i)\ P(T_{i+1}; t) = 0.$$

Thus

$$V_{\text{swap}}(t) = \sum_{i=1}^{n-1} (L(T_i, T_{i+1}; T_i) - S_i) \ (T_{i+1} - T_i) \ P(T_{i+1}; t)$$

$$= \sum_{i=1}^{n-1} \left(L(T_i, T_{i+1}; T_i) - S_{\text{par}}(t) \right) \ (T_{i+1} - T_i) \ P(T_{i+1}; t)$$

$$+ \sum_{i=1}^{n-1} \left(S_{\text{par}}(t) - S_i \right) \ (T_{i+1} - T_i) \ P(T_{i+1}; t)$$

$$= \sum_{i=1}^{n-1} \left(S_{\text{par}}(t) - S_i \right) \ (T_{i+1} - T_i) \ P(T_{i+1}; t).$$

For a swap with $S_i = K$ this implies

$$V_{\text{swap}}(t) = (S_{\text{par}}(t) - K) \sum_{i=1}^{n-1} (T_{i+1} - T_i) \ P(T_{i+1}; t)$$

$$= (S_{\text{par}}(t) - K) \ A(t),$$

where $A(t) := \sum_{i=1}^{n-1} (T_{i+1} - T_i) \ P(T_{i+1}; t)$ is called *swap annuity*. Since $A(t) > 0$ we have for the value of the option on this swap

$$V_{\text{swaption}}(t) = \max(S_{\text{par}}(t) - K, 0) \ A(t).$$

Now chose A as numéraire. Under the corresponding martingale measure $\mathbb{Q}^A$, S is an A-relative price, thus a martingale and thus $dS(t) = \sigma(t) S(t) \, dW^{\mathbb{Q}^A}(t)$. The evaluation formula for a swaption now follows as in the derivation of the Black formula for a caplet.

APPENDIX D

Java™ Source Code (Selection)

D.1 Java™ Classes for Chapter 30

Listing D.1. `BinomialDistributedRandomVariable`: *A toy sample class to illustrate the concepts of "classes", "data" and "methods".*

```java
package com.christianfries.finmath.tutorial.example1;

/**
 * @author Christian Fries
 * @version 1.0
 */
class BinomialDistributedRandomVariable {

  private double value1;
  private double value2;
  private double probabilityOfState1;

  /**
   * This class implements a binominal distributed random variable
   * @param value1 The value in state 1.
   * @param value2 The value in state 2.
   * @param probabilityOfState1 The probability of state 1.
   */
  public BinomialDistributedRandomVariable(double value1, double value2,
    double probabilityOfState1) {
    this.value1 = value1;
    this.value2 = value2;
    this.probabilityOfState1 = probabilityOfState1;
  }

  public static void main( String args[] ) {
    /*
     * Test of class BinominalDistributedRandomVariable
     */

    System.out.println("Creating random variable.");
    BinomialDistributedRandomVariable randomVariable =
```

```java
    new BinomialDistributedRandomVariable(1.0,-1.0,0.5);   // Create object

    double expectation = randomVariable.getExpectation();
    System.out.println("Expectation is: " + expectation);

    System.out.println("Changing value of state 1 to 0.4.");
    randomVariable.setValue1(0.4);

    expectation = randomVariable.getExpectation();
    System.out.println("Expectation is: " + expectation);
  }

  /**
   * @param value1 The value of state 1.
   */
  public void setValue1(double value1) {
    this.value1 = value1;
  }

  /**
   * @return Returns the probability of state 1.
   */
  public double getProbabilityOfState1() {
    return probabilityOfState1;
  }

  /**
   * @param probabilityOfState1 The probability of state 1.
   */
  public void setProbabilityOfState1(double probabilityOfState1) {
    this.probabilityOfState1 = probabilityOfState1;
  }

  /**
   * @return Returns the probability of state 2.
   */
  public double getProbabilityOfState2() {
    return 1.0 - probabilityOfState1;
  }

  /**
   * @param probabilityOfState2 The probability of state 2.
   */
  public void setProbabilityOfState2(double probabilityOfState2) {
    this.probabilityOfState1 = 1.0 - probabilityOfState2;
  }


  /**
   * @return The expectation of the random variable
   */
  public double getExpectation() {
    double expectation = probabilityOfState1 * value1
      + (1-probabilityOfState1) * value2;
    return expectation;
  }
}
```

488

Listing D.2. `BisectionSearch:` *Root finder implementing the RootFinder-interface using the bisection method.*

```
/*
 * Created on 16.02.2004
 *
 * (c) Copyright Christian P. Fries, Germany.
 * All rights reserved. Contact: email@christian-fries.de.
 */
package net.finmath.rootFinder;

/**
 * This class implements a Bisection search algorithm,
 * implemented as a question-and-answer search algorithm.
 *
 * @author Christian Fries
 * @version 1.1
 */
public class BisectionSearch implements RootFinder {

    // We store the left and right end point of the intervall
    double[] points = new double[2]; // left, right
    double[] values = new double[2]; // left, right

    /*
     * State of solver
     */

    double  nextPoint;            // Stores the next point to return by getPoint
        ()

    int     numberOfIterations = 0;    // Number of iterations
    double  accuracy = Double.MAX_VALUE; // Current accuracy of solution
    boolean isDone   = false;      // True, if machine accuracy has been reached

    /**
     * @param leftPoint left point of search interval
     * @param rightPoint right point of search interval
     */
    public BisectionSearch(double leftPoint, double rightPoint) {
        super();
        points[0] = leftPoint;
        points[1] = rightPoint;

        nextPoint = points[0];
        accuracy  = points[1]-points[0];
    }

    /**
     * @return Best point optained so far
     */
    public double getBestPoint() {
        // Lazy: we always return the middle point as best point
        return (points[1] + points[0]) / 2.0;
    }

    /**
     * @return Next point for which a value should be set using <code>setValue</
     *     code>.
     */
```

489

```java
public double getNextPoint() {
  return nextPoint;
}

/**
 * @param value Value corresponding to point returned by previous <code>
 *     getNextPoint</code> call.
 */
public void setValue(double value) {
  if (numberOfIterations < 2) {
    /**
     * Initially fill values
     */
    values[numberOfIterations] = value;

    if (numberOfIterations < 1) {
      nextPoint = points[numberOfIterations + 1];
    } else {
      nextPoint = (points[1] + points[0]) / 2.0;
      /**
       * @todo Check if values[0]*values[1] < 0 here
       */
    }
  }
  else {
    /**
     * Bisection search update rule
     */

    if (values[1] * value > 0) {
      /*
       * Throw away right point (nextPoint is the point corresponding to
       *     value)
       */
      points[1] = nextPoint;    // This is not yet the nextPoint.
      values[1] = value;
    } else {
      /*
       * Throw away left point
       */
      points[0] = nextPoint;    // This is not yet the nextPoint.
      values[0] = value;
    }

    // Calculate next point (bisection)
    nextPoint = (points[1] + points[0]) / 2.0;

    // Savety belt: check if still improve or if we have reached machine
    //     accuracy
    if(points[1]-points[0] >= accuracy) isDone = true;

    // Update accuracy
    accuracy = points[1]-points[0];
  }

  numberOfIterations++;
  return;
}

/**
```

```
 * @return Returns the numberOfIterations.
 */
public int getNumberOfIterations() {
  return numberOfIterations;
}

/**
 * @return Returns the accuracy.
 */
public double getAccuracy() {
  return accuracy;
}

/**
 * @return Returns the isDone.
 */
public boolean isDone() {
  return isDone;
}
}
```

List of Symbols

Symbol	Interpretation
$\emptyset$	Empty set.
$\mathbf{1}$	Indicator function; $\mathbf{1}(x) = \begin{cases} 1 & \text{for } x > 0, \\ 0 & \text{else.} \end{cases}$
$\mathbf{1}_{(a,b]}(x)$	Indicator function; $\mathbf{1}_{(a,b]}(x) = \begin{cases} 1 & \text{for } x \in (a,b], \\ 0 & \text{else.} \end{cases}$
$x^\top$	Transposed (of a vectors or a matrix x).
$\mathcal{N}(\mu, \sigma^2)$	Normal distribution with mean μ and variance σ^2.
W	Brownian motion. See Definition 29.
$\mathbb{P}$	Real measure.
$\mathbb{Q}^N$	Martingale measure corresponding to the numéraire N. N-relative price processes $\frac{V}{N}$ of traded assets V are $\mathbb{Q}^N$-martingales. Exists (as a measure equivalent to $\mathbb{P}$) under certain assumptions.
$E^{\mathbb{Q}^N}$	Expectation operator with respect to the measure $\mathbb{Q}^N$.
$[x]$	Gauß bracket. Largest integer, being less than or equal to x. $[x] := \max\{n \in \{0,1,2,\ldots\} \mid n \le x\}$
$\|x\|_1$	ℓ_1-norm of a vector $x = (x_1, \ldots, x_n)$. $\|x\|_1 = \sum_{i=1}^n |x_i|$.
$\|x\|_2$	ℓ_2-norm of a vector $x = (x_1, \ldots, x_n)$. $\|x\|_2^2 = \sum_{i=1}^n |x_i|^2$.
$\mathrm{diag}(x_1, \ldots, x_n)$	Diagonal matrix. $\mathrm{diag}(x_1, \ldots, x_n)_{i,j} = \begin{cases} x_i & \text{for } i = j, \\ 0 & \text{else.} \end{cases}$

$P(T)$	Zero-coupon bond with maturity T. $P(T)$ (in general) is a stochastic process. Evaluated at time t on path ω we write $P(T; t, \omega)$. See Definition 97.
$L(T_1, T_2)$	Forward rate for the period $[T_1, T_2]$. See Definition 99.
$S_{i,j}$	$S_{i,j} := S(T_i, \ldots, T_j)$. Swap rate for the tenor structure $T_i, \ldots, T_j$. See Definition 122.
B	Money market account. See Equation (9.6).
$m(t)$	$m(t) := \max\{i \; : \; T_i \leq t\}$. Projection to last fixing in tenor structure. See Definition 124.

List of Figures

498

List of Tables

List of Listings

Bibliography

Books

[1] BAUER, HEINZ: Maßtheorie und Integrationstheorie. 2. Auflage. de Gruyter, Berlin, 1992. ISBN 3-11-013625-2.

[2] BAUER, HEINZ: Wahrscheinlichkeitstheorie. de Gruyter, Berlin, 2001. ISBN 3-11-017236-4.

[3] BAXTER, MARTIN W.; RENNIE, ANDREW J.O.: Financial Calculus: An Introduction to Derivative Pricing. Cambridge University Press, Cambridge, 2001. ISBN 0-521-55289-3.

[4] BIERMANN, BERND: Die Mathematik von Zinsinstrumenten. Oldenbourg Verlag, Munich, 2002. ISBN 3-486-25976-8.

[5] BINGHAM, NICHOLAS H.; KIESEL, RÜDIGER: Risk-Neutral Valuation: Pricing and Hedging of Financial Derivatives. (Springer Finance). Springer, London, 1998. ISBN 1-852-33458-4.

[6] BJÖRK, THOMAS: Arbitrage Theory in Continuous Time. Oxford University Press, New York, 1999. ISBN 0-198-77518-0.

[7] BRIGO, DAMIANO; MERCURIO, FABIO: Interest Rate Models—Theory and Practice. Springer, Berlin, 2001. ISBN 3-540-41772-9.

[8] CONT, RAMA; TANKOV, PETER: Financial Modelling with Jump Processes. CRC Press, Boca Raton, 2003. ISBN 1-584-88413-4.

[9] DRAPER, NORMAN R.; SMITH, HARRY: Applied Regression Analysis. 3rd edition. Wiley-Interscience, Hoboken, 1998. ISBN 0-471-02995-5.

[10] DUFFY, DANIEL J.: Finite Difference Methods in Financial Engineering: A Partial Differential Equation Approach. Wiley, Hoboken, 2006. ISBN 0-470-85882-6.

[11] ECKEL, BRUCE: Thinking in Java. 4th edition. Prentice Hall, Boston, 2002. ISBN 0-131-87248-6.

[12] EBMEYER, DIRK: Essays on Incomplete Financial Markets. Doctoral Thesis. University of Bielefeld, Bielefeld.

[13] HUNT, PHIL J.; KENNEDY, JOANNE E.: Financial Derivatives in Theory and Practice. Revised edition. Wiley, Chichester, 2004. ISBN 0-470-86359-5.

[14] GAMMA, ERICH; HELM, RICHARD; JOHNSON, RALPH E.: Design Patterns. Addison-Wesley Professional, 1997. ISBN 0-2-016-3361-2.

[15] GATHERAL, JIM: The Volatility Surface: A Practitioner's Guide. Wiley, Hoboken, 2006. ISBN 0-471-79251-9.

[16] GLASSERMAN, PAUL: Monte-Carlo Methods in Financial Engineering. Springer, New York, 2003. ISBN 0-387-00451-3.

[17] GÜNTHER, MICHAEL; JÜNGEL, ANSGAR: Finanzderivate mit MATLAB. Mathematische Modellierung und numerische Simulation. Vieweg, 2003. ISBN 3-528-03204-9.

[18] JÄCKEL, PETER: Monte-Carlo Methods in Finance. 238 Seiten. Wiley, Chichester, 2002. ISBN 0-471-49741-X.

[19] JOSHI, MARK S.: The Concepts and Practice of Mathematical Finance. Cambridge University Press, Cambridge, 2003. ISBN 0-521-82355-2.

[20] KARATZAS, IOANNIS; SHREVE, STEVEN E.: Brownian Motion and Stochastic Calculus. 2nd edition. Springer, New York, 1991. ISBN 0-387-97655-8.

[21] KLOEDEN, PETER E.; PLATEN, ECKHARD: Numerical Solution of Stochastic Differential Equations (Applications of Mathematics. Stochastic Modelling and Applied Probability, Vol. 23). Springer, Berlin, 1999. ISBN 3-540-54062-8.

[22] MALLIAVIN, PAUL: Stochastic Analysis (Grundlehren Der Mathematischen Wissenschaften). Springer, Berlin, 1997. ISBN 3-540-57024-1.

[23] MEISTER, MARKUS: Smile Modeling in the LIBOR Market Model. Diploma Thesis. University of Karlsruhe, Karlsruhe, 2004.

[24] MUSIELA, MAREK; RUTKOWSKI, MAREK: Martingale Methods in Financial Modeling: Theory and Applications. Springer, Berlin, 1997. ISBN 3-540-61477-X.

[25] PAUL, WOLFGANG; BASCHNAGEL JÖRG: Stochastic Processes. From Physics to Finance. Springer, Berlin, 2000. ISBN 3-540-66560-9.

[26] PELSSER, ANTOON: Efficient Methods for Valuing Interest Rate Derivatives. Springer, London, 2000. ISBN 1-852-33304-9.

[27] ØKSENDAL, BERNT K.: Stochastic Differential Equations: An Introduction with Applications. Springer, Berlin, 2000. ISBN 3-540-63720-6.

[28] OESTEREICH, BERND: Objektorientierte Softwareentwicklung. Oldenburg, 2004. ISBN 3-486-27266-7.

[29] PROTTER, PHILIP E.: Stochastic Integration and Differential Equations. Springer, Berlin, 2003. ISBN 3-540-00313-4.

[30] REBONATO, RICCARDO: Modern Pricing of Interest-Rate Derivatives: The LIBOR Market Model and Beyond. Princeton University Press, Princeton, 2002. ISBN 0-691-08973-6.

[31] ROGERS, L. C. G.; WILLIAMS, DAVID: Diffusions, Markov Processes and Martingales: Volume 2, Ito Calculus. 2nd Edition. Cambridge University Press, Cambridge, 2000. ISBN 0-521-77593-0.

[32] SEYDEL, RÜDIGER: Tools for Computational Finance. Springer, Berlin, 2003. ISBN 3-540-40604-2.

[33] SHIRYAEV, ALBERT N.: Essentials of Stochastic Finance: Facts, Models, Theory. World Scientific, Singapore, 1999. ISBN 9-810-23605-0.

[34] STEELE, J. MICHAEL: Stochastic Calculus and Financial Applications. Springer-Verlag, New York, 2001. ISBN 0-387-95016-8.

[35] TAVELLA, DOMINGO; RANDALL, CURT: Pricing Financial Instruments: The Finite Difference Method. Wiley, Hoboken, 2000. ISBN 0-471-19760-2

[36] ULLENBOOM, CHRISTIAN: Java ist auch eine Insel. 5. Auflage. Galileo Press, 2004. ISBN 3-898-42526-6.

[37] YOUNG, JEFFREY S.; SIMON, WILLIAM L.: iCon Steve Jobs. Wiley, Hoboken, 2005. ISBN 0-471-72083-6.

[38] WILDE, OSCAR: The Picture of Dorian Gray. ISBN 0-679-60001-9.

[39] WILDE, OSCAR: The Importance of Being Earnest and Other Plays. Oxford University Press (Reprint). ISBN 0-198-12167-9.

[40] WILMOTT, PAUL: Paul Wilmott on Quantitative Finance. Wiley, Chichester, 2006. ISBN 0-470-01870-4.

[41] WILLIAMS, DAVID: Probability with Martingales. Cambridge University Press, Cambridge, 1991. ISBN 0-521-40605-6.

[42] WILLIAMS, DAVID: Weighing the Odds: A Course in Probability and Statistics. Cambridge University Press, Cambridge, 2001. ISBN 0-521-00618-X.

[43] ZHANG, PETER: Exotic Options: A Guide to Second Generation Options. World Scientific, Singapore, 1998. ISBN 9-810-23521-6.

Papers

[44] ANDERSEN, LEIF: A Simple Approach to the Pricing of Bermudan Swaptions in the Multi-Factor LIBOR Market Model. Working paper. General Re Financial Products, 1999.

[45] ANDERSEN, LEIF; BROADIE, MARK: A Primal-Dual Simulation Algorithm for Pricing Multi-Dimensional American Options. Working paper. General Re Financial Products, 1999.

[46] ANDERSEN, LEIF; SIDENIUS, JAKOB; BASU, SUSANTA: All Your Hedges in One Basket. Risk Magazine 11, 67–72, 2003.

[47] BENHAMOU, ERIC: Optimal Malliavin Weighting Function for the Computation of the Greeks. 2001.

[48] BLACK, FISCHER: The Pricing of Commodity Contracts. *Journal of Financial Economics* 3, 167–179, 1976.

[49] BOUCHAUD, JEAN-PHILIPPE; SORNETTE, DIDIER: The BlackScholes Option Pricing Problem in Mathematical Finance: Generalizations and Extensions for a Large Class of Stochastic Processes. Journal de Physique I, 4, 863, 1994.

[50] BRACE, ALAN; GATAREK, DARIUSZ; MUSIELA, MAREK: The Market Model of Interest Rate Dynamics. Mathematical Finance 7, 127, 1997.

[51] BRASCH, HANS-JÜRGEN: A Note on Efficient Pricing and Risk Calculation of Credit Basket Products. Preprint, 2005.
http://defaultrisk.com/pp_crdrv_54.htm.

[52] BREEDEN, D. T.; LITZENBERGER, R. H.: Prices of State-Contingent Claims Implicit in Option Prices. *Journal of Business* 51(4), 621–651, 1978.

[53] BOYLE, PHELIM; BOADIE, MARK; GLASSERMAN, PAUL: Monte-Carlo Methods for Security Pricing. Journal of Economic Dynamics and Control, 21, 1267–1321, 1997.

[54] BRIGO, DAMIANO; MERCURIO, FABIO; RAPISARDA, FRANCESCO: Lognormal-Mixture Dynamics and Calibration to Market Volatility Smiles. 2000.
http://www.damianobrigo.it.

[55] BROADIE, MARK; GLASSERMAN, PAUL: Estimating Security Price Derivatives using Simulation. *Management Science*, 42(2), 269–285, 1996.

[56] BROADIE, MARK; GLASSERMAN, PAUL: Pricing American-Style Securities by Simulation. *Journal of Economic Dynamics and Control*, 21, 1323–1352, 1997.

[57] BROCKHAUS, OLIVER: Implied Monte-Carlo. Bachelier Conference Crete, June 2002.

[58] CARR, PETER; THALELA, M.; ZARIPHOPOULOU, T.: Closed Form Option Valuation with Smiles. Working Paper, 1999.

[59] CARRIERE, JACQUES F.: Valuation of Early-Exercise Price of Options Using Simulations and Nonparametric Regression. *Insurance: Mathematics and Economics* 19, 19–30, 1996.

[60] CLÉMENT, EMMANUELLE; LAMBERTON, DAMIEN; PROTTER, PHILIP: An Analysis of a Least Squares Regression Method for American Option Pricing. *Finance and Stochastics* 6, 449–471, 2002.

[61] CHEYETTE, O.: Markov Representation of the Heath-Jarrow-Morton Model. Working Paper. BARRA Inc.

[62] DAVIS, MARK; KARATZAS, IOANNIS: A Deterministic Approach to Optimal Stopping, with Applications. In Whittle, Peter (Ed.): *Probability, Statistics and Optimization: A Tribute to Peter Whittle*, Wiley, New York and Chichester, 1994, pp. 455–466.

[63] ECKSTÄDT, FABIAN: The Valuation of Hybrid Options with a Two Dimensional Markov-Functional Model. Diploma Thesis. Bielefeld University, 2006. http://www.fabian-eckstaedt.de/.

[64] FENGLER, MATTHIAS R.: Arbitrage-Free Smoothing of the Implied Volatility Surface. SFB 649 Discussion Paper 2005-019. Berlin. ISSN 1860-5664. http://sfb649.wiwi.hu-berlin.de/.

[65] FOURNIÉ, ERIC; LASRY JEAN-MICHEL; LEBUCHOUX, JÉRÔME; LIONS, PIERRE-LOUIS; TOUZI, NIZAR: Applications of Malliavin Calculus to Monte-Carlo Methods in Finance. Finance Stochastics. 3, 391–412, 1999.

[66] FRIES, CHRISTIAN P.: Localized Proxy Simulation Schemes for Generic and Robust Monte-Carlo Greeks. 2007. http://www.christian-fries.de/finmath/proxyscheme

[67] FRIES, CHRISTIAN P.: The Foresight Bias in Monte-Carlo Pricing of Options with Early Exercise: Classification, Calculation and Removal. 2005. http://www.christian-fries.de/finmath/foresightbias.

[68] FRIES, CHRISTIAN P.; ECKSTÄDT, FABIAN: A Hybrid Markov-Functional Model with Simultaneous Calibration to Interest Rate and FX Smile. 2006.

[69] FRIES, CHRISTIAN P.; JOSHI, MARK S.: Partial Proxy Simulation Schemes for Generic and Robust Monte-Carlo Greeks. 2006. http://www.christian-fries.de/finmath/proxyscheme.

[70] FRIES, CHRISTIAN P.; KAMPEN, JÖRG: Proxy Simulation Schemes for Generic Robust Monte-Carlo Sensitivities, Process Oriented Importance Sampling and High Accuracy Drift Approximation. 2005. http://www.christian-fries.de/finmath/proxyscheme.

[71] FRIES, CHRISTIAN P.; ROTT, MARIUS G.: Cross Currency and Hybrid Markov Functional Models. Preprint, 2004. http://www.christian-fries.de/finmath/markovfunctional.

[72] GLASSERMAN, PAUL; LI, JINGYI: Importance Sampling for Portfolio Credit Risk. Working paper, Columbia University, 2003. http://www2.gsb.columbia.edu/faculty/pglasserman/Other/is_credit.pdf.

[73] GLASSERMAN, PAUL; ZHAO, XIAOLIANG: Arbitrage-Free Discretization of Log-normal Forward LIBOR and Swap Rate Models. *Finance and Stochastics* 4, 35–68, 2000.

[74] GLASSERMAN, PAUL; ZHAO, XIAOLING: Fast Greeks in Forward LIBOR Models. *Journal of Computational Finance*, 3, 5–39, 1999.

[75] HAGAN, PATRIK S.; KUMAR, DEEP; LESNIEWSKI, ANDREW S.; WOODWARD, DIANA E.: Managing Smile Risk (SABR Model), *Wilmott Magazine*, September 2002.

[76] HAGAN, PATRIK S.; WEST, GRAEME: Interpolation Methods for Curve Construction. Preprint. 2005.

[77] HAUGH, MARTIN; KOGAN, LEONIS: Pricing American Options: A Duality Approach, MIT Sloan Working Paper No. 4340-01, 2001.

[78] HEATH, DAVID; JARROW, ROBERT; MORTON,ANDREW: Bond Pricing and the Term Structure of Interest Rates: A New Methodology for Contingent Claims Valuation. *Econometrica* 60(1) (January), 77–105, 1992.

[79] HUNT, PHIL J.; KENNEDY, JOANNE E.; PELSSER, ANTOON: Markov-Functional Interest Rate Models. *Finance and Stochastics*, 4(4), 391–408, 2000.

[80] HUNTER, CHRISTOPHER J.; JÄCKEL, PETER; JOSHI, MARK S.: Drift Approximations in a Forward-Rate-Based LIBOR Market Model. Getting the Drift. *Risk*, 14, 81–84, July, 2001.

[81] JAMSHIDIAN, FARSHID: LIBOR and Swap Market Models and Measures. *Finance and Stochastics* 1, 293–330, 1997.

[82] JOSHI, MARK S.: Applying Importance Sampling to Pricing Single Tranches of CDOs in a One-Factor Li Model. QUARC, Group Risk Management, Royal Bank of Scotland. Working paper, 2004.

[83] JOSHI, MARK S.; KAINTH, DHERMINDER: Rapid Computation of Prices and Deltas of Nth to Default Swaps in the Li Model. *Quantitative Finance*, 4(3), 266–275, 2004.
http://www.quarchome.org/.

[84] KOHL-LANDGRAF, PETER: A PDE Approach to the Valuation of Interest Rate Products under Markovian Yield Curve Dynamics. Diploma-Thesis. University of Bayreuth. Bayreuth, 2007.

[85] LI,D.: On Default Correlation: A Copula Approach. *Journal of Fixed Income* 9, 43–45, 2000.

[86] LONGSTAFF, FRANCIS A.; SCHWARTZ EDUARDO S.: Valuing American Options by Simulation: A Simple Least-Square Approach. *Review of Financial Studies* 14(1), 113–147, 2001.

[87] MATSUMOTO, M.;NISHIMURA, T.: Mersenne Twister: A 623-dimensionally Equidistributed Uniform Pseudorandom Number Generator. *ACM Transactions on Modeling and Computer Simulations*, 1998.

[88] MILTERSEN, KRISTIAN R.; SANDMANN, KLAUS; SONDERMANN, DIETER: Closed Form Solutions for Term Structure Derivatives with Lognormal Interest Rates. *Journal of Finance* 52, 409–430, 1997.

[89] PITERBARG, VLADIMIR V.: A Practitioner's Guide to Pricing and Hedging Callable LIBOR Exotics in Forward LIBOR Models, Preprint. 2003.

[90] PITERBARG, VLADIMIR V.: Computing deltas of callable LIBOR exotics in forward LIBOR models. *Journal of Computational Finance*, 7, 2003.

[91] PITERBARG, VLADIMIR V.: TARNs: Models, Valuation, Risk Sensitivities. *Wilmott Magazine*, 2004.

[92] RITCHKEN, P.; SANAKARASUBRAMANIAN, L.: Volatility structures of forward rates and the dynamics of the term structure. *Mathematical Finance*, 5

[93] ROGERS, L. C. G.: Monte-Carlo Valuation of American Options, Preprint. 2001.

[94] ROTT, MARIUS G.; FRIES, CHRISTIAN P.: Fast and Robust Monte-Carlo CDO Sensitivities and their Efficient Object Oriented Implementation. 2005. http://www.christian-fries.de/finmath/cdogreeks

[95] SCHLÖGL, ERIK: A Multicurrency Extension of the Lognormal Interest Rate Market Models. *Finance and Stochastics* 6(2), 173–196, 2002. Springer-Verlag, 2002.

[96] SCHLÖGL, ERIK: Arbitrage-Free Interpolation in Models of Market Observable Interest Rates. Preprint. 2002.

[97] WICHURA, MICHAEL J.: Algorithm AS 241: The Percentage Points of the Normal Distribution. *Applied Statistics*, 37, 477–484, 1988.

Index

Q

R

zero-coupon bond